THE
COOL WAR

Titles in the Series

Transforming War

Paul J. Springer, editor

To ensure success, the conduct of war requires rapid and effective adaptation to changing circumstances. While every conflict involves a degree of flexibility and innovation, there are certain changes that have occurred throughout history that stand out because they fundamentally altered the conduct of warfare. The most prominent of these changes have been labeled "Revolutions in Military Affairs" (RMA). These so-called revolutions include technological innovations as well as entirely new approaches to strategy. Revolutionary ideas in military theory, doctrine, and operations have also permanently changed the methods, means, and objectives of warfare.

This series examines fundamental transformations that have occurred in warfare. It places particular emphasis upon RMAs to examine how the development of a new idea or device can alter not only the conduct of wars but their effect upon participants, supporters, and uninvolved parties. The unifying concept of the series is not geographical or temporal; rather, it is the notion of change in conflict and its subsequent impact. This has allowed the incorporation of a wide variety of scholars, approaches, disciplines, and conclusions to be brought under the umbrella of the series. The works include biographies, examinations of transformative events, and analyses of key technological innovations that provide a greater understanding of how and why modern conflict is carried out, and how it may change the battlefields of the future.

THE
COOL WAR

NUCLEAR FORCES, CRISIS SIGNALING, AND
THE RUSSO-UKRAINE WAR, 2014–2022

SEAN M. MALONEY

Naval Institute Press
Annapolis, Maryland

Naval Institute Press
291 Wood Road
Annapolis, MD 21402

Library of Congress Cataloging-in-Publication Data

Names: Maloney, Sean M., author
Title: The cool war : nuclear forces, crisis signaling, and the Russo-Ukraine War, 2014–2022 / Sean M. Maloney.
Description: Annapolis, Maryland : Naval Institute Press, [2025] | Series: Transforming war | Includes bibliographical references and index.
Identifiers: LCCN 2025017334 (print) | LCCN 2025017335 (ebook) | ISBN 9781682476895 hardback | ISBN 9781682476925 ebook
Subjects: LCSH: Information warfare—Russia (Federation) | Russo-Ukrainian War, 2014— Propaganda | Nuclear weapons—Russia (Federation)—Strategy | Russia (Federation)— Foreign relations—Ukraine | Ukraine—Foreign relations—Russia (Federation)
Classification: LCC U163 .M263 2025 (print) | LCC U163 (ebook) | DDC 355.3/4309477—dc23/ eng/20250819
LC record available at https://lccn.loc.gov/2025017334
LC ebook record available at https://lccn.loc.gov/2025017335

9 8 7 6 5 4 3 2 1

For Fritz Heinzen—Agent and Friend

Auch aus Steinen,
die einem in den Weg gelegt werden,
kann man Schönes bauen.

—Johann Wolfgang von Goethe

CONTENTS

MAPS AND CHARTS

MAPS

CHARTS

PHOTOS

FOREWORD

Millions of people around the globe have been consistently following the daily details of Russia's unprovoked and brutal second invasion of Ukraine. Whether you get your news from social media feeds, television, or newspapers, these daily snippets of information only give you part of the story on this violent war. Behind the scenes, another story has been taking place—nuclear messaging between the world's two largest nuclear powers. In this remarkable historical analysis, Dr. Sean Maloney delves deep into the intricate web of nuclear messaging between the United States, key NATO nuclear powers, and the Russian Federation that so far has been out of the public's eye.

The story you're about to read provides an unparalleled peak behind the curtain on the untold aspects of war in Ukraine. Dr. Maloney lays bare the hidden strategies, calculated moves, and subtle nuances that have shaped the nuclear aspect of this conflict. Drawing upon an extensive evaluation of senior leader comments, social media postings, and observing the uncharacteristic broadcasting or "squawking" of positional data from strike and nuclear command and control aircraft military aircraft, Dr. Maloney has put this complicated puzzle together that shows the nuclear aspect of this conflict. While nuclear signaling isn't new to this rivalry, it will highlight, as the author describes, a new "cool war" between Putin and the West.

As you read this historical analysis, you will all be members of the jury determining if the actions observed were preplanned, coincidence, or both. But what isn't in doubt is the movement of aircraft, corresponding flight profiles, and the days that these events occurred—none of this is speculation; it is fact. At the heart of this groundbreaking investigation lies the enigmatic case of Ukraine—an embattled nation reeling from the vicious attacks from a leader undaunted by the war crimes he's committing. As tensions escalated between the United States, the West, and Russia, the author reveals the pivotal role that nuclear messaging has played in shaping the course of events. With a keen eye for detail and an insatiable appetite for uncovering the truth, Dr. Maloney has unearthed a treasure trove of insights, offering readers a unique vantage point from which to view the highly guarded world of nuclear deterrence.

By unraveling the intricacies of U.S.-Russian nuclear messaging, Dr. Maloney unveils the strategies, doctrines, and psychological dynamics that have been obscured from public view. In doing so, he provides us with a fresh and profound understanding of the complex interplay between nuclear powers and the crucial role of communication in maintaining strategic stability. Dr. Maloney's rigorous research and penetrating analysis not only sheds light on the past but also provides us with a critical framework for understanding the present and shaping the future of nuclear deterrence.

Prepare to embark on a journey that will challenge your preconceptions, broaden your horizons, and deepen your understanding of the intricate dance of nuclear messaging between the United States and Russia. This book is destined to become a seminal work, guiding scholars, policymakers, and concerned citizens alike in navigating the complexities of deterrence. Make no mistake, the war in Ukraine is a nuclear conflict.

Lt. Gen. Jack Weinstein, USAF (Ret.)
Professor of the Practice of International Security
Pardee School of Global Studies, Boston University

ACKNOWLEDGMENTS

This book started as a series of three lectures I put together to categorize and explain nuclear signaling I observed in the very early stages of the crisis. I gave the first one of these to the Royal Alberta United Service Institute, by complete coincidence, on the night of 24 February 2022. As the crisis evolved, the demand for more was insatiable and I wound up putting together a three-lecture program that I gave in multiple venues as people struggled to deal with the nuclear aspects of the crisis. Lieutenant General Jack Weinstein saw the lecture series and encouraged me to do a full treatment, as did alte kamaraden Quentin Innes and Thomas Bradley. My good friend and agent Fritz Heinzen and his protégé Frannie Dove said the same thing. And Dr. P. J. Springer promised to harangue Padraic (Pat) Carlin and Adam Kane at Naval Institute Press until they relented and published *The Cool War*, which they ultimately did. Thank you very much, everybody.

Writing contemporary history brings with it challenges that are different from other forms of history. *The Cool War* is in many ways an experimental and risky early attempt at examining these phenomena using new methods. I would particularly like to thank Drs. Melvin Deaile, Ed Redman, and John Terino from the Department of Airpower, Air University for not only commenting on the draft but for their encouragement as well. The same goes for Dr. David Charters, whose student I was thirty-five years ago; I'd especially like to thank him for retaining an interest in my work over those long years. The incomparable Richard Martin was superb at helping me "red team" crucial portions of the narrative and as a philosophical sounding board through the writing process. I'd also like to thank Rob Silliman for his assistance with aerospace industry, SSBN, and space operations, Matt Larson for his experience with cyber operations, and Larissa Reise for expertise in space and communications operations. Also, Bernie DeGagne on NORAD and air operations, and Allen Jones for all things aeronautical. Dr. Greg Liedtki assisted with things Germanic. My colleague at RMC Dr. Steven Lukits insisted that the production of a broader narrative was a crucial contribution to our understanding these matters.

There are numerous observers of the scene I would like to recognize as well. Ambassador Michael McFaul's constant input on the matter of Russia and Ukraine is crucial. Julia "I watch Russian state TV so you don't have to" Davis kept us informed about the bewildering output from Rossiya1 and RT, as did Dmitri @wartranslated. Sam Ramani and his take was particularly valuable as well as Kevin Rothrock, Andrei Soldatov, Max Seddon, Sergej Sumlenny, Alex Kokcharov, Sergy Radchenko, Francis Scarr, Viktor Kovalenko, Christo Grozov, and Bellingcat. Aircraft watchers TheIntelFrog, Evergreen Intel, MeNMyRC, RivetJoint, #haveglass, Theneware51, and SkyScanWorld and CivMilAir produce significant situational awareness, as well as Ryan Chan's ability to ferret out obscure but useful air, ship, and nuclear weapons information. Similarly, H. I. Hutton, Saturnax, Sheila Weir, Iain Cameron, and Dougie Coull on the naval front. Thomas Nilsen, the staff of *The Barents Observer*, The Lookout, and Aki Heikkinen keep a snow hawk's–eye view on the northern flank. Thanks also to GlasnostGone and Def Mon for their observations.

Anna Clara Arndt and Liviu Horovitz's work on tracking the diplomatic and informational give-and-take was particularly valuable and deserves a wider audience. At RUSI, Jack Watling and Nick Reynolds deserve recognition for their efforts. Thanks also to George Allison at UK Defence Journal. On cyber, there is the Anonymous Collective and their efforts. The Hams also helped: Radioskaner, From the Static, Radio & Nukes, TJ, Spy Stations, redandblackattack, and particularly The Sky Kings. On the IO and influence fronts, I'd recommend reading everything by Peter Pomerantsev and Anton Shekhovtsov. I would particularly like to thank Trish Lejano at JETNET, Lori O'Donley at Flight Aware, and Liliia at Fightradar 24 support for their kind and helpful assistance.

Finally, I would like to recognize my friend Gordon Ohlke who assisted in key areas but passed away while I was writing *The Cool War*. Gord was a lifelong observer of the Soviet, then Russian, scene while serving as an intelligence officer in the 1980s, 1990s, and 2000s. His sudden death was the equivalent of a library burning down and I never had the opportunity to thank him properly for his decades of support and friendship.

Introduction

*O*hio-class ballistic missile submarines have been characterized as the most efficient killing machines ever devised. Capable of remaining concealed in the ocean's depths, these submarines carry twenty Trident II D5 missiles. Each missile can carry up to twelve nuclear weapons capable of yielding up to 475 kt each. Though the load-outs of the submarines vary per deployment, the combat potential of an *Ohio* ballistic missile submarine consists of the ability to strike up to 240 separate desired ground zeros at a range of 7,000 miles or 11,300 km from their launch point. That is one submarine. There are fourteen of these ships serving in the United States Navy, of which about two-thirds are on patrol at any one time.

At 1633 hours Eastern Standard Time on 19 October 2022, U.S. Central Command (CENTCOM), the military organization responsible for American operations in the Middle East, used the social media platform Twitter to inform its audience that the CENTCOM commander, General Erik Kurilla, made a visit to the *Ohio*-class submarine USS *West Virginia* "at an undisclosed location at sea in the international waters of the Arabian Sea." At 1714 hours, the U.S. Strategic Command (STRATCOM), responsible for controlling American nuclear forces, retweeted the CENTCOM tweet noting, "These submarines are the crown jewel of the nuclear triad, and the *West Virginia* demonstrates the flexibility, survivability, readiness & capacity of USCENTCOM & USSTRATCOM forces at sea."

Identifying the location of a ballistic missile submarine, even in such vague terms, let alone having its parent command do so over social media, is highly unusual. In fact, unmasking American ballistic missile submarines like this has only occurred on a handful of occasions in recent history. Such a move is only done under careful consideration: when the leadership of the United States wants to signal an adversary that they should alter the trajectory of their present course of action.

Three weeks earlier, Russia mounted a naval exercise in the arctic off Alaska called Umka 2022. The operation involved two cruise missile submarines, a ballistic missile submarine, and a forward command element consisting of nuclear command post aircraft. Conventional cruise missiles were fired in the direction of Alaska, but the Russian ballistic missile submarine ran silent and eventually revealed itself on entry to the base at Petropavlovsk. The same day as Umka 2022, President Biden warned Vladimir Putin in a CNN interview not to use nuclear weapons in the war in Ukraine. Two days later Putin announced that Russia was mobilizing to backfill losses in Ukraine and implied nuclear forces could be used if he deemed them necessary. The next day, former Russian president and now Security Council member Dmitry Medvedev threatened the use of nuclear weapons against the United States and Europe. He repeated those assertions on 27 September. Three days later, Putin announced the annexation of occupied parts of Ukraine and made a threat that he reserved the right to use nuclear weapons to defend them. A flurry of messages on social media triggered widespread public speculation and fear throughout the world that nuclear war was a real possibility. On 11 October, President Biden reiterated his stance on nuclear weapons use: the United States would respond as it deemed appropriate. The public fear level as expressed through mainstream and social media increased yet again.

At some point, U.S. authorities made the decision to unmask the *West Virginia*. The relative position of the *West Virginia* in the Arabian Sea to its potential targets is instrumental in understanding this signal. From Vladimir, east of Moscow, to Irkutsk on the shores of Lake Baikal, Russia's Strategic Missile Forces deployed ten missile divisions. Most of these formations consist of Topol-M or YARS mobile intercontinental ballistic missiles (ICBM), mounted on very large trucks. These missiles, loaded with nuclear warheads and capable of reaching North America, are kept at central bases until they are alerted and deployed into nearby forests. This complicates targeting, as the launch positions are neither constant nor consistent. The *West Virginia*'s geographic position in the Arabian Sea permits up to 288 nuclear weapons to rain down on these Russian bases. More important, the flight time from the *West Virginia* to those targets would be around fifteen minutes.

The Russian detection system consists of early warning satellites and ground-based ballistic missile radar. ROSCOSMOS (State Corporation for Space Activities) complained early in the crisis that their satellite constellations had been hacked. Indeed, then–ROSCOSMOS director Dmitry Rogozin publicly asserted that such a move was a casus belli, implying that war could result from this event alone. So how reliable were Russian early warning satellites at this point on 19 October? That is unclear. As for ground-based ballistic missile early warning radars, there is a gap in their coverage areas that corresponds to a slice of ocean in the Arabian Sea. Could the Konteyner over-the-horizon radar site at Vasilevka designed to cover this gap detect a Trident II D5 launch? That would depend on the state of maintenance of the equipment at the site and the proficiency, or even the sobriety, of its crew. Indeed, there were reports that the whole system was dysfunctional due to corrupt construction practices. And could it detect a depressed trajectory missile launch? The U.S. Navy demonstrated in November 2015 that it could launch Trident II D5's with a reduced angle that permitted shorter flight time and a lower ballistic path that in turn made the missile much more difficult to detect.

There would be little or no warning if the decision to launch was made. Russian defense minister Sergey Shoigu contacted his American opposite number, Lloyd Austin. This was followed by a flurry of phone calls between Shoigu and other ministers of defense, and between General Valery Gerasimov, the Russian chief of the general staff and his opposite numbers in the West over the next forty-eight hours. No missiles flew. There was no nuclear war.

This is signaling using strategic nuclear forces. And this is only one of many episodes that occurred from the winter of 2021 to the winter of 2023, the latest period of the "Cool War."

What Is the Cool War?

We are in the Cool War now. Like the war on terrorism and the Cold War, our opponents declared war on the West before we even realized it. The Soviet Union declared war on the West in 1921, twenty-five years before the Cold War was recognized in 1946–47. Bin Laden declared his war on the West in 1996, five years before the 9/11 attacks and Operation Enduring Freedom to counter Al Qaeda. Vladimir Putin essentially declared war on the West in 2007. Misguided constructs, denial, and fear based on obsolete constraints and "oldthink" of certain Cold War–era theories of realism and escalation prevented the United States from ascertaining the full nature of the threat until the Cool War was thrust upon it. Fortunately, the United States had the Cold War experience to fall back on and this is where strategic signaling with nuclear forces was revitalized to positive effect in 2022. This time, information operations using new forms of communication played a key role when used to augment strategic signaling were employed.

The Cool War is a period of protracted maneuvering by Vladimir Putin's Russia using strategic nuclear forces to underpin or otherwise support Russian methods and objectives in its quest to subjugate Ukraine and intimidate its allies, specifically during the 2007 to 2023 period. It also consists of the maneuvering of American, French, and British nuclear and conventional forces in response to Russian activities. The Cool War appears similar to, but very different from, the Cold War and the nuclear crises that regularly populated that conflict. Though many have sought to compare events within it to the 1962 Cuban Missile Crisis, or even the Second World War, this current confrontation is unique in human history. The Cool War involves signaling with nuclear forces and conventional forces intertwined with information operations and what is colloquially referred to as "hybrid warfare." The protracted nature of nuclear force signaling in this crisis is unparalleled, as is its visibility to a broad public audience due to new forms of communications technology. In effect, the Cool War is objectively something that has become superimposed over the Russo-Ukrainian conflict, with the understanding that the Russian leadership views both holistically.

The role of strategic signaling using nuclear forces has been and remains integral to the Cool War. In this new environment, it is possible to discern four types of this activity. There is *general deterrence signaling*, the ongoing maintenance of the day-to-day deterrence posture. There is *intimidation signaling* where forces are used on an ongoing basis specifically to generate pervasive fear and lower resistance to a course of action. Its corollary is *assurance signaling*, whereby the intent is to confront the effects of intimidation. Using forces to underscore diplomatic activities constitute *backstop signaling*. Finally, there is *crisis-specific signaling*, in which signaling moves are designed to react in the context of a designated crisis itself. It is important to differentiate between these activities simply for the reason that the messaging sent by the signal may not reach its audience and have its intended effect, which in turn has serious ramifications in a nuclear-armed world. They also provide the language of the conflict the United States is now engaged in; this includes the movements of military forces, to a great extent aircraft but also other nuclear forces, that give us a sense of the rhythm of signaling and how this shifted over time.

Then there is the message the signal conveys. That is open to interpretation. There may be multiple messages: what do they mean and who are the intended recipients? Without access to official materials, this is a highly subjective process but contextualizing them can increase the probability of analyzing these phenomena successfully. In such cases, this book will necessarily use tentative language, with the caveat that new and better information in the future may provide sharper, or perhaps differing, definitions.

Important questions emerge in this inquiry into signaling. Could signaling have prevented or forestalled aggressive activities at any time during the period of the Cool War? How viable was it in a so-called gray-zone warfare or hybrid warfare environment? How well did signaling work in a transition from the ongoing day-to-day situation to a crisis situation and then within it? Was there any point to signaling if one party had its mind made up on war?

Structure

It is essential that we examine Russian ideology, policy objectives, strategy, and the role of the information environment as these aspects evolved significantly and in an overlapping fashion during post–Cold War period. These were not disconnected spheres or stovepipes; ideology drove declaratory policy and strategy, and the subjugation of Ukraine was a precondition for Russian success. This is the subject of chapter 1.

The Cool War buildup took place after 2008, and both sides had different signaling styles that used similar tools but in different combinations for different purposes. Chapters 2 and 3 examine the Russian, and then American and NATO developments in these areas up to 2020, with an emphasis on force deployments, structures, and exercises. This establishes the signaling language that we will see employed in 2021 and 2022. Chapter 4 sets us up to understand the events of 2021 in which all the strands established thus far came together, and we blend Russian and Western signaling as the system's volume increased throughout 2021.

Chapters 5, 6, 7, and 8 handle the move–countermove rhythm of the crisis as it evolved throughout 2022 in relationship to the war in Ukraine. Chapter 6 examines the Russian buildup in Belarus, the Kazakhstan crisis, the run-up to the February invasion, and the American attempts to throw off the Russian program. Chapter 6 deals with the invasion and the immediate aftermath in which the Putin regime got frustrated and resorted to nuclear intimidation. This in turn generated significant pushback by NATO members. Chapter 7 handles the Putin regime's desperate attempt to restore freedom of action on the international stage by applying pressure outside Eastern Europe; it also discusses NATO countermoves. Finally, chapter 8 examines the Russian response to the Kharkhiv counteroffensive in the fall and the dramatically ramped-up intimidation and signaling that occurred the rest of the year.

The decision to end this narrative in December 2022 was a tough one. The Cool War has not ended, nor has the Russo-Ukrainian war. The chess-like move–countermove in the realm of nuclear forces and the realm of information operations reached a comparatively repetitive steady state in the first six months of 2023, consonant with the lack of significant movement in the ground in Ukraine, the Battle of Bakhmut, and the buildup for the summer counteroffensive.

All history starts with a narrative trajectory and this work will establish a first cut at it, especially for the role of nuclear force signaling. Examining the Cool War demands that a unique and new methodology be identified, explained, and employed so that the parameters of the conflict can be understood. This methodology, which permits us to track and observe certain aspects of strategic nuclear force activity in near real time, is crucial to understanding how conflicts of the future will build on and exploit the methods by which the Cool War was fought. No matter what happens to Putin's Russia, there will be utility to understanding this dangerous period. These future crises—be they with Communist China, Iran, nuclear-armed fragments of Russia, or emergent competitors—will employ derivatives of the Cool War methodology, so it is in the best interests of the United States and its allies to grasp this early. Human conflict is not going away, and neither are nuclear weapons. Misunderstanding these phenomena can potentially lead future policymakers and practitioners down paths that will be inimical for effective crises response and de-escalation.

◆ ◆ ◆

A note on definitions, as they do not necessarily conform to current doctrinal constructs. In this book, "information warfare" is defined as the ongoing calculated manipulation of an opponent's perception using a variety of means. The term "information operations" refers to targeting an opponent with a specific, discrete objective in mind related to manipulating that opponent's perception. Information warfare is itself a subset of "influence activities," which have as their objective the manipulation of an opponent's policies and actions. Cyberwarfare is one of several tools that can be employed in all three due to the broad reach of computer-facilitated communications.

CHAPTER 1

• • •

Getting Here from There

Russian Policy Objectives, Strategy, and Application

The present conflict over Ukraine is rooted in the breakup of the Soviet Union and subsequent social currents, particularly two important aspects. The first is what some refer to as the "yesterday forever" group. In effect, KGB chairman Vladimir Kryuchkov and those around him hedged their bets as the Soviet Union collapsed. This took several forms. First, overseas financial support mechanisms for KGB and GRU covert action remained in place for possible future employment. Second, narcotics networks and associated financial mechanisms controlled by KGB and GRU personnel remained in place. Third, new political parties and social movements that emerged after the collapse were infiltrated and funded. This included organizations across the political spectrum, but particularly what the West would call the extreme right and extreme left. In essence, elements of the KGB established a shadow parallel power to the Yeltsin government. When the 1991 coup collapsed and Kryuchkov was removed, the "yesterday forever" group remained in place.[1] These elements firmly believed that the USSR and the coup collapsed due to a conspiracy orchestrated by the United States, and there was a strong revanchist tone among them.[2]

The second aspect was the emergence in the late 1980s during glasnost of a "soup" of philosophies. A variety of shades of Russian nationalism—including neo-Nazism, esotericism, and even the neo-Nazi esotericism that emerged in the wake of the events of 1991—interacted with tremendous societal paranoia and

8

conspiracism that flourished at that time. These included the works of Ivan Ilyin, Alexander Prokhanov, Alexander Dugin, Lev Gumilev, and Sergei Kurginyan. An unhealthy dose of anti-Semitism which had a long basis in Russian society, was also injected into the mix through some of these authors. The Russian Orthodox Church with its nationalistic doctrines was resurgent, and Pan-Slavism and 1930s-era Eurasianism also became staple products in this morass.[3]

At the same time, a new politics of victimization built on an established cult of the Great Patriotic War was also active in this mix. This revolved around the idea that breakaway portions of the former Soviet Union were "ungrateful" and indeed antagonistic to the sacrifice of the Russian peoples against fascism, particularly with the removal of Soviet symbols (specifically statues) by newly independent states. The possibility that these regions needed to be "liberated" from a new fascism emanating from the West (and that this could be the basis of a new Russian unity as well as a recovered global legitimacy) was also operational at this time.[4]

Putin's centrality in events did not occur until 2000: he was at this point the face man to the international community representing the city-state of St. Petersburg, having initially been placed there by yesterday forever elements to monitor emergent liberals running the city. By 1993 he had formed his views on the Soviet breakup and the future direction that the country should take in its wake, and publicly expounded on them in various circles. He ranted about the alleged ill treatment of Russian minorities in Estonia during the Narva affair and refused to accept Estonian independence. Putin railed to foreign representatives that Ukrainian desires for independence constituted "a case study in ingratitude," that Russia had "lost Crimea" and the combination of alleged maltreatment of Russian minorities in the Baltic states reflected Russian national "humiliation." Finally, he asserted that "Russia should not be a buffer zone, a barrier between East and West [but] first and foremost a connecting link between the two." The role of the former Soviet and Warsaw Pact states should "serve as a go-between, linking Russia with Western Europe and Asia," but under Russian dominance. In a 1995 diatribe to a British delegation, Putin railed against the West because NATO allegedly broke a pledge to not accept members from Eastern Europe. He was unable to accept or understand that democratically elected governments had freedom of choice and association.[5]

When Putin ascended to power in 2000, he brought with him a collection of trusted personalities known as the *siloviki* ("people of force"), many former KGB personnel. Each of these men was influenced, to one extent or another, by various elements of the 1990s philosophic strains that were being imbibed by the various nationalist groups. It is important, however, that we distinguish between "instrumental" philosophies used to gain and retain power and what these men actually believed. There is substantial literature on "Russia-as-mafia-state," and the term "kleptocratic authoritarianism" has been applied to describe Putin's *sistema*. One view is that "Putin's ultimate aim is not the revival of Russia, restoration of the Russian empire, or the wellbeing of the Russian nation, but rather the preservation of the existing patrimonial regime at any cost."[6] The fundamental question as to whether authoritarian leaders actually believe the ideologies they promulgate as a means of justification and social control is a complex one. It is possible that Putin and his *siloviki* both believe in what they say and are monetarily greedy. One can also justify monetary acquisition as legitimate compensation for "saving the people."

It is possible to discern common ideas that did influence Putin and his *siloviki*. Putin's economic advisor, Sergy Glazyev, and future foreign minister Sergy Lavrov were heavily influenced by Lev Gumilev. Distilling Gumilev's work is difficult. He believes in the sociopolitical movement labeled "Eurasianism," that is, there is a definable geographic area that corresponds to where Russian is spoken, where Orthodox Christianity is practiced, and where Russia has historically dominated—and should dominate. His political philosophy is Darwinist and anti-Semitic; it asserts the individual plays no role in history and is subordinated to "national collectivity" and that there are immutable laws when it comes to gene pools, ethnicity, and superiority. Europe is older and therefore less dynamic than the "younger" Russia: "harm has always come to us from the West." Gumilev has a "fear of contact" and essentially expresses the belief that Jews are some kind of Western vanguard to destroy Russia. This is not fringe academia in the Russian context: "His ethnic theories are immensely popular; many post-Soviet scholars consider them to be scientifically proven."[7] Gumilev is also influenced by American conspiracist and anti-Semite, Lyndon Larouche.[8]

Alexander Dugin is another influence. Dugin is a philosophical omnivore and opportunist, writing in a variety of fora from youth journals to political journals throughout the 1990s to achieve social and political influence. In the 1990s Dugin was introduced to the leadership of the Russian Academy of the General Staff by hard-line nationalist and anti-Semite Alexander Prokhanov. Dugin lectured in the Department of Strategy and provided closed lectures to Russian's military leadership. This lecture series blended elements of nationalism, communism, Gumilev's writings, and the writings of various Russian fascists into what he called a "hybrid ideology" later dubbed Eurasianism. These lectures formed the basis of the book *Foundations of Geopolitics* (1997), which was then used as a textbook at the General Staff and other military academies before its introduction into the public domain in the 2000s.[9]

In essence, *Foundations* is based on a plethora of cherry-picked ideas and philosophies. These include Carl Schmitt and Karl Haushofer, who directly influenced Hitler's national socialist ideology, specifically lebensraum and "blood and soil." Dugin employs a variation of Halford Mackinder's Heartland thesis, where Russia is the geographic-economic and resource center that the world revolves around. The theories of Alfred Thayer Mahan also make an appearance. The racial and cultural superiority of *Russkiy Mir* (the Russian World) and orthodoxy are in effect the "software" to control the Eurasian zone that Russia is destined to occupy. There is even an underpinning of occultism.[10]

The "Atlanticists," seafaring societies led by the United States, constitute a significant threat to this project. This threat can be defeated by using subversion and covert action—specifically by influencing France and Germany to reject Atlanticism and ally with Russia as vassal states and by extending Russian covert methods to generate instability to Latin and South America to pin down the Atlanticists. Allying with Iran serves the same purpose to the south of Eurasia, and Japan in the Pacific. The People's Republic of China (PRC) is to be dismantled.[11]

For Dugin, the primary obstacle to this project is Ukraine, which he considers to have no intrinsic value except as a buffer between Russia and Europe; he considers such "sanitary cordons" as unworthy of independent status.[12] In basic terms, Dugin's Russia-dominated Eurasia extends from Dublin to Vladivostok.

Dugin was widely read by the bulk of the *siloviki*, including Gleb Pavlovsky, Putin's main "political technologist"; Vladimir Yakunin, "orthodox Checkist," purveyor of anti-LGBTQ conspiracism and controller of the Russian railway system; Igor Sechin, leader of the Rosneft oil concern; and Putin's right arm,

Nikolai Patrushev.[13] Patrushev, has been described as "a visionary, an ideologist for the rebuilding of the Russian empire" who wanted "the Soviet Union, only with capitalism."[14] After departing the FSB in 2008, he ran the Security Council. Patruschev has, since at least 2009, publicly and repeatedly articulated Halford Mackinder's geostrategic theories in relation to Russia, which were introduced to him through Dugin's work. Mackinder assumed that dominance of central Eurasia naturally led to global hegemony, and Patruschev sees Russia in that position.[15]

After 2004, Putin dramatically expanded state recognition and endorsement of the philosophies of expatriate Russian professor Ivan Ilyin, a figure who died in obscurity in 1954.[16] Ilyin's overall philosophy called for a monarchy in Russia and an acceptance of the social order by Russian citizens, mixing dictatorship and theocracy with Russian culture.[17]

Ilyin adopted Judeo-Bolshevik conspiracy theories that predominated in the interwar years, especially in the émigré communities.[18] His philosophy had "three core features: it celebrated will and violence over law; it proposed a leader with a mystical connection to his people; and it characterized globalization as a conspiracy rather than a set of problems."[19]

A revitalized Ilyin thus became part of a coordinated multimedia offensive unleashed in 2005 by Vladislav Surkov, Putin's public relations guru. This period is crucial to understanding the present circumstances. After ascension to power, Putin and his *siloviki* realized that the various Russian nationalisms that emerged in the 1990s could be as much a threat to the regime as the "color revolution" conspiracies allegedly cooked up by the West to repeat the events of 1991. Consequently, Surkov, supported by Gleb Pavlovsky, was placed in charge of handling "managed nationalism." In effect, the various forms of nationalist energies from across the Russian political spectrum needed to be controlled. These measures included the creation of youth groups and the suppression of Russian media oligarchs to ensure the media landscape was "coordinated."[20] Surkov, of course, is best known for changing the harsher sounding concept of "managed democracy" into the more anodyne "sovereign democracy."[21] More important, Surkov "relocated [anti-Western conspiracy theories] from the margins of Russian political discourse to its centre and made criticism of the West—when framed within the conspiratorial narrative—a legitimate part off official political and media discourse."[22] Summed up, the message was, the West is a perpetual threat and there are conspiratorial dark forces trying to destroy Russia from all quarters.[23]

All of these ideas and more came together in 2011–12 in the Izborsk Club, a think tank/forum headed by Putin interlocutor Alexander Prokhanov, who saw "Eurasianism [as] a vehicle to return to Soviet power in fascist form." Dugin and Gumilev were not only required reading, but were also members. In addition to Eurasianism, Izborsk's core beliefs have been characterized as anti-LGBTQ, in that AIDS is considered to be a weapon introduced into Russia by gay ideology to destroy it; anti–Barack Obama, because he is Black; anti-Semitic for all the usual Russian reasons; and anti-Zionist in the sense there is an active ongoing Jewish conspiracy that facilitates infiltration and penetration of Russia. The only way to combat this was to unify Russia, Belarus, and Ukraine and then take over Western Europe to "protect" Russia as a stepping-stone to Dugin-esque Eurasianism. As Timothy Snyder notes, Putin was so enthralled by its output that he even permitted a TU-95 bomber to be named after the Izborsk Club.[24]

Once reelected, Putin was able to reveal his designs for Russian ideology and its relationship to policy and strategy. The writings of Dugin, Prokhanov, and Glazyev were activity promoted by the government, and there were direct connections between Glazyev and Sergy Lavrov in the formulation of the 2013 Foreign Policy Concept. The future would be a period of atavistic competition for resources as the legal order that protected weaker states succumbed and they were absorbed by empire-like powers. This state of affairs would affect the entire planet. The European Union (EU) would be replaced with a Eurasian state, so it behooved Eastern and Western Europe to abandon the EU and get on board with this inevitable process. Not coincidentally, the Foreign Policy Concept made reference to Ukraine, which was just embarking on negotiations with the EU—better to give up on that because Russia will dominate anyway. The Foreign Policy Concept was to be backed up with an information warfare and diplomatic onslaught targeting Western European leaders and societies in order to undermine the EU.[25]

In fundamental terms, there was a Russian ideology and it had Putin's imprimatur. No matter how dressed up it was in pseudo-academic or intellectual garb, in retrospect it amounted to little more than *Mein Kampf* in Cyrillic, with the direction of effort pointing west instead of east. It told its consumers what they wanted to hear. The ideology was multidimensional in that it was taught in the school system, promoted by state-controlled media, distributed within the bureaucracy, and taught in the armed forces higher staff colleges. The ideology also progressively underpinned Russian foreign policy starting in 2005 but then "came out," as it were, in 2013. Going after Ukraine was a means to retaining power and constructing a new future for Russia.

You Say You Don't Want a Color Revolution

The expressed concern by the regime of what were termed "color revolutions" became central to the narrative of Russian aggression against Ukraine in that it built on the conspiracism established by the collection of philosophers drawn upon by the regime. The use of the term "color revolutions" in Russian discourse emerged after election disturbances in Ukraine. Russian interference with the 2004 election, including the poisoning of a presidential candidate and a rigged vote, resulted in mass demonstrations leading to a new vote that defeated the Russian-backed candidate. The colors of the winning party were orange and the event became known as the Orange Revolution. Prior to this, however, was a similar series of events in Georgia in 2003 whereby economic mismanagement and a rigged election were rebuffed with large-scale demonstrations and opposition leaders carrying bundles of roses with them as they demonstrated. This was called the Rose Revolution by the media. In 2005 Kyrgyzstan's autocratic leader, Askar Akayev, told a Russian media influencer during an interview "that he was aware of the West's plans to incite a 'tulip revolution' in his country."[26]

At the same time, Russian FSB chief Nikolai Patrushev made similar statements in the State Duma in early 2005 using the term "velvet revolutions" to describe "foreign special services using non-traditional methods, using non-governmental organizations." Patrushev claimed that millions of dollars were allocated for a "velvet revolution in Belarus" similar to the Orange Revolution in Ukraine and he was coordinating efforts with his counterparts in the Commonwealth of Independent States to stop this phenomenon.[27] Of note, the 1968 Czech uprising was known as the "Velvet Revolution" and was part of men like Patrushev's zeitgeist who continue to believe it was some form of CIA plot rather than a spontaneous internal event. By the end of 2005, Russian foreign minister Sergy Lavrov was using the term in speeches, as did foreign policy guru Alexy Arbatov in his writings.[28]

In parallel with these statements, the official Union State website published five pieces on the phenomenon throughout 2005. The first announced security forces had intercepted agitators and asserted that "orange revolutions will not happen in either Belarus or Russia."[29] The second made specific reference to color revolutions: "A series of 'color revolutions' in the post-Soviet space inevitably raises the question, 'who is next?'"[30] The third and fourth were anonymous essays on the subject which went into some detail defining the phenomenon. They examined the conditions underlying the Georgian, Ukrainian, and Kyrgyzstan revolutions and concluded that the "Western liberal model of economic thought" which is "alien to local conditions" raised expectations in the population who expressed "public dissatisfaction with the ruling groups." All three "color revolutions . . . were the result of the delegitimization of elections."[31] The underlying reasons were not legitimate dissatisfaction with politics; they were the result of American manipulation. An official Union State news website features an anonymous fifteen-page article, "Color Revolutions as a Consequence of the Network War." In essence, this essay stitches together disparate 1990s-era American academic theories on network-centric warfare, connecting them to nongovernmental organizations, soft power theory, and the writings of Russian subversion war theorist Evgeny Messner. The article posited that the solution, or part of it, is to stave off revolutions by indoctrinating youth to the point where they don't want to rebel and cut off foreign human rights groups to curtail their influence.[32] The ultimate solution, apparently, is the Union State, which amounted to Belarus allowing itself to be absorbed by Russia again.[33]

Through 2022, the term "color revolution" was used by the Putin regime as a pejorative term to describe any popular opposition that expressed itself against governments in the "post-Soviet space," driven by nefarious American intent. Color revolutions directed against Russia were a tool of subversion war, network-centric warfare, or both. No evidence was ever presented to substantiate these assertions, even within the Russian academic analysis of the subject, of which there was substantial discussion.[34]

Declaratory Defense, Foreign Policy, and Nuclear Strategy, 2000–2020

Understanding nuclear strategy entails separating declaratory policy and strategy, and academic and media discussion of it from actual strategy formulated by military organizations (along with the weapons systems assigned to carry it out). We have no means in the public domain to ascertain the specifics of contemporary Russian nuclear strategy. What we can see—that is, what the Russian leadership wants us to see—is their declaratory policy which is periodically issued in the form of National Security Concept/Strategy, Foreign Policy Concept, and Military Doctrine documents. The ensuing academic and media discussions augment the release of these documents, which we can conclude are signaling and information warfare tools of the highest order. They cannot be taken at face value.

There are three clusters of declaratory documents: one from 2000, when Putin became leader; one from 2009–10 in the wake of the invasion of Georgia; and one from 2014–16 in the wake of the invasion of Ukraine. The 2000 cluster was led by the National Security Concept, which Putin signed on 10 January. He is seen as the main driver behind the document and specifically the nuclear aspects of it. This and the Military Doctrine document were extensively modified from their 1999 drafts, apparently at Putin's instigation.[35]

In many ways, the 2000 declaratory policy documents were a specific reaction to the NATO-led humanitarian intervention in Kosovo to stop Serbian genocide against the Kosovar Albanians. This was viewed as a unilateral move by either NATO, the United States, or both and deemed by the Russian leadership and others as an infringement on their Pan-Slavic sphere of interest as well as the prerogatives of the United Nations Security Council, which they clung to as if it were their only identifiable stake in the post-Soviet world order. Militarily, the speed and effectiveness of the NATO-led air campaign alarmed Russian national security specialists. The National Security Concept essentially had a skewed view that the world was "unipolar," that is, coded language that it was run by the United States, and that reestablishing "multipolarity" was key to protecting Russian interests. Reestablishing a strong Russia was central to this, and the basis of a strong Russia was its "centuries of history and rich cultural traditions" as well as "its unique strategic location on the Eurasian continent."[36]

"Fundamental threats" include "above all, NATO's eastward expansion" and the "possible emergence of foreign military bases and major military presences in the immediate proximity of Russian borders." Another was attempts to "weaken [the Russian] position in Europe, the Middle East, Transcaucasus, Central Asia, and the Asia-Pacific region." This was not a purely military threat. The document immediately links the "increased threat . . . in the information sphere." The periphery is problematic because of "the development by a number of states of 'information warfare' concepts." When combined with "NATO's transition to the practice of using military force outside its zone of responsibility and without UN Security Council sanction," this could "destabilize the entire global strategic situation" if it were applied to the Russian sphere of influence, through "economic, demographic and cultural-religious expansion by neighbouring states into Russian territory." Thus, ideas and economic systems not controlled by Russia in peripheral areas were a threat. They could influence the Russian "constitutional system" by generating "political parties and public associations that pursue separatist and anti-constitutional goals."[37]

Ideas that were not in line with Russia's emergent ideology were also a threat to it. Those ideas were transmitted from the periphery to Russia proper. The Russian sphere of influence was a buffer zone, but it was an informational, idea-based buffer zone as much as a military one. And in this, Putin was recasting Stalin's postwar strategic outlook in modern terms. Russia was the core and there were peripheral entities that needed to be influenced by Russia to protect the core.

The 2000 National Security Concept explicitly noted that "deterrence to prevent aggression on any scale . . . nuclear or otherwise" was a "vital task." The Russians' concept of deterrence differs from Western thought. The basic word is *sderzhivanie*, more akin to compulsion than deterrence but incorporating both concepts. In the context of Russian declaratory policies, we are talking about strategic *sderzhivanie*.[38] Strategic *sderzhivanie* was not designed to "focus on the perceptions of the decision-makers of the adversary" but "is aimed at the society of the potential aggressor." It "seeks to force specific changes in behaviour."[39]

The 2000 Military Doctrine document built on the National Security Concept document in many ways, but importantly it reflected alarm over "the possibility of achieving military-political goals through indirect, non-close quarters operations," in this case couching them as "humanitarian intervention" (again referencing Kosovo) or "the utilization by entities in international relations of information and other (including nontraditional) means and technologies for aggressive (expansionist) purposes." Fundamental to the Military Doctrine document is the explicit statements that "maintenance of Russia's military potential—primarily its nuclear deterrent potential" was crucial to and underpinned all other activity.

Note again, this is the Russian definition of *sderzhivanie*, not the Western concept. The nuclear forces exist to compel and thus facilitate combinations of state power, supporting conventional forces and preventing an overthrow of the state.[40]

Notably, "the Russian Federation … maintains the status of nuclear power to deter aggression against it or its allies," of which Belarus is specified, and that "the Russian Federation proceeds in the basis of the need to have a nuclear potential capable of guaranteeing a set level of damage to any aggressor (states or collation of states) under any circumstances."[41] A senior officer clarified nuclear weapons would be used in response to weapons of mass destruction (WMD) attacks upon Russia or its allies, or conventional attacks threatening Russian national security.[42]

That is, the use of non-nuclear WMD and conventional systems that threatened "the end of Russia" could be countered using nuclear weapons. As an aside, in 1998 Russia unleashed a disinformation campaign branding HIV/AIDS as a biological weapon employed by the West.[43] Later on this was looped in with Russian anti-LGBTQ policies related to the "health" of the Russian body politic. The idea that this could be used as a pretext for nuclear weapons use emerged in the 2022 period with the disinformation campaign against "U.S. bioweapons labs" in Ukraine.

The Foreign Policy Concept, released much later in April 2000, reinforced and amplified Putin's concerns and preoccupations. The paper asserts that the Russian objective was "to form a good-neighbor belt along the perimeter of Russia's borders, to promote elimination of the existing and prevent the emergence of potential hotbeds of tension" and "to uphold in every possible way the rights and interests of Russian citizens and fellow countrymen abroad." "Globalization" was bad because it disadvantaged Russia. Importantly, and reflecting the influences on the Russian establishment, Russia's place in the world was "predetermined by the geopolitical position of Russia as one of the largest Eurasian powers." Strangely, the Concept paper noted that "the main aim of Russian foreign policy in Europe is the creation of a stable and democratic system of European security and cooperation." The Russian definition of democracy is, like deterrence, not the same. Democracy in Russia is "managed" or a "sovereign" democracy which is a democracy simulacrum. The implication in this statement is that Russia will stabilize (manipulate) Europe to conform to its interests, and Europeans will have little or nothing to say about it: "Russia retains its negative attitude towards the expansion of NATO." In regard to Russian-speaking minorities in adjacent states, the paper noted that this was of interest "including the key question of respect for the rights of the Russian-speaking population."[44] The repeated emphasis in the concept papers about Russian minority rights cannot help but echo similar statements from Berlin in the 1930s regarding the Sudetenland.

Each subsequent declaratory policy paper was modified depending on the international situation. The 2008 Foreign Policy Concept prioritized the "protection of rights and legitimate interests of Russian citizens and compatriots abroad ... to promote an objective image of the Russian Federation [and] promote and propagate, in foreign States, the Russian language and Russian people's culture," statements clearly aimed at the Baltic states and Ukraine. There was a new emphasis on combating "a selective approach to history ... as regards the World War II and post-war period." And for the first time, Russia's objectives included "firmly counter[ing] manifestations of neofascism, any form of racial discrimination, aggressive nationalism, anti-Semitism and xenophobia, attempts to rewrite history, use it for instigating confrontation and revanchism in world politics, and revise the outcome of World War Two." Clearly, the truth about the history of this period was seen as a significant threat to the Russian agenda to be countered. Importantly, the democratization project (Russian "managed" or "sovereign" democracy) that Russia was implementing in Europe was to be extended to include "the unity of the Euro-Atlantic region, from Vancouver to Vladivostok." Canada, which was "traditionally stable and almost immune to the effects of the political environment, is an important element of the North American dimension [of Russian policy]."[45] That is, Canada was to be neutralized and used as a lever against the United States as part of the effort to create this new Russian-influenced space. The Russian leadership were fully aware that Canada had the largest Ukrainian diaspora in the world.

The supporting Military Doctrine document did not come out until 2010. During its formulation in 2009, aspects of it supposedly leaked out. The nuclear aspects of the new doctrine were in a classified annex, something that obviously limited the deterrent (Western definition) utility of the document.[46] Nikolai Patrushev stated that Russia intended to use nuclear weapons even in regional and local wars, depending upon the situation and enemy intent.[47]

The commander of the Strategic Rocket Forces (RVSN—*Raketnye voyska strategicheskogo naznacheniya*), Lieutenant General Andrey Shvaychenko, enhanced Patrushev's statement: "In a conventional war, [the nuclear ICBMs] ensure that the opponent is forced to cease hostilities, on advantageous conditions for Russia, by means of single or multiple preventive strikes." This became known in Western analytical circles as "escalation to de-escalate."[48] There is no Russian military definition for this per se, but it was seen as a dramatic and significant departure from the 2000 Military Doctrine.

The 2010 Military Doctrine was a significant amplification in many ways of the 2000 doctrine and clearly responded to the post–Georgian War environment. The document distinguished between military dangers and military threats. Military dangers could become military threats, prompting the implementation of military conflict, which could take the form of armed conflict, local war, regional war, or large-scale war. As of 2009–10, the movement of NATO "military infrastructure . . . closer to the borders of the Russian Federation, including by expanding the bloc" was the primary danger. Related to this was "the build-up of troop contingents . . . on the territories of states contiguous with the Russian Federation and its allies [read: Belarus] and also in adjacent waters."

The Military Doctrine demonstrated Russian sensitivities toward precision guided systems; in this case it was not made clear whether they meant conventional, nuclear, or both in the wake of Prompt Global Strike tests in the United States. The implication was that nuclear weapons would be used in response to precision guided attacks. Importantly, the doctrine emphasized that Russia's main task was to deter and prevent military conflict. This involved "[neutralizing] possible military dangers and military threats using political, diplomatic, and non-military means," that is, information warfare and active measures methodologies were to be used on an ongoing basis to achieve this, not just in a crisis situation. This was backed up by maintaining "strategic stability and the nuclear deterrence potential at an adequate level." Thus Russia "reserves the right to utilize nuclear weapons in response to the utilization of nuclear and other types of weapons of mass destruc-tion . . . and also in an aggression against the Russian Federation involving the use of conventional weapons when the very existence of the state is under threat."[49]

The Russian Foreign Policy Concept for 2013, likewise, built on its predecessor document. The emphasis remained of enhancing Russian culture globally and crowed that the Western financial crisis now permitted "for the first time in modern history, global competition [to] take place on a civilizational level."[50] The document warned that there was "a risk of [the] destructive and unlawful use of 'soft power' and human rights concepts to exert political pressure on sovereign states, interfere in their internal affairs, destabilize their political situation, manipulate public opinion, including under the pretext of financing cultural and human rights projects abroad."[51]

This clearly reflects Putin regime's fear of "color revolutions" and the Arab Spring. The updated concept continued to note fears of historical examination of the Stalinist past, noting that a Russia policy objective was to "contribute to the depoliticization of historical discussions to ensure their exclusively academic character," a remarkable statement that appears to be laying information operations groundwork for future Russian activities. Pan-Slavism was invoked twice, supporting a new emphasis in the Balkans.[52]

The Military Doctrine released in December 2014 was mostly a rehash of the 2010 doctrine. The differences in the military risk/military threat priorities were similar except there was a greater emphasis on concern over the use of "information and communication technologies for military political purposes . . . aimed against sovereignty, political independence, territorial integrity of states, and posing [a] threat to international peace." Another new risk was the "establishment of regimes, which policies threaten [Russian] interests in the states contiguous to the Russian Federation, including by overthrowing legitimate state administrative bodies." The 2014 statement added a new risk: "subversive information activities against the population, especially young citizens of the State, aimed at undermining historical, spiritual and patriotic traditions related to the defense of the Motherland." The idea that promulgating history was categorized as a "military risk" is extraordinary and reveals the fear endemic in Putin and those surrounding him about the power of that discipline and the truth it could bring to bear on their enterprise. The 2014 Military Doctrine went on to warn its readers what Russia considered military threats. These did not significantly change except a slight modification: "demonstration of military force in the course of exercises in the territories of states contiguous with the Russian Federation and its allies" was now categorized as a military threat. Military risks and threats were to be "neutralized . . . through political, diplomatic, and other non-military means."[53]

On the nuclear front, declaratory policy remained the same, though retired general Yury Baluyevsky was trotted out to note that "the conditions of preemptive nuclear strikes were contained in classified policy documents," which led to another Western analytical frenzy over how low the threshold was for Russian nuclear weapons use under the supposed new policy.[54] One analysis was that the 2014 Doctrine statement was in part directed at NATO to deter moving resources to or holding exercises in Ukraine in the wake of the Wales summit.[55]

In December 2015, the Putin regime released its updated National Security Strategy. The bulk of the document was a list of deficiencies in Russian society that needed to be corrected in order to ensure national security. It was a long list and amounting to 75 percent of the document. The situation after the invasion of Ukraine was elaborated on for the first time, with the document asserting that Western support of an unconstitutional coup in Ukraine split Ukrainian society and triggered the conflict, creating an unstable and dangerous situation near Russia's borders.[56] The most disturbing aspect of the document, looking at it after the 2022 phase of the Russian invasion, was the insertion in 2015 of what would become one of several bogus justifications for the 2022 operation, the suggestion that the U.S. military was building biological weapons laboratories near the Russian border.[57]

Incredibly, it also decries that there is "an intensifying confrontation in the global information arena caused by some countries' aspiration to utilize the informational and communication technologies to achieve their geopolitical objectives, including by manipulating public awareness and falsifying history." To combat this, Russian national security had to "[focus] efforts on strengthening the internal unity of Russian society . . . [by] preserving and developing culture and traditional Russian spiritual values." In effect, Russian national security was defined as defending itself in a cultural war with the West.

The 2015 document is more explicit in articulating Russian objectives. The continued existence of Russian culture and its ability to be dominant in the Eurasian space is under threat from a number of azimuths, but the most important one comes from Ukraine. By implication, the pro-Western alignment of Ukraine exemplified by the Maidan Revolution allowed it to become a conduit for all the things the Putin regime saw as threats to its power base, not the least of which was the challenge to the manufactured history of the Soviet Union and Russia's attempts to develop an ideology based on some equally manufactured history of Russian nationalism. The Ukrainians do not fit into that schema and must be, as the documents assert, neutralized.

The Foreign Policy Concept of the Russian Federation in 2016 retained all of the provisos established in previous concept documents. The main changes or augmentations emphasized establishment of "equal and indivisible security . . . for the Euro-Atlantic, Eurasian and Asia-Pacific regions" to Russian advantage. It emphasized as a key objective "the strengthening and expanding integration within the Eurasian Economic Union" former Soviet-dominated states, specifically Belarus and Kazakhstan." It unleashed a broadside against NATO and the EU "with their refusal to begin implementation of political statements regarding the creation of a common European security and cooperation framework" that has "produced a serious crisis in the relations between Russia and the Western states." This was specifically the result of "geopolitical expansion pursued by NATO." The U.S.-led Western response to Russian activity over Ukraine was described as a "containment policy" placing "pressure" on Russia that "undermine[s] regional and global stability." This situation could be mitigated by "establish[ing] a common economic and humanitarian space from Atlantic to Pacific by harmonizing and aligning interests of European and Eurasian integration processes," processes presumably dominated by Russian interests. NATO expansion was the key blockage as it violated "the principle of equal and indivisible security." Unlike previous concept papers, this one placed emphasis on Arctic cooperation, especially in the Barents Sea.[58]

This was the public face of the Russian ideology presented to those who might get in the way of Russia achieving its objectives, in this case by Putin himself because he contributed to and signed off on all of these documents. Common to all of them was the idea that Russia, with its people and culture, was the dominant player in Eurasia. Western Europe should be part of this space in cooperation with Russia, but there were organizations in the way to realizing this objective, specifically NATO and the EU. Attempts to parse nuclear strategy by Western analysts missed the point. Russia consistently laid out the idea that nuclear weapons could be used under any condition they deemed to be a threat, which should be interpreted broadly. This was a signal: Russian nuclear weapons backstopped any activities undertaken by Russia in the pursuit of its objectives. Its purpose was not to deter in the classic sense, but to compel.

Russia increasingly alluded to excuses they could or would use to further their objectives. This included the "Russian minority" issue in former Soviet areas. It included countering soft power and color revolutions. It included the alleged existence of fascists and neo-Nazis in adjacent countries. It hinted that alleged biological warfare laboratories were operating in peripheral states. These were threats to the Russian people writ large. None of this can be decoupled from statements about the conditions under which nuclear weapons might be employed. Notably, Ukraine is only mentioned in documents *after* the initial 2014 invasion. After that, NATO and Ukraine are identified explicitly as impediments to achieving the larger Russian objectives that, when coupled with the ideology behind the mask, are not different from those espoused by Hitler in *Mein Kampf*. And having a neutralized Western Europe with the permanent threat of the Red Army backing up Soviet whims was exactly what Stalin wanted in the late 1940s, which led to the creation of NATO in the first place.

Into the Cyber Hybrid Gray Zone?

Russian policy concept documents repeatedly made reference to "[neutralizing] possible military dangers and military threats using political, diplomatic, and non-military means." In retrospect, this refers to what an army of analysts have interpreted as "hybrid warfare" or "gray-zone warfare" in which "the rapid spread of gray-zone warfare as a concept led to institutional change and the development of a dedicated research industry."[59] This in turn has muddied the waters in an attempt to grasp the nature of the Russian origins of the phenomenon. Let us for a moment dispense with Western interpretations and instead examine the context and basis of the Russian policy concept statements' phrasing.

In a general sense there is nothing new in the Soviet Union or the Russian Federation employing coordinated political, diplomatic, and nonmilitary means to accomplish Soviet or Russian policy objectives, just as there is nothing new generally in employing signaling using strategic nuclear forces as part of those efforts. The Comintern, active measures, centralized propaganda, and virulent speeches in the UN tied foreign aid and support to wars of national liberation; these were all earlier manifestations of nonmilitary means.[60] Similarly, there is nothing new to the idea of the Soviet Union or Russia producing energy dependency in Europe, then turning energy supplies on and off as a policy objective tool in concert with other methods.[61]

The difference is that the Russian policy elite in the 2000s consisted of men who believed some novel, nefarious means had been employed by the "Main Enemy" to collapse the Soviet Union. And they wanted to determine what those means were and use them for revenge, an effort that furthered emergent Russian ideology as a codependent effort. This conspiracism underpins the basis of the entire Russian enterprise: external dark forces did this, not the complex failure of an unsustainable economic model. The literature laying out aspects of American Cold War strategy during the Reagan era—which did include economic warfare, covert support to Solidarity in Poland, and the psychological aspects of the intermediate nuclear force deployment—likely was a ready source of fodder for this Russian outlook.[62] What came to be known as hybrid or gray-zone warfare had at its basis conspiracism and revenge that significantly predated the 2013 writings of General Valery Gerasimov, who is generally credited with creating so-called hybrid warfare.

There were several relevant key elements that intertwined after Putin attained power in 2000. First was the progressive consolidation of the Russian information space under government control. Oligarch-controlled media found itself under attack on multiple fronts until they were no longer independent entities. To firmly establish the government salient in the information space, Gleb Pavlovsky, whose Fund for Effective Politics played a key role in harnessing "political technologies" to get Putin elected, established volume-based news aggregators to compete with what was left of independent Russian media. By 2002 this "new patriotic media" was ascendent, by 2006 "online media was no longer considered a threat by the Putin regime," and in 2007 it became a crime to criticize Russian state officials. By 2009, the "Kremlin School of Bloggers" was in operation, generating a chill in the Russian blogosphere.[63] This consolidation allowed for two things: suppression of internal dissent and amplification of Russian ideology in the international information space. After the Orange Revolution, Russia Today (later rebranded RT) was set up in 2005 specifically for these purposes.[64]

These new techniques were then tested by the Putin regime externally. The first apparent use of Russian cyber operations in conjunction with other methods as a destabilization tool was in Estonia in 2007. One report concluded that the cyberattack used in conjunction with diplomatic and economic measures "demonstrated the ability of the Russian Federation to inflict expensive, disruptive costs onto the state of Estonia without needing to draw on conventional and escalatory forms of military and political force. It sent a message that even a few hackers could cause asymmetric damage and weaken the Estonian population's confidence in their military and security structures."[65] During the invasion of Georgia in 2008, the RT news apparatus was heavily engaged as part of the Russian war effort, while Russian hackers interfered with Georgian government computer operations in what looked like a supporting psychological operations campaign but was in fact what is today called "strategic communications."[66] There were other Russian experiments that involved social media interference during unrest in Brazil, Venezuela, and Spain after 2009.[67]

Control of media and the 2005 push by Putin for a national ideology was to act as a prophylactic against an Orange Revolution in Russia. It was the phenomenon of the Arab Spring throughout 2011, however, that revitalized Russian concern over so-called color revolutions. For example, Belarus's Lukashenko expressed concerns in September 2011 at a Collective Security Treaty Organization (CSTO) meeting that a "constitutional coup" could pose a threat to his regime in Minsk. The existing Regional Group of Forces (a joint Russian-Belarusian command) had conducted exercises against "bandit formations" ("and what kind of bandit formations in the east of Europe is not easy to guess"). A compromised Belarus was seen as a direct threat to Moscow, given its geographic proximity.[68]

After discussion came a controversial public briefing on 18 November 2011 by General Nikolai Makarov, chief of the General Staff, to the Public Chamber. In effect, the West was using "the technologies of color revolutions to achieve their strategic interests. . . . These technologies can be projected onto Russia and its allies." Makarov was quoted as saying, "Under certain conditions, I do not rule out that local and regional armed conflicts can escalate into a large-scale war, including the use of nuclear weapons." To what extent Makarov was suggesting that nuclear weapons might be a response to those supporting a "color revolution" in Russia or among its allies was deliberately ambiguous, but he did allude to the Military Doctrine section dealing with nuclear weapons.[69] Makarov's comments were amplified throughout the Russian media ecosystem, which strongly suggests this was signaling.[70]

Makarov had in fact been lambasting the Russian military establishment throughout 2011 for not learning lessons from Desert Storm and Iraqi Freedom, and defense scientists for "instead of developing a theory of network-centric wars, they still continue to defend dissertations on the significance of the partisan movement during the war years."[71] Makarov handed off to Valery Gerasimov in November 2012 after stimulating a vast outpouring of Russian military analytical literature on countering the color revolution threat.[72] This was fashionable after the Putin crew amped up the fear factor of a color revolution during the 2011 election campaign.

In effect, what Western analysis called hybrid warfare or gray-zone warfare was, at one level, the application of what the Russians *thought* "Western color revolution technology" was to support their policy objectives. The counter to this emerged as a series of tools Russia created in 2012–13. These included the establishment of "Russian clubs" in Russian-speaking communities in eastern Ukraine, specifically ones with a paramilitary orientation like paintball and airsoft clubs, Russian language clubs, and those that pumped up the prestige and social image of the Russian Orthodox Church. This was supported with political agitation, social media, and other information operations. It was eventually labeled by those in the know as "the Russian Spring."[73] Russia Today was completely refocused away from domestic Russian issues to emphasize "Western social problems and the flaws of democracy."[74] The year 2012 was also identified as the period when Facebook "turned itself into the world's most powerful propaganda machine."[75] The CIA also identified social media as a threat, specifically the Russian-controlled platform Telegram.[76]

It was also at this time that the Slavonic Corps, a Russian private military corporation that later became the Wagner Group, was established. It was first deployed to Syria to assist the Assad regime and was probably a covert and deniable mechanism of the GRU.[77] In parallel to this was the creation of the Internet Research Agency (IRA), also known as "the trolls from Olgino," in St. Petersburg. The IRA was also associated with the GRU.[78]

On the military front, Gerasimov penned "The Value of Science in Foresight," widely assessed as the blueprint for what Western analysts would call hybrid warfare. In it, Gerasimov identifies the problem of "the so-called color revolutions" in the wake of the Arab Spring. He noted that "the rules of war have changed significantly. The role of non-military methods in achieving political and strategic goals has increased, which in a number of cases have significantly surpassed the force of arms in effectiveness." He specifically noted that the "methods of confrontation" have shifted to the use of "political, economic, informational, humanitarian, and other non-military measures implemented with the use of protest potential of the population." He felt that peacekeeping and humanitarian operations used by the West were merely cover for furthering objectives. He specifically used the Libya operations as an example. Russia needed to "create an integral theory of such actions" which among other things should be used "to reduce the combat potential of the enemy." Gerasimov suggested that there is no more declared war.[79]

Those who call this the "Gerasimov Doctrine," based on his 2013 article, completely miss the complex background that influenced Gerasimov. In the early 2000s, Alexander Dugin and Igor Panarin at the Military Academy of Science of the Russian Federation had independent, though similar, ideas when it came to information warfare. Dugin saw net-centric warfare (NCW) as an extension of his Eurasian ideas and made connections between American conceptualizations of NCW and the collapse of the Soviet Union. As Dugin noted in 2007, Putin had removed American-dominated networks from Russian space. Panarin, who embraced what he called "information war," saw control of the target's public opinion as the prime objective of any conflict. This was to be done using tools including information manipulation, disinformation, fabrication of information, lobbying, blackmail, and denial of information. Both Dugin's and Panarin's theories are, unsurprisingly, "based on a narrative of western aggression against Russia."[80]

Sitting on the periphery of this morass were two retired Russian colonels, Sergey Chekino and Sergy Bogdanov, both of whom served as faculty at the Centre for Military-Strategic Research of the General Staff of the Russian Federation. Their ideas (which traced their pedigree to the 1990s and before) included "new generation warfare," which focused on nonmilitary methods, some of which were based on *gibridnaya voyna*. In essence, *gibridnaya voyna* was "a mix of political, diplomatic, economic, information and other non-military means intended to subvert and undermine an adversary." The concept of new-generation warfare was "a full-scale military operation preceded and accompanied by different non-military actions intended to weaken an adversary's military power and political resilience."[81]

Gerasimov's 2013 article was not published in an official publication, which suggests it was a trial balloon. He published a follow-up piece in 2016, after the Western analytical community drove itself crazy trying to understand Russian hybrid warfare in the wake of the 2014 invasion of Ukraine. Gerasimov uses the term *gibridnaya metody*, or hybrid methods, in discussing nonmilitary means contributing to supporting military power.[82] Importantly, both Putin and Lavrov frequently speak using language that is clearly derived from the ideas of Dugin, Panarin, Chekino, Bogdanov, and Gerasimov.[83]

Through the Russian strategic and policy lens, Maidan was a color revolution taking place in the Russian sphere of influence and Russian operations constituted a preemptive move to attenuate what they asserted was a Western-controlled incursion. This allowed them to insulate Russia from a domestic color revolution, which in real terms permitted them to accomplish the objectives of their ideology as they were informed by Eurasianism. That meant taking down Ukraine and firmly incorporating it into the Russian sphere of influence.

The 2014 Invasion of Ukraine

As part of his Eurasianist policy, Putin created the Eurasian Economic Community in 2000, which Ukraine refused to join. Putin then formulated the Single Economic Space, another Russian mechanism for economic dominance, which initially included Russia, Belarus, and Kazakhstan. Ukraine joined after being strong-armed using the "gas weapon" in 2003.[84] The military counterpart to these moves apparently included very early contingency planning that took place in 2003 but was formalized in 2004 after the Orange Revolution.[85] At some point the Belarusian leader claims to have seen contingency plans for the seizure of Crimea.[86] In testimony to the Senate Foreign Relations Committee, Stephen Blank noted that the Ukraine operation "had been planned by Moscow since 2005."[87] Mikhail Koval, who served in the Ukrainian security apparatus, believed that operations were planned at least by 2006 by the General Staff of the Armed Forces of the Russian Federation.[88]

Starting in 2005 Russian popular culture was mobilized to introduce the idea of a war to "liberate" Ukraine. This included the production of mass market speculative fiction released on the internet, followed by mass market publishing. These were written in some cases by Russian ex-military personnel living in eastern Ukraine and included scenarios that involved "'Orange' Nazis" who provoke a civil war in Ukraine and unleash genocide against the Russian-speaking population, "wiping entire cities off the face of the earth"—aided by NATO "peacekeeping" troops and American airpower. This was decried by Kharkiv governor Arsen Avakov, who believed that the books were part of a deliberate Kremlin strategy to build up popular support for war against Ukraine by playing to Soviet nostalgia among older readers and ignorance among younger ones.[89]

The next attempt by the Putin regime to economically dominate Ukraine included plans to create a Customs Union announced in 2009 to include Russia, Belarus, Kazakhstan, and Ukraine. Intense diplomatic and economic pressure was employed in this move.[90] In February 2010 pro-Russian Viktor Yanukovych was elected president in Ukraine, with a voting base dominated by eastern Ukrainian jurisdictions. Over the next two years, Yanukovych attempted to appease pro-EU elements and pro-Russian elements. By the summer of 2013, however, Putin had had enough and generated direct economic pressure including use of a variety of protectionist measures and blockades. Yanukovych backed away from dealings with the EU.[91]

This pressure was supported with military means. Zapad 2013 and its "parallel exercises" involved 80,000 troops, most of them deployed to Belarus. Ukrainian intelligence was alarmed by the dramatic increase in Russian overflights and aerial harassment that was estimated to be one hundred times the norm.[92] There are allegations that Yanukovych learned about Russian invasion plans and this may have played a role in his 21 November shift away from the EU and toward the Customs Union.[93] At this point, the Maidan demonstrations started, as did a review of Russian contingency plans for the seizure of Crimea.[94] By January "Moscow had already lost hope" in using Yanukovych to control Ukraine.[95]

The Revolution of Dignity (or the Maidan revolution) was decried by Russian propagandists and their enablers as a "fascist coup d'état," in keeping with the Kremlin's perception of so-called color revolutions. The reality was far more complex and the employment of historical grievances and associated paraphernalia generated a situation which was readily exploited by the Russian machine. No matter what spin was put on it, the majority of Ukrainians did not want to be a Russian vassal state. The deliberate introduction of inflammatory discourse on genocide by Russian outlets and comparisons to ethnic cleansing in the Balkans did not change this state of affairs. Yanukovych was deposed and a provisional government formed. He was later "rescued" by Russian special forces.

Four days after Yanukovych departed, on 22 February 2014, Putin ordered a "snap inspection" that involved 150,000 troops in both the western and central military districts. This was believed by some to be a diversion to distract intelligence resources from observing the movements of Russian units tasked with the Crimea operation.[96]

Consider the Crimean operation and the larger Ukraine operation: the literature on the subject tends to focus on Crimea and the "Little Green Men/Polite People" phenomenon in the context of what analysts call "hybrid warfare," while the uprisings in eastern Ukraine are treated as follow-on effects of the Crimea seizure. It is more likely that there was a Crimean operation nested within a larger plan or that there were two plans mounted simultaneously. One study notes that in late February the GRU "dispatched several hundred members of the 45th Spetsnaz Regiment to Crimea to create a 'popular uprising' aimed at facilitating Russia's annexation of the region. Simultaneously, GRU agents used bribery among the ethnic Russian population to win support for annexation."[97] At the same time, however, Russian elements were dispatched to Odesa and other Ukrainian cities to create organizations and generate agitation for exploitation in the uncertain political environment. For example, the organization in Odesa was led by Russian neo-Nazis from the Slavonic Unity organization. Their leaders' nom de guerre was "Maidanek."[98]

It is clear that the Crimean operation was important, given the need to secure the Russian Black Sea Fleet and its logistics apparatus. Not coincidentally, the geographic position of Crimea gave Russia a central base of operations against the rest of Ukraine. The rapid annexation of Crimea using a referendum enforced by the Night Wolves Motorcycle Club and special forces personnel dressed as police occurred on 16 March, and Putin signed it into law on 19 March.[99]

Concurrent with this was the seizure of seats of government in Donetsk, Luhansk, Slovyansk, Kharkiv, Kramatorsk, Horlivka, and Mariupol by pro-Russian elements.[100] Notably, "the main core of the protesters was coordinated by Russia's secret services."[101] Ukrainian security forces temporarily regained control, but then a more hard-line leadership emerged in the anti-Kyiv forces that had clear links to Moscow. For example, Putin's economic advisor, Sergy Glazyev, was apparently in communication with the Odesa component of the uprising for motivational and monetary purposes, and there were implications that Crimea could be repeated in Odesa with "little green men."[102] Not coincidentally, the Ukrainian government internet system was taken down at the start of the first day of the Crimean operation and the communication systems of the members of parliament were blocked.[103]

Throughout April pro-Russian hard-liners retook seats of government in Donetsk and Luhansk. At this time, local pro-Kyiv elements coalesced and successfully confronted pro-Russian elements in Odesa, Dnipro, and Kharkiv and blocked any further attempts to seize power. At the same time, Russia conducted another "snap inspection" on the Ukraine border, presumably to pin down Ukrainian security forces while the pro-Russian elements in the Donbas consolidated. Man-portable surface-to-air missiles appeared in the hands of the pro-Russian elements inflicting serious damage to the Ukrainian air force during the "anti-terrorist operation," as the operations to retake the Donbas region were called. Russian "volunteers" started to pour in by late May.[104]

During this period, the Putin regime exercised its strategic nuclear forces on several occasions. There was a large four-day RVSN exercise, with the usual public denials that it was related to the Ukrainian situation. There was the deployment of Tu-95MS Bears to the North Sea. And then there was another major strategic nuclear forces exercise on 8–9 May involving movements and launches from all three legs of the nuclear triad. Throughout this period there were stepped-up Long Range Aviation (LRA) operations off California, Alaska, and Guam, as well as "snap inspections" of mobile ICBM units. This culminated with Sergy Lavrov's veiled nuclear threat of 11 July.

The conflict continued in 2014, and eventually regular Russian ground forces were employed in eastern Ukraine. In retrospect, Russian planning in early 2014 envisioned the seizure of something much more than Crimea and the Donbas, using nonmilitary tools supported by information operations and cyberwarfare. Nuclear signaling was employed to compel the Western Europeans and the United States not to intervene. When it became apparent that the non-Crimean aspects of the plan were not going to work, however, Russia transitioned the situation into a "frozen conflict," neither peace nor war. But a country at war cannot join NATO, so Ukraine was neutralized, at least for the time being. Planning for the next lunge to seize Ukraine started in 2014 as the situation solidified, and Putin declared the "liberated" parts of eastern Ukraine to be "Novorossiya."[105] The supposed Ukrainian color revolution was thwarted even though the "Russian Spring" did not fully materialize.

Aftermath

The methodology that the Putin regime employed during the Ukrainian invasion of 2014 kept the Western powers off-balance and generated tremendous uncertainty in formulating responses to the situation. The Russian activities operated on the margins of accepted ideas as to what "war" was and was not, which inhibited assessments and responses. The question for the Putin regime to address was, how do we forestall a "color revolution" in Ukraine that is inimical to Russian interests without provoking Western intervention? Answer: Generate a "color revolution" that was backed by real force, with nuclear weapons looming in the background. Ukraine had already given up its nuclear weapons back in the 1990s and there was no NATO nuclear umbrella, so this state of affairs facilitated Russian activities. There is every reason to believe there was a calculated synergy between Russian conventional forces buildup, covert activities, information warfare, and the manipulation of strategic nuclear forces in the pursuit of their objectives. Clearly not everything was well coordinated, and not everything went according to plan. If it had, Odesa, Dnipro, Mariupol, and Kharkiv would have succumbed like Donetsk, Luhansk, and Crimea did, which would have left the rest of Ukraine bereft of economic tools and maritime access.

CHAPTER 2

• • •

Red Forces

Russian Nuclear Signaling in Context Before 2021

There was day-to-day general deterrence signaling with nuclear forces, and then there was signaling using nuclear forces during crises as they developed during the course of the Cold War. These two states of affairs were set against a backdrop of varying levels of readiness that the military forces of both sides maintained during the 1945 to 1991 period. For the most part, this system waned and faded into the background after 1991 as concerns over ethnic conflicts and failed states in a de-communized world took center stage. Yes, military forces exercised, trained, and maintained but nowhere near the levels attained during the Cold War and they certainly never achieved anywhere near the same levels of readiness.

This changed in 2001, when the United States and its allies removed the Taliban regime in Afghanistan and, later, the Hussein regime in Iraq. These wars provided focus for the United States, its military systems, and its allies which put paid to the remnants of the Cold War system. Not so in Russia. Putin progressively set the stage for a dramatic shift in Russia's relationship with the international community, starting with his State of the Union Address in April 2005 and culminating with his aggressive speech at Munich in February 2007. The Munich speech "for many in the audience . . . sounded remarkably like the declaration of a new Cold War."[1] And in retrospect that is exactly what it was. The policy was, Russia is back. And that policy expressed itself across the board—from invading Georgia in 2008 and Ukraine in 2014, increasing large-scale exercises and intimidating Eastern Europe, improving strategic nuclear forces, and, yes, to signaling using those nuclear forces. The Cool War started here. And what happened back then contextualizes our present circumstances in the Russo-Ukrainian war and particularly the role of strategic nuclear forces in this crisis.

Elephants in the Room: Large-Scale Exercises Before 2014

When it comes to signaling, large-scale exercises overlap with general deterrence signaling, declaratory policy/strategy, and combat potential. The evolution of Russian large-scale exercises after 2008 underpins both general deterrence signaling and nuclear force specific exercises. The Russian exercise cycle before 2008 consisted of four geographically based exercises: Zapad (West), Kavkaz (Caucasus), Tsentr (Center), and Vostok (East). After the mediocre performance of Russian forces in the 2008 Georgian invasion, the tenor and size of these exercises changed in response to reforms initiated by Defense Minister Anatoly Serdyukov and his successor, Sergey Shoigu.[2]

Consequently, the 2009–10 exercise season was a significant departure from those of the 1990s and early 2000s. Zapad, held in concert with Belarusian forces, was ostensibly a Kosovo-like interventionist scenario that included the use of Tu-95MS, Tu-160, and Tu-22M strategic bombers launching presumably conventional standoff weapons in support of a conventional frontal assault force, followed by FSB and interior ministry troops to suppress "bandits." Analysis suggested that it was set up to provoke confrontation with former Warsaw Pact countries that had not yet fully "NATO-ized" their forces. Exercise Ladoga 2009 was run concurrently with Zapad 2009 in the St. Petersburg area, probably to get around Vienna Document agreements on limiting exercise size. Its focus was on the use of military forces, FSB, interior ministry troops, and VDV (airborne) forces in countering so-called bandits to protect the logistics rear area in the region. The possibility that Zapad exercised the seizure and occupation of some or all of the Baltic states, or that it was designed to intimidate them, exists.[3] It is not a stretch to see early preparations for the Ukraine operation in 2014 in Zapad/Ladoga 2009.

Kavkaz 2009 in the Caucasus consisted of a counterinsurgency operation backed up with conventional forces. The real motive behind the exercise was to deter Georgia from conducting any activities on the spectrum of conflict inside

Russian-occupied Ossetia and Abkhazia. As for Vostok 2010, the exercise play had Russian forces delay the functional equivalent of a People's Liberation Army assault in the Russian Far East, in which Russian forces delay the enemy, mobilize forces from western Russia, and use technologically superior forces to defeat the enemy. Vostok 2010 was the only exercise of the four in which nuclear weapons were employed, in this case, nuclear land mines.[4]

After 2010, the same four exercises were run annually: Tsentr 2011, Kavkaz 2012, Zapad 2013, and Vostok 2014. However, during "parallel exercises" that started in 2011 and 2013, the new chief of the General Staff, Valery Gerasimov, brought in "combat readiness inspection exercises" also known as "snap inspections."[5] These were mechanisms employed to conceal the nature and extent of Russian military training and capabilities while at the same time to deceive any observers as to the intent of the training in relation to national strategy and policy. In effect, the unnamed "parallel exercises" grew to be the same size as the regular named exercise; they just took place in a different geographic area.

Tsentr 2011 was an example of this. Held in central Russia in September, the exercise ostensibly had 12,000 military personnel to keep it under Vienna Document numbers. These numbers did not include the thousands of CSTO, FSB, Emergency Ministry, Federal Drug Control, and Federal Penitentiary Service personnel that also participated. The parallel exercise, Union Shield, held from Nizhny Novgorod to the Caspian Sea, with 7,000 Russian and 5,000 Belarusian troops. The possibility that Union Shield represented the assault echelon and that Tsentr was the second echelon and occupation force was noted by observers at the time.[6]

Kavkaz 2012 exercised the Southern Military District as well as Black Sea and Caspian Sea forces. It had 8,000 personnel and focused on complex command and control aspects related to operations in the Caucasus. The parallel exercise, simply referred to as the "joint inter-service staff exercise," was mounted on the Kola Peninsula after Kavkaz 2012 was completed. Analysis of this exercise suggests that the Russian navy had renewed its focus on nuclear submarines.[7]

When put together, Kavkaz 2012 and the Kola exercise suggest a larger scenario whereby Russia first protects its seagoing deterrent forces, as a signal, then launches an intervention, or launches the intervention and then signals with its strategic nuclear forces to those outsiders who might intervene to confront

Russian forces in the crisis area. In other words, this was essentially the nucleariza-
tion of the Kavkaz 2009 exercise.

Zapad 2013 was conducted in Belarus and Kaliningrad, with 11,920 personnel. It
also involved as many as 20,000 other personnel from the FSB, interior ministry,
and so on. A further 25,000 troops were mobilized elsewhere in Russia at the
same time, some of which were civil military cooperation units. The Russian
rail network and commercial air in support of operations was also manipulated
as part of Zapad 2013. To outside observers, this looked like an occupation force
being exercised.[8]

There were two parallel exercises to Zapad 2013. The first involved CSTO's
Collective Operational Reaction Forces (CORF), based around VDV divisions.
The second was an unnamed exercise of the Northern Fleet's ground defense
formations. This exercise included nuclear submarines in a scenario testing
escalation with nuclear weapons.[9]

The implementation of "surprise combat readiness inspection" exercises under
the Gerasimov tenure started in 2013. There were many different types and scales
of forces for these exercises (there were twelve held in 2013 alone), but the most
important was held on 22 July 2013. This involved the RVSN's 31st Missile Army
in Orenberg, which included the 42nd Missile Division that consisted of silo-
launched ICBMs and mobile ICBMs. It was held immediately after two snap
inspections in the Eastern Military District and Central Military District and was
judged to be an exercise "in escalation from conventional conflict to nuclear."[10] In
October another snap inspection exercise included "air, sea, and land-based nuclear
weapons forces as well as the Aerospace Defense Forces." Putin commanded this
exercise.[11] When compared with NATO exercises held from the 2008 to 2013
period, none approached the scale or complexity of any of the Russian exercises.[12]

Russian Signaling: Understanding Combat Potential

"Combat potential" has a very specific meaning in the Russian context. It is
"the totality of available means, as well as the material and spiritual capabilities
of the armed forces . . . to carry out combat missions facing them." However,
"the material basis of combat potential . . . is the presence of a rational weapons
system" and its supporting personnel and infrastructure. Combat potential "can
be used as a force to deter" (that is, the Russian definition of deterrence, which

is "compellence" to us) as well as a means of carrying out combat missions. The "main component of [Russian] combat potential" is the nuclear forces.[13] Combat potential is a mix of what we would call a "strategic force in being" and something that can be employed or maneuvered to support policy on an ongoing basis. On the flip side, it is also something that Russia's opponents possess and must be taken into account.

During the Cold War, the ability of a bomber aircraft to reach and bomb its target was the measure of credibility. Those aircraft carried one or two megaton-yield gravity bombs; later models carried one or two standoff missiles equipped with nuclear warheads. Combat potential was measured in sheer numbers: seventy-two B-47s in the Arctic, each possibly carrying two nuclear bombs, was a significant signal when observed by radar and signals intelligence and transmitted back to Soviet leadership. By the 2000s, the measurement of combat potential changed significantly. Bombers were now like cruise missile Pez dispensers. Those cruise missiles had significant range, so the bomber did not have to fly anywhere near its target.[14] One aircraft could do the job of ten or twenty Cold War–era bombers, if it was positioned carefully for optimal dispersal of missiles. The cruise missile itself, essentially an unmanned bomber, could use a variety of tactics to reach its target on its own. Placing potential targets at risk, therefore, has become a new measurement for signaling.

Some cruise missiles are carried internally, while some can be carried externally. And given the fact that cruise missiles are dual-purpose systems—that is, some can carry conventional high-explosive warheads as well as nuclear warheads—there is significant ambiguity when using these systems to signal. Is the plane loaded with missiles? Are they conventional or nuclear missiles? Flying in international airspace with an ambiguous load is potentially destabilizing. A ballistic missile, by definition, leaves a ballistic track that can be seen by satellites and radar and thus provides early warning. What if the opening move in a war is to spit out twelve cruise missiles less than fifty miles from their targets, and those targets consist of the decision-making bodies for a country, or the ballistic missile radar systems? Warning and reaction time is tremendously reduced. Surprise is achieved. Attempts to downplay aerial activity by Russia (LRA) by asserting that such activities are not threats to sovereignty because they do not enter sovereign airspace or are merely proficiency flights are absurd when nuclear-armed cruise missiles can range out to 4,000 km and there is no way of knowing for sure what is aboard the aircraft.[15] Every aircraft is a potential threat and must be treated as such.

Less overt in the signaling game are the conduct of major exercises involving strategic nuclear forces with a focus on command and control. It is one thing to see bombers moving about, but when the activity is confined to underground command posts, command post aircraft, and the means of communication between those assets and the deployed forces, it is difficult to see in the public domain. The purpose of such exercises is to establish and improve readiness and to demonstrate that a given country can efficiently conduct a sequence as follows: warning is received, a decision is made to use nuclear forces, the transmission of that decision is made, it is authenticated by the deployed forces, and the launch is executed. Conveying that this can be done to an adversary is dependent on knowing an adversary can listen in and observe the exercise using intelligence methods. Announcing in mainstream media and social media that such an exercise is ongoing or has just taken place can augment the nonpublic messaging or at the very least confirm for an adversary that what they were listening to was in fact an exercise.

Signaling using silo-based and submarine-based missiles is much more difficult than it is, say, with bomber aircraft. The silo-based missile force's inert, protected status mitigates against its use as a signaling mechanism, though creative means have emerged to change this. Similarly, ballistic missile submarines have as their primary goal to remain hidden in the ocean depths until called up to be used. Of course, footage of ICBM and SSBN activities disseminated through media contributed to general deterrence during the Cold War, but for the most part their operational activities remained concealed. This changed dramatically in 2022 during the Cool War.

The Invasion of Ukraine and Russian Signaling Activities: 2014

Decisions in late 2013 made by Ukrainian president Viktor Yanukovych to move away from Ukrainian aspirations of joining the EU and strengthen ties with Russia led to massive popular unrest in January 2014. As violence increased into February, Russian officialdom and media outlets made unsubstantiated accusations that the United States was staging a coup d'état. On 22 February, Yanukovych was impeached by Parliament and fled Ukraine. Five days later, organized armed groups seized government buildings and airports in Crimea, unfurling Russian flags as they did so. Immediately after, the Putin regime received "permission" from its legislature to use military force if necessary in Crimea. Both Ukrainian and Russian forces were alerted. On 22–23 February, Russian special operations forces supported with airborne units and amphibious units landed, and by 9 March they had secured Crimea. A joint statement by Western leaders condemned these moves.[16]

Russian activities were backstopped by three snap inspection exercises. The first in January consisted of a snap inspection that exercised the Russian strategic nuclear forces command and control apparatus that involved "evaluating the fallout from the use of weapons of mass destruction."[17] The second, held immediately after, exercised the 27th Missile Army, consisting of two missile divisions, the 7th and the 54th.[18] The third, held in February, involved the Western Military District bordering Ukraine, the Baltic Fleet, air defense forces, and LRA. This snap inspection exercise was retroactively seen as "a diversionary manoeuvre against Ukraine" and was mounted "just days before Russian elite forces and Special Forces invaded Crimea."[19]

In what would come to be seen as Russian modus operandi later in 2014 and in 2022, its occupation forces supervised a sham "referendum" by the population to legitimize the intervention and military occupation on 16 March. This was backed up with public statements by Rossiya 1 television media personality Dimitry Kiselyov, a close ally of Putin. Kiselyov asserted that "Russia is the only country capable of turning the United States into radioactive ashes."[20] As we would see later in 2022, statements from Kiselyov should be taken as direct, but deniable, threats from the Putin regime leadership. Indeed, it was a signal for the United States as well as other NATO powers to stay out of events in Crimea.

In late March, a combination of staged demonstrations of Russian surrogate forces supported by sophisticated information operations and active measures attempted to seize control of Slovyansk, Kramatorsk, Kharkhiv, Donetsk, and Luhansk. Ukraine responded by establishing the Anti-Terrorism Operation (ATO) in April to take back control of these areas. Thrown back initially by "separatist" forces, a more deliberate operation was mounted in May. After Putin gained "permission" from the legislature in June, "volunteers" and weapons, including theater-level air defense equipment, materialized from Russia to support the irregular separatist units. An attempt to generate a similar "separatist" environment in Odesa led to violence as ATO forces staved off this move. The ATO forces pushed back the "separatist" units in August, which produced the introduction of regular Russian units to combat Ukrainian forces.[21]

Russian strategic missile forces conducted a four-day exercise starting 29 March. The reported scope of the activity was massive in that it involved two of the three RVSN missile armies and seven out of twelve missile divisions. It is unclear to what extent mobile ICBM units deployed from their bases, but coordination and communications procedures were tested at all levels. The existence of the exercise was reported in Russian media and then picked up and retransmitted by Western mainstream media. Russian outlets repeatedly insisted that the exercise had "nothing to do with events in Ukraine." Western media, however, interpreted the exercise as a signal of Russian "resolve" in the Ukraine situation. The timing of the exercise in relation to Kiselyov's statement and Russian activities was, of course, too coincidental. Of note is the specific use of Russian media as a conduit for signaling on nuclear matters.[22] In early April, there was the deployment to the North Sea via the Norwegian Sea of a pair of Tu-95MS bombers. These were tracked by RAF Typhoons and intercepted by Dutch F-16s as they entered Dutch airspace. Capable of carrying at least twenty-four cruise missiles, their position in the North Sea posed a threat to all NATO countries in range—in effect, all NATO countries surrounding the North Sea. The message was clear: NATO, do not get involved in Ukraine.[23] Throughout April, Russia emphasized its position and stepped up aerial activity in both the Baltic and Black Sea areas. This included the buzzing of the guided missile destroyer USS *Cook* by nuclear-capable Su-24 strike aircraft that made multiple runs on the ship.[24] Russian aircraft conducted reconnaissance of Swedish installations and NATO aircraft shadowed multiple Russian aircraft in the Baltic region.[25]

Russia held another major strategic nuclear exercise on 8–9 May, this one led by Putin. Russian media noted that "the exercises simulated dealing a massive retaliatory strike in response to an enemy attack."[26] Russian propaganda outlet RT asserted that "the planned drills come ahead of the May 9 celebrations dedicated to victory in World War II." The exercise included a live ballistic missile interception, cruise missile launches from Tu-95MS aircraft, and importantly, the launch of SLBMs from the Delta IV–class *Tula* assigned to the Northern Fleet, and the Delta III–class *Podolsk* from the Pacific Fleet. The launch of a mobile Topol-M ICBM from Plesetsk was also part of this exercise. One assessment concluded that the exercise was "structured around a 'launch on warning' scenario."[27]

It is difficult to determine whether some Russian LRA activity overlapped with or was independent of the exercise. In May, several Russian bombers launched from their Far East bases, flew to Guam which hosts American air and naval forces, circumnavigated the island twice, and returned to base. On 4 June, four Tu-95MS bombers flew to within fifty miles of California, placing a possible forty-eight targets within cruise missile range.[28] Concurrent with the LRA activity, a snap inspection exercise was held employing the 29th Missile Division equipped with mobile ICBMs.[29] The 29th Missile Division is geographically positioned to fire against targets in North America, probably the ICBM fields in Montana and Wyoming, and not targets in Western Europe.

On 11 July, Russian foreign minister Sergy Lavrov made a veiled nuclear threat: "[Russia] has the doctrine of national security, and it very clearly regulates the actions that will be taken in this case," with reference to Russian officials suggesting that "Russia reserves the option of nuclear usage to retain new territory."[30] To repeatedly underscore this statement, Tu-95MS aircraft simulated strikes with cruise missiles in the Black Sea in July, presumably to deter outside naval interference in that region or to intimidate Turkey.[31]

From here, Russian signaling activity broadened out regionally. It is clear that there was a campaign conducted from June to November designed to intimidate the Baltic countries. Russian aircraft conducted reconnaissance of Swedish military installations in April, intercepted Swedish surveillance aircraft in July, and entered Swedish airspace with nuclear-capable Su-24 Fencer bombers in September. Russian aircraft moved from shadowing Swedish surveillance aircraft to outright harassment with aircraft armed for combat operations. Finally, in October Sweden was forced to mount its largest antisubmarine warfare operation since the Cold War to track suspected Russian submarine activity in Swedish waters. More harassment of Swedish surveillance aircraft occurred concurrently. On one occasion an American surveillance aircraft was chased into Swedish airspace by Russian fighters.[32]

Other Baltic states were subjected to Russian pressure. In June, Russian aircraft made repeated runs at the Danish island of Bornholm, and Latvian airspace was probed. Finnish research vessels were harassed by Russian air and naval forces. In September a Lithuanian ship was boarded by Russian forces and taken to Murmansk. Right after an American presidential visit to the Baltics, Russian operatives kidnapped an Estonian intelligence officer from Estonian territory in an operation that involved jamming regional communications, while Russian warships tacked back and forth off Latvia.[33]

In the United States, the North American Aerospace Defense Command (NORAD) forces were confronted with sixteen bomber flights in the NORAD air defense identification zone over a ten-day period in August.[34] This appears to have been a prelude for an early September operation whereby Tu-95MS Bear bombers were tracked on a route from Iceland to Greenland and then Labrador. Apparently "analysis of the flight indicated the aircraft were conducting practice runs to a predetermined 'launch box'—an optimum point for firing nuclear-armed cruise missiles at U.S. targets." A Russian Ministry of Defence (MOD) spokesman publicly noted that Russian nuclear doctrine was under revision: "It is necessary to hash out the conditions under which Russia could carry out a pre-emptive strike with the Russian Strategic Rocket Forces."[35] This suggests that the exercise was related to sorting out timings between the arrival of Russian ICBMs and then penetration of the air defense system by bomber-launched cruise missiles. That operation was conducted the same day as the annual NATO summit, which was underway in Wales.

The August operations occurred while the Russians surged in troops to prop up the "separatists" in Ukraine. The LRA flights, both in August and September, were clearly designed to intimidate the United States, and Putin used circumlocutory language to express this on 29 August to an audience at a youth camp in Russia.[36]

LRA operations against NORAD did not let up. This was followed by a complex operation mounted against the Alaskan air defense identification zone fifty-five miles from the Alaskan coast involving Tu-95MS bombers escorted by MiG-31 fighters supported by Il-78 tankers. A supplementary operation involved another pair of Russian bombers being intercepted by Canadian CF-18s. One interpretation was that these moves were related to Ukrainian president Petro Poroshenko's state visit to Canada and the United States, thus messaging all three countries: Canada and the United States to stay out, and Ukraine to stop looking for North American support.[37]

Immediately after the NATO Wales summit, Russian mounted Vostok 2014, a mass joint exercise involving all forces in the Russian Far East. The exercise also included Tu-95MS and Tu-22M3 activities, the 35th Missile Division near Barnaul, the 39th Missile Division near Novosibirsk, and the 62nd Missile Division near Uzhur with their mobile ICBMs. The RVSN component was assessed as "the biggest and probably most complex exercise for the Strategic Missile Forces" up to 2014.[38] The exercising of more than one missile division at once does not appear to have occurred previously. The fact they were all in mid-eastern Russia and are geographically positioned to strike targets in North America indicated that significant signaling was in progress.

The NATO European area was subjected to a sophisticated two-day Russian operation in October. Four waves of Russian aircraft flew into the NATO area over a two-day period on 28–29 October. Transponders were shut off and none of the aircraft filed flight plans, which forced air traffic controllers to rapidly reroute civilian air traffic. The first wave involved four Tu95MS and four Il-78 tankers. These proceeded from the Kola Peninsula over the Norwegian Sea, where they were intercepted by RNAF F-16s. Six aircraft returned to base, but two Tu-95MSs continued west of the British Isles, where they were intercepted by the RAF. They continued to a point off Portugal, where they were intercepted by Portuguese F-16s. At the same time, another aerial excursion consisting of six Russian fighters were intercepted by Luftwaffe Typhoons over the Baltic. The next day, two Tu-95MS Bears and two Su-27 Flanker fighters were detected over the Black Sea and intercepted by Turkish F-16s, while another Baltic excursion was mounted by Russian aircraft.[39] This large multiday operation was significant in that it demonstrated coordination with intent to intimidate. The presence of Tu-95MS cruise missile carriers and the deliberate disruption of civilian air traffic underscored this. Russian LRA activity directed at the United Kingdom also took place in the September–October time frame. On three occasions, RAF Typhoons scrambled to intercept pairs of Tu-95MS bombers operating in close proximity to British airspace.[40]

PHOTO 1 ✦ After its invasion of Ukraine in 2014, Russia regularly conducted deep penetrations of Western European–adjacent airspace using Tu-95MS Bears to points off the Iberian Peninsula and Tu-160 Blackjacks over the North Sea to intimidate and support information operations. These disruptive operations interfered with commercial air traffic and forced NATO members to intercept them. *Belgian Defense*

In the Black Sea, NATO committed a standing naval task group in September to assert freedom of the seas. The Russians selected the Canadian frigate HMCS *Toronto* for special attention. A Russian reconnaissance aircraft located the *Toronto* and a pair of Russian fighters buzzed the ship at three hundred meters separation. The *Toronto* locked on with its air defense systems, but the Russian aircraft departed.[41] When taken as a whole, all of this Russian aerial activity in the fall of 2014 may also have been used to backstop Russian diplomacy designed to generate a "frozen conflict" in Ukraine, that is, consolidate areas held by Russian forces using insincere ceasefire and peace talks to set the stage for future operations.[42] There was, however, a larger game. All of this activity was not just signaling Russian displeasure. It appears to have been part of a sophisticated strategy to undermine, intimidate, and possibly decouple the United States and Canada from western NATO, and then decouple western NATO from eastern NATO so that Putin's Russia could have a free hand in Ukraine.

Russia: Continuing to Push the Boundaries, 2015–2020

In terms of what was being called "bomber diplomacy," RAF fighters intercepted Russian Tu-95MSs operating close to the United Kingdom on four occasions from 28 January to 20 February, some of them in sight of Cornwall and off Bournemouth. These flights led to increased public concern.[43] There were Tu-95MS flights in April and May, and it appears as though the LRA was reaching a steady state of at least one flight per month. It should be emphasized again that these flights and the RAF Quick Reaction Force interceptions were hazardous for civilian air traffic, in part because in some cases the Russian aircraft did not squawk with transponders and this forced deft actions by air control staff.[44]

There was some LRA activity off Norway in February, but it was likely part of the reconnaissance for a major unannounced Russian naval exercise involving fifteen submarines and forty-one other ships. Opinion is split on the nature of this exercise. One argument is that the Putin regime increased its economic interest in the region, particularly oil, in the wake of sanctions imposed during the Ukrainian invasion. Reestablishing new infrastructure on Cold War–era bases was a sine qua non to economic exploitation, and presence was a precondition for that.[45]

LRA activity was limited in the first half of 2015. There was one probe against the Alaskan air defense identification zone (ADIZ), which was not intercepted, and an explanation was not provided by NORAD. Bombers were observed by Japanese forces over the Rykuku Islands.[46] The most spectacular event was a one-off. On 4 July 2015, American Independence Day, four Tu-95MS bombers transited through the Alaskan ADIZ, and two proceeded to a point forty miles off California, where they were turned back by USAF F-15 fighters. Rumors abounded in the mainstream media that Putin called the American president to wish him a happy Fourth of July, but apparently it was the crew of the Bears that did this over the radio.[47]

Russian LRA activity in 2015 was attenuated by the crash of two Tu-95s, one that skidded off the runway at Ukrainka Air Base on 9 June and another that crashed and burned near Khabarovsk on 14 July. During this period, four other

Russian military aircraft crashed. Older aircraft and inexperienced air and ground crews were blamed, but there was some form of operational pause in Tu-95MS activity.[48] Operations were resumed against the United Kingdom, Japan, and NORAD using Tu-142 maritime patrol variants and Tu-160 Blackjacks on four occasions in September, October, and November.[49]

By this point Syria was undergoing its version of the Arab Spring, with the Assad regime refusing to succumb to those attempting to unseat it. The decision by the Putin regime to enter the conflict in 2015 was multifaceted. The most obvious reason was support for a longtime, Cold War–era ally that was undergoing just the sort of color revolution the Putin regime feared. Syria was a base to project Russian naval and airpower into the "NATO lake" known as the Mediterranean Sea. It outflanked Turkey. And it permitted Russia to interfere with increasingly successful American operations in the region against the Islamic State, particularly in Syria and Iraq. Significantly, Russian military power was used against the Syrian rebel groups receiving American aid. An additional purpose of the military operations was to have Russia participate in a war on par with Afghanistan and Iraq, targeting "radical Islamic terrorism," in a conflict that was legitimate under international law and acted as a moral equivalent to feed Russian information operations lines in the global information environment. It also provided a distraction from events in Ukraine. To be seen to be succeeding somewhere was deemed important if one was not succeeding elsewhere.

An incident in which a Russian commercial Airbus crashed in the Sinai with the death of 212 people on 31 October 2015 was portrayed by ISIS media as a successful terrorist operation, though some doubt remains as to why the plane went down. Nevertheless, the Putin regime used it as an excuse to mount Operation Retaliation, which expressly showcased LRA forces and capabilities. On 17 November, twelve Tu-22M3 Backfires operating from Mozdok Air Base in the Caucasus, struck targets in Syria. They were preceded by two Tu-160 Blackjack firing sixteen Kh-101 cruise missiles over the eastern Mediterranean Sea, while three Tu-95MS fired Kh-555 cruise missiles from Iranian airspace. Three days later, eight Su-34 Fullback bombers departed occupied Crimea and hit targets in Syria, while twenty Tu-22M3 conducted ninety-six sorties from Mozdok. Sixteen sorties of Tu-160 and Tu-95MS from Engels Air Base fired eighty-five cruise missiles from Iranian airspace. Russian submarines in the Mediterranean launched an additional eight cruise missiles.[50] The audience for this show was conceivably everybody: the Russian domestic audience, Ukraine, NATO, the United Kingdom, the United States, and Israel.[51]

These aircraft were dual-capable systems, and in some cases the same ones that were probing NORAD and the United Kingdom. And it neatly underscored what small numbers of Russian LRA aircraft were able to do; on the same day as Op Retaliation, the RAF intercepted two Tu-160s Blackjacks over the North Sea. What made this event notable was that it was also reported over Russian propaganda outlet RT's English language service. In this report, a former RAF air marshal was quoted by the Russian propaganda outlet as saying, "They fly in these regions to check our air defenses and have probably worked out we are not as sharp as we were." And that was the message Russian propagandists wanted its international and domestic audiences to hear.[52]

How Russian messaging related to 2016 deliberations over Brexit is murky, and the degree and effectiveness of Russian interference in assisting Brexit remains under discussion. Russian embassy staff and Russian media outlets were engaged in supporting Brexit. Brexit disrupted the EU, which supported sanctions against Russia in the wake of its invasion of Ukraine, and it reduced British influence while aggravating domestic turmoil.[53] There appears to have been a single probe by Tu-160 Blackjacks on 17 February, or roughly when the debate got underway, but then there are no interceptions until late September and throughout October when pairs of Tu-95MS Bears and Tu-160 Blackjacks were intercepted on 22 September, 5, 12, and 16 October, and 23 November.[54] The idea that the United Kingdom could be reminded of its strategic vulnerability while not having it rubbed in its face publicly during an event that benefited Russian global objectives, and then amping up LRA operations again after those objectives had been achieved appears reasonable.[55]

It is equally curious to correlate Russian LRA activity and the American presidential election campaign, which was in full swing throughout 2016. No interceptions of Russian LRA aircraft in the NORAD ADIZs are recorded in the public domain after 6 May 2016. Unlike previous Tu-96MS operations, this one was covered by Russian pro-military media, an information channel that expanded in importance starting with the Syria operations. In this case, Russian military public affairs filmed the operation and distributed edited footage to interested parties via the Ministry of Defence Facebook page. The Su-35 pilot is seen to adjust his rearview mirror to show the audience he is carrying live air-to-air missiles.[56]

In contrast to previous years, however, the blatant harassment of American surveillance aircraft took place three times. In January an RC-135U Combat Sent signals intelligence collection aircraft was intercepted by an Su-27 Flanker fighter, which used its exhaust to generate wake turbulence to destabilize the Combat Sent. In April, a Russian Su-27 Flanker, incredibly, did a barrel roll over a U.S. Air Force RC-135 Rivet Joint surveillance aircraft while it was over the Baltic. In September, a U.S. Navy P-8 maritime patrol aircraft had an intercepting Su-27 Flanker come within ten feet of it over the Black Sea.[57]

Similarly, there are no publicly recorded interceptions of LRA aircraft in the NORAD ADIZs for the first four months of 2017. It was not a question of a post-crash Tu-95MS stand-down. It was not a question of sortie rates, as they were maintained against the United Kingdom. There was a deliberate decision made not to conduct operations against North America, while retaining pressure on the United Kingdom. It appears as though there was a shift to harassing surveillance aircraft. These were doing secretive things close to Russia and fair game. By harassing them, and with this information going public, it would embarrass the United States to some degree, especially if there were no responses to the Russian activities which would confirm the nefarious nature of the activity to Russian domestic audiences. Conversely, by backing off on LRA operations against North America, there could be no media coverage if there was no activity. Hillary Clinton, viewed with fear by the Russian establishment during her presidential campaign, would have nothing to use in the electoral arena besides Russian actions in Syria, something distant from the American populace and its concerns. The American people would not feel pressurized or threatened directly if they were not regularly subjected to "Russian bombers entered the Alaskan Air Defense Identification Zone today" in the media.

At the same time, the Russian Ministry of Defence ensured that it was publicly known through social and mainstream media that the new mobile ballistic missile systems and their command and control apparatus were functional. Indeed, there was a large-scale RVSN exercise in which the missile divisions were reported in Russian pro-military media to have been on alert for more than a month, that is, October. The Moscow-area RVSN command and control units were exercised on

14 December, and the media was treated to footage of a YARS ICBM being loaded into a silo launcher at Kozelsk. Viewers were also informed that a command post pod had been installed in a silo in the Kozelsk ICBM field back in 2014.[58]

The Russian honeymoon with newly elected president Donald Trump ended in April 2017. Images of a Syrian sarin attack against civilian targets shown to him by his daughter, Ivanka, moved Trump to lash out. Trump and Secretary of Defense Jim Mattis debated options. Trump favored a large-scale aerial campaign, but Mattis was concerned about escalation due to the presence of Russian forces in Syria. In the end, U.S. Navy ships fired sixty cruise missiles at the base which the Su-22 Fitter that launched the chemical attack came from, while the Assad regime moved as many political prisoners as possible and packed them into hangars so they would be killed by American airstrikes.[59]

The Putin regime expressed its displeasure by ramping up LRA activity against NORAD. The first flight was a pair of Tu-95MS, flying one hundred miles south of the Kodiak Island on the night of 17 April. These were supported by an Il-38 Coot maritime patrol aircraft that was probably being used to collect signals intelligence. The Bears were intercepted by a pair of USAF F-22s supported by an E-3B Sentry AWACS aircraft and a KC-135 tanker. Lt. Gen. Kenneth Wilsbach told the media that this "was the first such flight seen near Alaska in nearly two years."[60] The next day, two Tu-95MS were tracked by an E-3B Sentry as they flew alongside the Aleutian chain. These were also supported by an Il-38 Coot. On 19 April, two IL-38 Coots proceeded halfway up the Aleutian chain but were only observed and not intercepted.[61] On the fourth day, RCAF CF-18s intercepted two Tu-96MS "off Canada's northern coast."[62] After a week's hiatus, two Tu-95MS and two Su-35 Flankers supported by tanker aircraft came within fifty miles of Point Hope after flying past the Alaskan North Slope before turning back. These were intercepted by F-22 Raptors.[63]

This was a signal on a whole new level. When taken together, this series of flights demonstrated that critical American facilities in Alaska could be put at risk in the early stages of a crisis that might involve nuclear weapons use. Specifically, the ballistic missile early warning system at Clear, Alaska; Eielson Air Force Base; and the anti-ballistic missile (aka "ground-based midcourse defense") silos at Ft. Greeley. Alternately, putting key oil and gas facilities at risk, as Russia had done with the Norwegians, was another possible message. Two pairs of Tu-95MS bombers, escorted by aerial-refueled fighters, could potentially overwhelm defenses in this area with air-launched cruise missiles if they were not closely observed with forces capable of destroying them.

Indeed, it was after this in 2017 that Russian authorities significantly expanded social media coverage of strategic nuclear forces, in part to support this new line on the Trump administration. There was a steady drumbeat focused mostly on RVSN developments deployed by the Russian Ministry of Defence through its tame supporting media outlets. In effect, most of this was general deterrence signaling. The exception was a Backfire exercise out of Anadyr in October. This is particularly significant, given the proximity of Anadyr to Alaska. One possible course of action in the event of conflict could be that the Tu-22M3s have a role in destroying the Alaskan air defense system with standoff missiles to permit Tu-95MS bombers with their cruise missiles free rein against critical strategic targets in Alaska, specifically, the much-maligned ballistic missile defense (BMD) facilities at Ft. Greeley and associated sensors at Clear Air Force Station. Another possible employment scenario is the reverse, with Tu-95MSs launching the first attacks against the air defense system and a follow-on attack using Tu-22M3 exploiting the breaches.

In 2018, this release of information by the Ministry of Defence ceased, and extensive coverage in Russian pro-military media of these topics abruptly stopped, as if the point had been made. The American intelligence apparatus was presumably aware of all of these activities through a variety of national technical means. Was this information campaign designed for an international audience, a domestic audience, or both? Outwardly, it collectively supported the general "Russia is back" line, with an emphasis on the ICBM forces. So the audience could have been Western specialists, both amateur and professional, in academia, the media, and social media. Domestically, and the fact that pro-military Russian media had priority over other outlets, also suggests it was designed to prop up national morale. Underpinning all of this, however, was a deeper message: Russia's strategic nuclear forces were now "back" and capable of blocking any threats by the West to escalate in Syria and even Ukraine. It could have been the functional equivalent of the Soviet Union achieving parity in the 1970s, which made the world safer for Soviet-initiated conventional warfare. The key statement was made by Shoigu in May 2017, that is, his forces were "solving nuclear deterrence problems," a Cold

War–era Soviet phrase used to confirm that the objectives had been achieved by the end of an exercise. In other words, Shoigu was expressing new confidence in the ability of the Russian strategic nuclear forces to deter.

From 2018 to 2020, the pattern of air intercepts conducted in the NORAD ADIZs and by the RAF around the United Kingdom remained relatively consistent with the 2015–17 period, so it is superfluous to list each and every intercept here. That said, Canada was concerned enough to expand the Canadian Air Defence Identification Zone (CADIZ) from the 50th to beyond the 80th parallel toward the North Pole in May 2018.[64] The Russian signaling tool of choice was a pair of Tu-96MSs or Tu-160s, and if they were operating off Alaska, they were accompanied by fighter aircraft and aerial refueled. On the north Norwegian, Baltic, and Black Sea fronts, the nature of the game revolved around interception and observation of surveillance aircraft by both sides with no apparent public flourishing of bomber aircraft. There was no noticeable change in LRA activity directed at the NORAD ADIZs during the U.S. 2020 presidential campaign, with one exception, and there was no letup in similar activity directed against the RAF after the Brexit implementation in January of that year.

The exception was a two-day operation conducted by the LRA against the NORAD ADIZ on 20 and 21 May 2019. On day one, four Tu-95MSs escorted by Su-35 fighters entered the Alaskan ADIZ and were intercepted by four F-22 Raptors supported by an E-3 AWACS. The next day two Tu-95MSs entered the Alaskan ADIZ, were intercepted by two F-22s, exited the ADIZ, reentered it escorted by Su-35s, and were intercepted again by F-22s.[65] The duration, size, and sophistication of the operation distinguished it from other operations, but there is no specific correlation to political or other activity. The open deployment of Tu-160 Blackjacks to Anadyr air base opposite Alaska was another bold move in August 2019.[66]

What did change after 2017, however, was the increased publicity given by the Russian authorities of the Grom ("Storm") series of nuclear force exercises and the increased use of these as signaling mechanisms. The 2018 exercise in October was similar in scope to the one held in 2017. This involved the ballistic missile submarine *Tula* which salvo-launched a number of SLBMs from the Barents Sea to the Kura test range in Kamchatka, and the launching of cruise missiles from LRA Tu-95MS, Tu-160, and Tu-22M3 at ranges at Pemboy and Teretka. Unlike

2017, there were no ICBM shots. Russian press releases specifically noted that land- and space-based ballistic missiles warning systems were used to detect the SLBMs, thus emphasizing that these systems were functional and capable. Putin participated in the exercise.[67]

Mention should also be made of an individual test conducted in May 2018 by the new *Borei*-class ballistic missile submarine *Yury Dolgoruky*, which volley fired four Bulava SLBMs from the Barents Sea to the Kura test range. It is likely this contributed to general deterrence signaling in that it was a test of a new system, and of importance, footage of the action was released by the Russian Ministry of Defence to interested media outlets. This was the first time since 1991 a Russian SSBN had volley fired its missiles.[68] During the Cold War, volley firing of Soviet SLBMs from the Barents Sea toward Kura was always of concern by NORAD in that until the trajectories of the missiles and the reentry vehicles could be determined, the launch had to be treated as though it was a possible attack against North America.[69] After nearly two decades of lowered intensity in this field, this may have come as a surprise to NORAD duty personnel.

The Grom exercise from 15–17 October 2019 went even further in that it was accompanied by a more intense information campaign that included a plethora of coverage and announcements by multiple "media" outlets and, notably, extensive coverage on TV Zvezda. Interestingly, Grom 2019 started with a briefing to foreign defense attachés. There was a specific scenario that the Russian establishment wanted to convey, of Western forces launching a nuclear attack against Russia and being met by a devastating response.[70]

Putin himself participated in Grom 2019 and there were images of him pushing buttons in an underground command post, itself a significant signal of increased interest in aspects of nuclear warfighting as well as in command and control. The main events of Grom 2019 included an SLBM launch from the Sea of Okhotsk to the Chizha test range east of the Kola Peninsula from the Delta III–class SSBN, the *Ryazan*; and another SLBM launch from the Barents Sea to the Kura test range by a Delta IV SSBN; and an RS-24 YARS ICBM. Of note, the *Ryazan* was supposed to launch a second SLBM that dated from the 1980s, but it misfired and the submarine returned to Petropavlovsk. Rather than conceal the failure, Russian commentators deflected and emphasized the command and control nature of the exercise.[71] They did not tell observers that the SLBM shots that did take place successfully released "mass simulators of warheads," that is, dummy MIRVs, though NORAD would have detected this.[72]

Additionally, Iskander cruise missiles and Kalibr sea-launched cruise missiles featured prominently, and imagery was made available to reinforce this.[73] The Kalibrs were launched from the Caspian Sea and the Barents Sea at land targets. This aspect differed from previous exercises in that Iskander and Kalibr are essentially theater-level nuclear weapons, not strategic, thus emphasizing the renewed flexibility of the Russian nuclear arsenal. Indeed, there is no analogue system in Western arsenals, with the Pershing II, GLCM, and TLAM-N having been removed from service and dismantled between twenty-five and thirty years ago.

Grom 2020, however, did not garner as much attention in Western media or specialist analysis as Grom 2019. The 9 December 2020 exercise featured Tu-160 and Tu-95MS bombers launching cruise missiles; a YARS ICBM launched from Pletsetsk; and an SLBM launch from the Delta IV–class ballistic missile submarine *Karelia*. These activities were duly reported by the Russian Ministry of Defence which provided footage of them. However, three days later on 12 December, there was an unannounced firing of four Bulava SLBMs from the *Vladimir Monomakh*, a new *Borei*-class ballistic missile submarine operating in the Sea of Okhotsk. These landed at the Chizha test range. There was virtually no Western media coverage of this event, with the exception of *The Barents Observer* in Norway, which retains a special interest in Russian Arctic activities.[74]

The Grom messaging was used to underpin all other Russian signaling efforts using its strategic forces. Indeed, maintaining Grom as an annual reminder to the West of Russia's nuclear potency was precisely its purpose, especially the SLBM portion of the Russian nuclear triad.

Elephants in the Room II: Large-Scale Exercises After 2014

Russian large-scale exercises continued to expand after 2014. The backbone consisted of the major regional exercises in rotation, but there were more parallel and snap exercises. Nuclear weapons use was increasingly featured in exercises other than Grom. Tsentr 2015 featured an "counter terrorism exercise"[75] in concert with Kazakhstan, but in reality it "focused on increasing strategic mobility in the context of full-scale intervention in a Central Asian state destabilized by civil war."

Russian information operations wanted observers to take away the possibility that the objective of the exercise "was to prepare troops for possible ground operations in Syria."[76] We may have to recast this assessment in light of the events of January 2022 with the CSTO intervention in Kazakhstan. Tsentr 2015 was accompanied by two parallel exercises: United Shield in Belarus and an unnamed exercise conducted by the Northern Fleet in the Barents and Laptev Seas. Two large snap exercises were also conducted, including a Northern Fleet exercise that involved nuclear weapons. Five "partial surprise combat readiness inspections" were conducted with strategic nuclear forces over the course of 2015, twice the number conducted in 2013 or 2014.[77]

In the past, the Kavkaz exercises handled operations in the Caucasus. In 2016, however, they involved "the redeployment of Russian troops near the border of Ukraine" and were "preceded by a series of smaller exercises that included defense of a Russian sea base in Sevastopol from submarine attack." Indeed, "most of the exercise took place in [occupied] Crimea." Other smaller exercises took place, notably in "Transnistria," which was interpreted as a Russian threat to Moldova. Another exercise involved the CSTO rapid response force where it was inserted to act as "peacekeepers enforcing the observance of a ceasefire between warring sides," which was interpreted as a possible option to insert CSTO "peacekeepers" between occupied Donbas and the rest of Ukraine to shield so-called "separatist" forces from the Ukrainian ATO.[78] These activities were conducted right on the heels of Russian protests over Ukraine allegedly staging "provocations" in Crimea.[79]

A parallel exercise was run by the Northern Fleet in the Barents Sea. One analysis suggests that the Northern Fleet would be used to attack reinforcement convoys from North America. This fleet hosts most of Russia's nuclear missile submarines, suggesting that nuclear escalation might occur very quickly after the commencement of a conventional war.[80] Exercises involving strategic nuclear forces in close proximity to conventional exercises are not coincidental. In each case, the Putin regime was reminding outside observers that this relationship exists. Strategic nuclear forces backstopped everything the Russian armed forces were planning and conducting.

Zapad 2017 was a dramatic departure from previous exercises, not only in terms of scope but in intent. Zapad 2017 was held in Belarus, but there were three parallel exercises involving the Northern Fleet as well as the Southern and Central Military Districts. No snap exercises were held before Zapad 2017. As for intent, there were significant Russian strategic information operations directed at NATO members.

The Belarus portion of Zapad 2017 had ulterior motives. Analysis suggested that "it was clear that Moscow wanted to test how Belarus would react to a potential deployment of Russian troops on its territory.... Thanks to Zapad 2017, Moscow could notice that the vast majority of Belarusians have either neutral or positive attitude to the Russian army."[81] In addition to forward-facing defensive exercise moves in Belarus, elements from three airborne divisions were deployed opposite Estonia and Latvia in response to "raids."[82] Another parallel exercise "repelled" a massive air attack from the West. This included the defense of Moscow "against a mass cruise missile attack."[83] Zapad 2017 also featured the exercising of so-called anti-access/area denial "bubbles" around Kaliningrad and Crimea using Bastion, Kalibr, and Iskander missile systems.[84] Russian Il-76 aircraft also violated Lithuanian airspace as they were flying into Kaliningrad.[85]

Meanwhile, there was a parallel exercise in the Northern Fleet area, which may have included an exercise that simulated an assault to take Svalbard Island.[86] Observers noted that it included a ballistic missile attack by a submarine and a missile strike against enemy surface combatants.[87]

This "allows Russia to test its escalation dominance in a future conflict." Indeed, another observer noted that SSBN "patrols are concentrated in areas of the Barents Sea, which is designated as a bastion. One of the prioritized tasks is to protect these bases and patrol areas.... In fact the bastion defense concept was at least partly tested during ZAPAD-2017."[88] It was evident that the Zapad 2017 scenario was not "counterterrorism" but simulated "a large scale state-on-state conflict."[89]

Vostok 2018, held in September, was preceded by several snap inspections in both the Central and Eastern Military Districts, as well as Northern Fleet, strategic transport aviation, airborne forces, and the LRA. Mobilization capability was employed and assessed. The number of troops involved was estimated to be "almost as large as the entire German or Polish armed forces." Vostok 2018 was not limited to the Far East: part of exercise play was using the Northern Sea route and Arctic operations.[90] The nuclear component was left deliberately ambiguous.

Tu-95MS bombers launched cruise missiles "over the Barents, East Siberian, Chukchi, and North Arctic Seas," while Iskanders and naval cruise missiles were live fired during the exercise to simulate the destruction of "groupings of forces."[91]

Tsentr 2019 involved formations from the Central Military District and was mounted in the Arctic, the Caucasus, the Caspian Sea, and central Russia. The ostensible purpose of Tsentr 2019 was to "fight against international extremism and terrorism." The parallel exercise was Union Shield held in the Western Military District (with no Belarusian participation), and its objective was to exercise against "illegal armed groups," conduct "rapid deployment," and ensure interagency coordination. Once again, this looked more like an occupation exercise than force-on-force kinetic action. The activities Northern Fleet component, which constituted another parallel exercise, was deliberately downplayed and blurred from public observation. A Northern Fleet landing exercise on Bolshevik Island was downplayed and Russian commentators insisted it had nothing to do with Tsentr 2019.[92] It didn't take a graduate degree in geography to see that Bolshevik Island was a Svalbard analogue. If nuclear weapons played a role in Tsentr or the parallel exercises, their simulated use was kept concealed in contrast to the other annual exercises. Finally, there was Kavkaz 2020. Held against the backdrop of the COVID-19 pandemic and instability in Belarus, the exercise was provocatively held in Armenia, South Ossetia, and Abkhazia, instead of the original location of Crimea, though units of the Black Sea Fleet and Caspian Sea Flotilla participated. Like Tsentr, there was no overt nuclear weapons play.[93] Why this was the case remains opaque, considering the activities and operations surrounding Zapad 2017.

It is appropriate to mention the Russian use of Notices to Air Missions (NOTAMS) as a mechanism that is a combination of political signaling and information operations, that is, intimidation. NOTAMS are notices "containing information essential to personnel concerned with flight operations but not known far enough in advance to be publicized by other means. It states the abnormal status of a components of the National Airspace System, not the normal status."[94]

In effect a country can temporarily declare a sea-air-subsurface space as a danger area and issue a warning that military or space activity will be occurring within it. Russia employs NOTAMS regularly in the Barents Sea, the Norwegian

Sea, the Baltic Sea, and the Black Sea. In some cases, military activity or space activity takes place, and in some cases it does not but still forces civilian air and surface traffic to avoid the declared areas. This has obvious economic ramifications for fishing, but NOTAMS can also be positioned to act as a virtual blockade. After 2014 and in the run-up to the events of 2022, Russia employed NOTAMS extensively to proactively disrupt NATO exercises and operations, particularly off Norway.[95]

Iskanders and Kaliningrad: Missile Games

An overlapping signaling channel during the period was the disposition of Russian Iskander missile systems. The Iskander (NATO reporting name SS-26 Stone) comes in two forms, a ballistic missile version (Iskander-M) and a cruise missile version (Iskander-K). Both versions are mobile and are carried by an eight-wheeled transporter erector launcher (TEL) that carries two ballistic missiles or four cruise missiles. The Iskander-M was taken into service by the Russian armed forces in 2006.[96]

There are several missile types and different load-outs which adds to the ambiguity of the Iskander system when it comes to signaling. Both the ballistic and cruise missiles types can use nuclear and conventional warheads. The ballistic missile is actually a MARV, like the Pershing II, in that it can maneuver and dispense countermeasures on the way to its target. The ranges of the missile systems have been subject to significant obfuscation by Russian authorities. In a general sense, the ballistic version is believed to have a range of 450 km, while the cruise missile versions are estimated to have a range of 1,500 to 2,500 km.[97]

The deployment of the Iskanders is wrapped around two issues. The first is the Cold War–era Intermediate Nuclear Forces Treaty; the second is the Anti-Ballistic Missile (ABM) Treaty. If the deployment of Ballistic Missile Defense (BMD) systems in the NATO area makes the ABM Treaty a dead letter, it might justify the deployment of Iskander even if it violates the Intermediate-Range Nuclear Forces (INF) Treaty. These matters were in play prior to the onset of Russian hostilities against Ukraine in 2014. Indeed, Dmitry Medvedev announced in 2008 that Iskanders would be deployed to the Russian enclave of Kaliningrad specifically because of American intransigence on BMD deployments to Poland.[98] Estonian sources indicated the first Iskanders were deployed between 2010 and

2011 opposite Estonia.[99] This was after the United States scaled back BMD deployment in favor of a more limited capability based on the U.S. Navy Aegis system.
In terms of combat potential, the Russian Iskander deployment put all three
Baltic states at risk, and there were no defensive systems capable of dealing with
the threat. Chief of the General Staff Nikolai Makarov noted that U.S. ABM
systems might lower the deterrent value of Russian nuclear weapons and their
ability to be used for coercion against Europe.[100]

Russian media sources announced in 2014 and 2015 that Russia was training the 152nd Missile Brigade on the Iskander system and implied it would be
deployed imminently to Kaliningrad. This was interpreted by Polish observers
as "a regular part of Russian psychological warfare stoking the sense of threat in
NATO countries."[101] It was also likely a response to the NATO Wales summit that
year where the decision was made to go ahead with an Aegis Ashore deployment
in Romania.[102] This was in response to American criticism that Russia had tested
a ground-launched cruise missile that violated the INF Treaty.[103]

The Iskander deployment to the Kaliningrad enclave took place during the first
week of October 2016. NATO responded with the deployment of an E-3 Sentry
AWACS aircraft to Siauliai air base. There was media speculation that the deployment was related to the collapse of yet another post–Cold War nuclear agreement
involving the disposal of surplus plutonium, but then Russian state media quoted
a senior official linking the deployment to NATO missile defense efforts.[104]

All countries in the Baltic Basin were at risk from the Kaliningrad Iskanders.
The weapons could reach Germany, Denmark, the Czech Republic, and Slovakia.
It was like West Berlin during the Cold War but in reverse. Kaliningrad was in
effect a missile launch pad to directly and specifically intimidate Scandinavia
and formally Soviet-occupied Eastern Europe.[105] The Russians stepped up the
fear factor in December 2017 by announcing through Russian media that the
missiles in Kaliningrad would be put on "combat alert" in 2018. Of note, combat
alert or combat duty in Soviet and now Russian lexicon is the ability to conduct
the primary mission, not a DEFCON equivalent. Whether the distinction was
known by the Russians or not is unclear.[106] And from 2018 to 2022, the Russian
leadership used the Iskanders to regularly poke at NATO. Russian media would
announce more Iskanders would be deployed to Kaliningrad. Then Western media
"discovered" Russia was upgrading its nuclear storage areas in Kaliningrad. After

that, media reports were disseminated that Iskander missile units in Kaliningrad were conducting "launch simulations."[107] This constant prodding on the Kaliningrad Iskander issue was a new form of signaling, something that sat between general deterrence signaling and a specific crisis move.

Conclusions

Signaling and the maneuver of nuclear forces as the means to express it is itself not new but has new dimensions in the 2000s and especially after 2014. Between 1999 and 2013 Russia's leadership resorted to nuclear force signaling, saw it got results, and then expanded its activities in this area to the point where conventional and unconventional operations against Ukraine became feasible, supported by threats to use nuclear weapons as a backstop against Western interference.

An examination of Russian actions from 2014 forward strongly supports the argument that the Russian leadership has embraced the doctrine that military actions short of war, and especially the manipulation of strategic nuclear forces, are actions that are meant to intimidate, achieve psychological ascendency, and thus dominance over opponents. These moves are not merely conducted to signal intent.

It is equally clear that the Russian leadership has manipulated strategic nuclear forces for two other mutually supporting reasons. First, the ongoing operations of the RVSN, LRA, and the naval deterrent forces were manipulated at an increasing pace from 2000 to 2014. There was clearly a plan to revitalize Russian strategic nuclear forces, but this was also meant to reestablish the credibility of Russian strategic nuclear capabilities in the eyes of its opponents and competitors. Second, there is every indication that all of this activity was part of a larger assertion of power for national restorative purposes by the Putin regime. A third purpose was to emphasize that whatever international action that Russia chose to embark upon, from counterterrorism, to counterinsurgency, to conventional warfare, would be backstopped by its strategic nuclear forces.

Finally, we start to see the progressive integration of nuclear force maneuver and what we will eventually call information operations, especially using new media. And with it, there is a rediscovery of the idea that an event can be employed simultaneously to influence an external audience but particularly a domestic audience.

CHAPTER 3

◆ ◆ ◆

Blue Forces

American and NATO Strategic Signaling in Context Before 2021

Russian strategic signaling using nuclear forces in the 2010s was designed to establish that "Russia is back" and to support an aggressive global policy first in the Caucasus, then in the Middle East, and finally against Ukraine and NATO. American strategic signaling during this period was significantly different in intent and content because, simply put, the United States never left. In the American case, the world reordering after the 9/11 attacks, the war on terrorism, and increased concern over non-peer WMD threats produced a shift in American declaratory nuclear strategy and how this was expressed through signaling. Similarly, the increased global assertiveness of both the PRC and North Korea posed dilemmas for American global and regional strategies with a subsequent effect on signaling. NATO and allied signaling, in contrast, had waned in the years before 2014 and required immediate revitalization given the events in Ukraine, the Baltics, the Black Sea, and Norway. These were complex challenges that were on an order of magnitude different from the Russian regime's situation and approach.

In the first decade of the twenty-first century, both the United States and NATO were passive, then reactive to Russia's aggressive moves, but by the late 2010s signaling picked up in quality and quantity. There were, however, limitations. The primary one was the inability to gain credible extended deterrence over Ukraine, a non-NATO member adjacent to NATO powers, that had been violently engaged by an aggressor contiguous to those same NATO powers. The second one was how to connect signaling to an inchoate policy vis-à-vis Ukraine and the region. The third was how to connect deterrence and signaling to Russia's increased use of so-called hybrid or gray-zone warfare, something that existed in different forms during the Cold War but tended to be completely detached from signaling involving Western forces. A new form of signaling had to be introduced between general deterrence and crisis signaling to counter Russia's intimidation signaling and that was assurance signaling.

Changing the Baseline: Nuclear Planning, Deterrence, and Signaling from the 1990s into the 2000s

There were several significant shifts in the American nuclear posture and thus the context for American nuclear weapons use thinking that formed the backdrop for American signaling in the 2000s and 2010s. U.S. Cold War plans included the Single Integrated Operational Plan (SIOP) which was generally focused on the Soviet Union. The regional commanders in chiefs (CinCs) had nuclear components of their regional plans and forces assigned to fulfill them. In 1992 the U.S. Joint Chiefs of Staff (JCS) reviewed its planning structures and process in light of the collapse of the Soviet Union and the new protracted conflict with Iraq. This review concluded that "a planning process was needed which could rapidly develop a flexible SIOP to meet new threats such as regional instability, the rise of hostile regional powers, proliferation of WMD, and the residual nuclear capability of the republics of the former Soviet Union."[1] An initial capability was achieved and under the rubric of OPLAN 8044. The basic plan was built in six months and notably included "adaptive planning options for developing new courses of action within 24 hours."[2] At this point, the JCS "maintained only three resourced and fully maintained numbered war plans; these are plans for two regional contingencies (Korea and Iraq), plus a supporting nuclear employment plan (SIOP)."[3]

The plans were called USCINCCENT OPLAN 1002, USCINCPAC OPLAN 5027, and USCINCSTRAT OPLAN 8044. The SIOP was a nuclear target list supporting military operational plans. In the post-Soviet era, U.S. Strategic Command (STRATCOM) responsibilities included support to regional strategies, contingency operations, and counterproliferation actions. The SIOP is no longer a stand-alone document prepared by the (former) Joint Strategic Target Planning

Staff; rather, it has been transformed into a numbered OPLAN prepared by the STRATCOM Plans and Policy Directorate, J-5, and is a target list integrated into the Joint Staff's Joint Strategic Capabilities Plan (JSCP).[4]

With the stand-down of Strategic Air Command (SAC) in 1992, the advent of the de-targeting agreement with Yeltsin's Russia in 1994, and the denuclearization of Ukraine, Belarus, and Kazakhstan in line with the Budapest agreements of 1994, the new era of de-emphasizing strategic nuclear forces as a part of international strategic discourse was well on its way.

However, new threats emerged in the 1990s. These were grouped under the term "proliferation" threats, but essentially it was the spread of nuclear weapons and delivery system technology to and from non-peer states that have problematic relations with the United States. In the 1990s this included Iraq, Iran, North Korea, Libya, and Pakistan. By 1998–99, American concern was taking form to include sanctions and an open discussion of the larger problem.[5] One product of this was a reassessment of strategic force planning that was completed in September 2000. Boiled down, the United States needed the ability to plan to use nuclear weapons to counter any emergent non-peer WMD capability. This included biological and chemical as well as nuclear weapons, while at the same time handling the more traditional deterrent tasks regarding peer or near-peer potential threats.[6]

This reassessment had several notable features, but the most important was the conduct of "deterrence framework analysis [to include] assessments of the WMD deterrence calculus or Iraq, North Korea, and China" as well as "non-state actor deterrence issues." The reassessment fed changes to Strategic Command's Alternative Employment Options "in both their SIOP and Theatre Support planning roles."[7] Procedures for "adaptive planning"—that is, quick response targeting in under twenty-four hours—were also part of the reassessment,[8] as were the development of measures against Iran, Iraq, and North Korea to deter them from proliferating WMD technologies or systems.[9] There was particular interest in deterrence effectiveness directed at Iraq regarding biological or chemical weapons use, and to what extent hard or buried targets in all appropriate countries could be "put at risk."[10] Equally interesting was a sub-study that examined "the impact of fallout effects in achieving denial or delay of enemy access to key installation as a result of US nuclear strikes, with a view to using fewer weapons or smaller yields than are required when prompt effects alone are considered."[11] Sub-studies were conducted to "develop fallout protection factor distribution for North Korea, Iraq, and the People's Republic of China."[12] Finally, there was an examination as to where all this fit into OPLAN 8044, and even if it should.[13]

While all of this was being processed, an OPLAN 8044 update was promulgated in January 2001, which had the SIOP embedded within it. OPLAN 8044–98's mission statement remained the same as it had from the beginning: "Deter major military attack on the UNITED STATES and its allies, and should deterrence fail, employ forces to achieve national objectives."[14] The plan was divided into five phases:

> Phase I: Pre-hostilities
> Phase II: Generation
> Phase III: Decisive Combat and Stabilization
> Phase IV: Recovery and Regeneration
> Phase V: Post-hostilities

Cold War–era SIOPs were to be implemented on order, and this was connected to Tactical and Strategic Warning criteria, as opposed to a series of potentially complex pre-hostilities moves and actions. In this sense, OPLAN 8044 was more like the 1950s Emergency War Plan that was connected to a series of movements intended to deter an adversary before the plan was unleashed.[15] Though the entire OPLAN 8044 is not within the public domain, there are enough pieces of information available to give us insight into how signaling fit into it. First, the DEFCON system was still in use, and it was understood that certain movements and activities conducted during DEFCON shifts could be seen by opponents to signal them.[16]

There were psychological operations (PSYOP) objectives built into the plan. These included[17]

- Deter a major attack, particularly a nuclear attack, on the U.S., U.S. forces, and its allies
- Promote attitudes that dissuade enemy leaders from continuing hostile courses of action and motivate them to accept U.S. objectives
- Gain enemy populace understanding and acceptance of U.S. objectives and operations

- Counter enemy propaganda targeting allies, neutrals, and adversaries
- Undermine enemy confidence in ultimate victory and weaken their will to fight
- Portray strong U.S./coalition solidarity and resolve

These were built into the pre-hostilities phase of OPLAN 8044. Exactly how force movements were to contribute to this is obscure. That said, several moves are suggested in the portions of the plan that are available. Dispersal of aircraft under Expanded Alert Operations was one. Second, the movement of tanker aircraft, especially Forward Located Alert Generated (FLAG) aircraft, was a component of maintaining "deterrence through posturing of strategic forces commensurate with strategic warning and as directed by the National Command Authorities" was another. Third, there was a specific requirement under Phase I – Pre-hostilities/deterrence to "conduct training, exercises, and monitor readiness of committed reconnaissance forces. Maintain forward presence in theatres deterring and monitoring regional conflict, and assessing strategic threats." Finally, the movements of command and control aircraft appears to have played a role here as well.[18]

After the 9/11 attacks, the reassessment accelerated and took several forms. The use of the term "SIOP" was dropped in 2003 to reflect the fact there was a family of plans and not a single nuclear war plan, though internally the term was informally used for aspects of it.[19] Accompanying this was an overhaul of exercises to improve readiness. Global Lightning was established as an annual STRATCOM command post exercise to test procedures and communications using "simulated forces in deterring a military attack and employing forces as directed." Global Thunder was the annual field training exercise in which deployed forces exercised their specific capabilities involving aircraft, ICBMs, and submarines. Finally, there was Global Storm, which was established specifically to address STRATCOM's activities "in the early stages of conflict, such as information operations and intelligence, surveillance, and reconnaissance."[20] These exercises contributed to general deterrence efforts, but the increased emphasis on the pre-hostilities and early stages of conflict before nuclear weapons were employed was relatively new.[21] Indeed, another new concept that emerged from the September 2000 reassessment overlapped with these changes.

In January 2003 a new mission was assigned to STRATCOM called Global Strike, defined as "a capability to deliver rapid, extended range, precision kinetic (nuclear and conventional) and non-kinetic (elements of space and information operations) effects in support of theatre and national objectives."[22] This was called CONPLAN 8022 and it went into effect in June 2004, specifically to provide "the President with a prompt, global strike capability."[23] Conditions for the implementation of CONPLAN 8022 were derived from the 2000 reassessment study.[24]

Global Strike focused on exposed ICBMs, underground structures, mobile missile launchers, port and airport facilities, and WMD storage facilities.[25] As for delivery systems, bombers with conventional precision strike capabilities were a feature. These were exercised in Global Lightning 05 held in October 2004, which was "the first STRATCOM sponsored GLOBAL STRIKE exercise" and handled nuclear strike options CONPLAN 8022 as well as OPLAN 8044. Another exercise nested within Global Lightning was Global Archer 04 that exercised the conventional options. During the exercise, thirteen B-52Hs from Barksdale AFB conducted a Cold War–style minimum interval takeoff.[26]

The real star of Global Strike, however, was the SSBN force. In March 2005 the USS *Tennessee* fired a Trident II D5 SLBM with a modified Mk. 4 reentry body carrying a simulated W79 warhead. This test demonstrated that the missile could be used with a "highly-compressed trajectory [with a distance] of only 1380 miles. Impact occurs only 12–13 minutes after launch." The reentry body was GPS-guided and steerable, with an accuracy of thirty feet.[27] For comparative purposes, the unclassified accuracy of the regular system is a three-hundred-feet Circular Area Probable.[28] By 2005, American SSBNs were "currently tasked under CONPLAN 8022. . . . The SSBNs do not deploy with a mixed [conventional–nuclear] payload and are only nuclear tasked under CONPLAN 8022."[29]

In effect, the procedures and capability existed by 2005–2006 for the president to order a precision, limited nuclear strike in under twenty-four hours against a target list of WMD capabilities outside of the context of a general or regional nuclear war. Indeed, the amount of press, congressional, and analytical coverage on the matter ensured that this capability was publicly known, thus contributing to general deterrence as well as, potentially, deterring non-peer potential adversaries possessing WMD programs. That said, comments made by a defense official noted that Global Strike operations could be conducted against targets "below the equator"; "large land masses of Asia [or] the Middle East; or "all the way up to the Baltics."[30]

From 2005 to 2007, all of the "Global" exercises took place in one form or another. In November 2005, Global Lightning and Global Strike were held simultaneously at the JCS level and dubbed Ex Positive Response, while on 9 December 2005 nine B-52Hs from Minot AFB conducted a Global Strike exercise.[31] The first Global Thunder FTX was held in April 2006, and in July load exercises were held to respond to OPLAN 8044 execution.[32] Global Storm was conducted in March 2007, while other elements of OPLAN 8044 were held in the summer.[33] These exercises should be seen within the context of improving STRATCOM readiness in the post-9/11 period after it had languished in the 1990s and within the context of general deterrence as oppose to overt signaling with a specific international audience in mind.

There was a significant change in 2008 when OPLAN 8044 was replaced with OPLAN 8010–08, Global Deterrence and Strike. Why this shift took place is unclear, but it was probably related to deteriorating relations with Russia throughout 2007, and by coincidence the plan went into effect the day before Russia invaded Georgia on 1 August.[34]

In effect, the Global Strike CONPLAN was subsumed into OPLAN 8044. OPLAN 8010–08 was designed to support the U.S. National Defense Strategy by "assuring allies and friends, dissuading potential adversaries, deterring aggression, and defeating adversaries." OPLAN 8010–08 responded to "a new concept of 'waging' deterrence, paired with revised joint force capabilities that provide a wider range of military deterrent options." Deterrence operations were designed to "convince adversaries not to take actions that threaten US vital interests by decisively influencing their decision-making." This was to be achieved by "credibly demonstrating the will and means to deny benefits and/or incur costs." Specifically, OPLAN 8010–08 was to provide the "capability to plan rapidly and deliver limited duration and extended-range attacks to achieve precision effects against highly valued adversary assets."[35]

Of critical importance, OPLAN 8010–08 was to operate "in concert with an integrated communications campaign to shape adversary perceptions, intentions, and actions across all campaign phases as well as influence other regional and global actors." This was to support deterrence that was based on "both the

perception and reality of its capabilities and political will." A critical component
of this was "forward presence."[36] Important phases of the plan that involved
signaling include

Phase 1: Deter, when the situation has "moved outside of normal day-to-day
          operations with the emergence of a crisis."
Phase 2: Seize the initiative, where "the emphasis of deterrent operations
          moves from diplomatic, informational, and economic efforts to
          military engagement."

The bulk of OPLAN 8010–08 is not available to the public, so what movements
the forces were to undertake are opaque. The DEFCON system remained the
main driver for preparatory moves and presumably assigned forces played a role
here as they had during the Cold War. It appears that the shift in the DEFCON
was assumed to become known by an adversary and thus constituted signaling.
The existence of OPLAN 8010–08 and when it was to supersede previous plans
also suggests that the existence and general content of the plan itself was also a
form of signaling. Significantly, peacetime operating areas for E-6B command
and control aircraft were to become unclassified, which strongly supports the
idea that their movements also constituted signaling. The existence, but not
the makeup, of a Secure Reserve Force was also unclassified. Reconnaissance
aircraft movements were classified, but it appears as though these also played a
role in Phase 1.[37]

Between 2008 and 2012, changes in the global situation produced an updated
STRATCOM operations plan, OPLAN 8010–12, which went into effect on 30
July 2012. Unlike previous plans, potential enemies were identified and broken
down into peer/near-peer states that could "compete economically and militarily,"
on the one hand, and regional states and non-state actors that could proliferate
WMD and generate regional instability on the other hand. Russia and China
constituted global threats in the peer/near-peer category, while the other three
states are redacted from the declassified document.[38] These are probably North
Korea, Iran, and Pakistan.

OPLAN 8010–12 genuflected to the same deterrence objectives its predecessor documents used. It employed different language, however, in its concept of
operations: "This plan follows a premise that to achieve escalation control, the

US military and other instruments of national power will effectively match an adversary on multiple levels of conflict." OPLAN 8010–12 "provides options to deter, dissuade, control escalation with, and if necessary, defeat adversaries who threaten U.S. interests. This plan also assures allies and partners of extended deterrence." This was to be done through "decisive influence."[39]

Unlike the previous OPLANs and more in line with Cold War–era deterrence as practiced in the late 1950s through the mid-1960s,[40] American movements were to be "aimed at the cognitive processes of an adversary, namely the key decision makers." Specifically, analysis was to be undertaken to "expose effects which can be generated to highlight the specific interests of the strategic decision makers which will restrain them from unacceptable actions." STRATCOM's forces were to conduct "routine deployment of strategic forces for planned training and exercises, and theatre cooperation activities (e.g., exercises) to ensure unified action with GCCs, allies and partners."[41] OPLAN 8010–12 followed its predecessor plans' phases: Deter, Seize the Initiative, Dominate, Stabilize, and Enable Civil Authority. The first two phases had signaling built into them.[42]

Signaling Tools Prior to 2014

The bulk of the U.S. signaling load during the 2000–2014 period was carried by conventional forces, with the strategic nuclear forces acting as a backstop in the background. In the 1970s the United States established Freedom of Navigation Operations (FONOPS) as a response to unregulated assertion of national control over international bodies of water, specifically the oceans, to ensure they can be used safely for commerce. From the 1960s on, the U.S. Navy conducted FONOPS using naval vessels in concert with diplomatic activity to assert this right globally, against friends as well as potential foes. FONOPS was not strictly a Cold War tool: it continues today.[43]

During the Cold War, vessels conducting FONOPS carried nuclear weapons as part of their normal load-out. This caused the Soviet Union some concern when U.S. Navy ships carrying Tomahawk cruise missiles conducted FONOPS in the Black Sea in the 1980s, given their proximity to certain sensitive facilities.[44] After the Cold War, however, nuclear weapons were removed from U.S. Navy surface vessels by 1992, and subsequent FONOPS were conducted with conventional capabilities only.[45] According to annual reports, FONOPS did not start up again against Russia until 2019, though they remained a staple to challenge PRC activity in the South China Sea over the past decades.[46]

Former secretary of state Henry Kissinger apocryphally referred to an aircraft carrier as 100,000 tons of diplomacy, and so it remains in the post–Cold War period. And like their other surface brethren, they no longer carry nuclear weapons. That said, U.S. Navy aircraft carriers, due to their flexibility and capabilities, remain a primary signaling tool and have been used by such by successive administrations throughout the 2000s and 2010s, particularly with Iran as the audience.[47]

The Ice Exercises (ICEX) were related to FONOPS, Cold War–era submarine exercises in the Arctic held every two years using pairs of nuclear attack submarines. These exercises had multiple purposes that included scientific information collection, but the primary objective was to maintain experience in maneuvering and surfacing a submarine in the ice and in the operations of passive and active sonar under Arctic conditions.[48] In 2003 the ICEX series was modified to include a British nuclear attack submarine, as Canada had no submarine force capable of operating in the Arctic. Royal Navy submarines had not conducted under-ice operations since 1996.[49] The ICEX operations remained a predictable staple presence activity throughout the 2000s and 2010s.

The USAF ICBM force was a general deterrence tool when it came to signaling. The Cold War ICBM force peaked at 450 Minuteman II, 500 Minuteman III, 54 Titan II, and 50 Peacekeepers but declined to 400 Minuteman III by 2022.[50]

Every year, Minuteman III missiles were randomly removed from their silos, had their W78 warheads replaced with a non-nuclear Joint Test Assembly, and launched from Vandenberg AFB at the Kwajalein test area. Code-named Glory Trip, "these tests warn us if there are any issues that need to be addressed with the weapon." Importantly, "they also show our adversaries that we're still quite capable of using our Minuteman III system, despite its age."[51] In essence, Glory Trip is designed to ensure the credibility of the system for all audiences. During the Cold War, there was an average of six Minuteman and three Peacekeeper Glory Trip exercises per year in the 1980s. In the 1990s this dropped to two to three a year. In the 2000s the average was two to three a year, plus five development launches over five years. From 2010 to just prior to the Russian invasion of Ukraine, the numbers were down to one or two a year. After 2014, however, the average was four Glory Trip exercises and one development launch per year.[52]

Notably, Glory Trip exercises were scheduled years in advance and were not keyed to specific crisis events.[53]

Like the ICBM force, the ballistic missile submarine force served as a general deterrence tool. In its original configuration, the *Ohio*-class SSBN carried twenty-four Trident II D5 missiles. By 2017 the number of operational missile tubes on each of the fourteen ships was limited to twenty instead of twenty-four.[54]

SLBM testing included live launches of SLBMs equipped with non-nuclear Joint Test Assemblies. These fell into three categories: operational test launches to ensure the reliability of the Trident II D5 missile itself; demonstration and shakedown operational launches, in which submarines and their crews coming out of refit recertified their procedures and ability to launch missiles; and commander in chief evaluation test launches that checked higher-level communications and procedures. After 1992 there were an average of six or seven of all types of launches per year. In the 2000s there was an average of five a year. From 2010 to 2014 the number fluctuated between four and seven launches.[55]

Starting in June 2010, the number of JTA reentry bodies that were deployed during launches increased to eight per launch, though the number of SLBMs tested during these launches is unclear. That number rose to fourteen reentry bodies in two 2015 tests.[56] Of importance were launches conducted in 2015 and 2019 off California in which the flight paths were observed by the public. These strongly suggest that these were depressed trajectory SLBM launches, whereby the missile's ballistic trajectory is significantly flatter than normal which reduces flight time and detectability with a consequent impact on warning.[57] Like the compressed trajectory test in 2005, proof of this new capability was likely intended for a wide audience. That said, U.S. Navy spokespeople emphasized that SLBM tests "are not conducted in response to specific world events."[58]

The reduction of the U.S. SSBN force from eighteen to fourteen in 1992 as a result of START II led to the modification of four *Ohio*-class SSBNs into guided missile submarines (SSGNs). Twenty-two missile tubes on each sub were converted to carry 7 cruise missile each for a total of 154 Tomahawk TLAM or Tactical Tomahawk (block IV). These two systems carry conventional warheads.[59] The Cold War–era TLAM-N, the nuclear version, was removed from deployed ships in 1992 and put in storage. In 2010 the Obama administration announced the missiles would be dismantled.[60] Warhead deconstruction is believed to have been completed at the Pantex facility by 2013.[61] The SSGN force was repeatedly employed for signaling purposes.

Finally, the USAF's bomber forces have repeatedly been used for signaling, though the composition of the mixed force of B-1, B-2, and B-52 bombers generates some ambiguity when it comes to whether the messaging is based on conventional or nuclear combat potential. Indeed, the capabilities of the three bomber types were in flux throughout the 2000–2021 period as they responded to arms control, maintenance, and political challenges.[62]

In terms of combat potential, the B-2A relies on its reputation for stealth and surprise, so its presence in a theater of operations generates a sense of the unknown. The B-52 is a known and visible quantity. In its 1950s configuration, it carried two high-yield thermonuclear bombs. In the 1960s it was loaded with two cruise missiles designed to destroy air defenses and four megaton-yield bombs. In the 1970s the B-52 carried ten SRAM for air defense suppression and four megaton-yield bombs. The progressive reduction of the number of available airframes from 744 in the 1960s to 46 today is notable, but with each of those capable of potentially carrying 20 AGM-86B cruise missiles, that is still a possible 920 desired ground zeros, with the AGM-86B ranging out to 1,500 miles (2,700 km), the system remains potent.

2014–2016: Reconfiguring Signaling

The seizure of Crimea in 2014 generated responses from NATO members in both bilateral and multinational capacities. The United States deployed a battalion of the 173rd Airborne stationed in Italy to the Baltic states and Poland for a series of joint exercises. Additional USAF F-16s and three C-130J transports deployed to Poland to augment air policing missions. Six F-15s were deployed to Estonia to augment the four already there, and six F-15Cs augmented the four F15Cs in Lithuania for the air policing role. U.S. Navy naval deployments included the destroyer USS *Truxton*, which carried a 90-cell vertical launch system capable of launching RIM-161 ABMs and BGM-109 Tomahawk cruise missiles: the *Truxton* conducted a Black Sea deployment in March, followed by the same-class USS *Donald Cook* in April, which was followed by the frigate USS *Taylor* that was part of the NATO Standing Maritime Group. Then the *Ticonderoga*-class guided-missile cruiser USS *Vella Gulf* with its 122-cell vertical launch systems that were similar in capability to the USS *Truxton*'s followed.[63] NATO's other transatlantic partner, Canada, deployed six CF-18 Hornet fighters to Romania to increase air policing capacity in April, followed by an airborne company contingent to Poland in May, and deployed a guided missile frigate HMCS *Regina* to the Mediterranean, also in May.[64] Russia expressed its displeasure with the Black Sea naval deployments. In one event, two nuclear-capable Su-24 Fencer strike aircraft buzzed the *Donald Cook* and a Polish helicopter twelve times while they were operating in international waters.[65] None of these assurance measures involved nuclear forces.

The Obama White House European Reassurance Initiative (ERI), released on 3 June 2014, grouped all American conventional deterrent activities in Europe under the umbrella Operation Atlantic Resolve.[66] The ERI essentially listed all of the ongoing military exercises intended to present a deterrent front to the Putin regime, except those involving strategic nuclear forces. For example, in May 2014 two B-2As flew over the Atlantic, rendezvoused with KC-135 tankers out of RAF Fairford, while maintaining communications with an E-6B over Rostock, Germany, and another E-6B over the Atlantic. This may or may not have been part of the annual Global Lightning exercise.[67] There was no official public visibility on this exercise, but Russian signals intelligence organizations would have been able to listen in. The overt messaging came in the form of three B-52Hs (call sign "Doom") and two B-2As that deployed to the United Kingdom on a Bomber Assurance and Deterrence (BAAD) exercise.[68] There was no explicit connection between NATO conventional deterrence exercises and U.S. nuclear deterrence exercises.

The origin and development of BAAD exercises is opaque, but the first ones do not appear to have been run before 2013.[69] USSTRATCOM's Global Strike command handles these in order to "assure audiences" in each of the Geographic Combatant Command's areas of responsibility. Notably, "requests for BAAD missions may be made by [Geographic Combatant Commands] or other states or USSTRATCOM may push the asset to theatre as part of its own planning cycle or reassurance requirements decided upon by the command."[70] BAAD missions are used "to create complex challenges to our adversaries' warfighting capability while simultaneously demonstrating our nation's commitment and resolve to our allies."[71]

NATO's recognition that the global and regional situation had dramatically shifted was expressed by the Wales summit of 4–5 September 2014. The declaration that emerged altered NATO's deterrent posture on all levels. The centerpiece was the Readiness Action Plan (RAP), which featured "assurance measures" and "adaptation measures."[72]

In reality, the assurance measures were established until NATO could reorganize itself and "adapt" to the new situation. Primarily this effort focused on expanding NATO exercises and establishing a Very High Readiness Joint Task Force (VJTF), which was really just reestablishing what had been called ACE Mobile Force (Land) and ACE Mobile Force (Air) during the Cold War. The VJTF's function was the same as the AMFs: deploy a brigade-sized force to an area of the Alliance that was under some form of threat below the threshold of hostilities in order to express NATO concern and signal the willingness to escalate, if necessary, through the spectrum of conflict.[73]

Regarding nuclear weapons, the declaration noted that NATO remained a nuclear alliance, with the U.S. arsenal the ultimate security guarantee for members, but the likelihood of nuclear usage remained remote. The Wales Declaration did not link signaling and other deterrent activities to nuclear weapons, nor was there any discussion of dual capable aircraft possessed by NATO members.[74]

The day after the Wales summit, Russian aircraft deliberately harassed HMCS *Toronto* while it was conducting operations in international waters in the Black Sea. Two Su-24 Fencers and an An-26 Curl intelligence collection aircraft were involved. Fearful Canadian commentators downplayed the event, and some even parroted a Russian statement that the aircraft posed no threat to the *Toronto*.[75] Other outlets, however, discovered that one Fencer came within three hundred meters of the ship and the *Toronto* had its fire control radar locked on to the Russian aircraft.[76] Canada is a weak link in NATO, and this kind of attention was probably produced for multiple purposes, one of which would have been to generate anti-military sentiment in the Canadian intelligentsia and degrade any Canadian support for Ukraine. Canada possesses the largest Ukrainian diaspora in the world.[77]

There were two significant American exercises held in the wake of the Wales summit. The first was Valiant Shield, the fifth in a series that started in 2006. This was a joint American exercise held around Guam and the Mariana Islands. It included two aircraft carrier strike groups as well as the deployed B-52s from the Continuous Bomber Presence (CBP) mission. In terms of signaling, there appears to have been no nuclear component in the exercise and it was geared toward Pacific Rim operations.[78] The second was Global Thunder 15, a USSTRATCOM exercise. There were few details released, though the blurred and darkened released imagery suggested B-2A and B-52H alert operations.[79] Global Thunder was part of the scheduled annual exercise series. Global Thunder 15 was conducted in conjunction with NORAD exercise Determined Dragon 14, in which NORAD forces "intercepted" B-52s "that had completed their GLOBAL THUNDER mission and were on their return leg" coming through the Goose Bay area, implying that the B-52s were operating in the Arctic, presumably against a notional Russian target system.[80]

The nature, extent, and messaging of the signaling game changed significantly when Ashton Carter took over as secretary of defense in February 2015. Given Russian aggression against Georgia and Ukraine, Carter "took a series of steps to deal with these challenges . . . including the creation of the first new war plan for responding to a Russian attack in twenty-five years."[81] Indeed, Carter said, "Not everyone in the White House shared my concerns about Russia. . . . The administration's less-than-wholehearted pushback against Russian aggression almost led to disaster in Syria."[82] The open-source record of U.S. signaling activity in 2015–16 reflects this in that the bulk of it occurred in the Asia-Pacific area, which reflected the Obama administration's tilt away from Europe. The administration authorized five FONOPS, two in 2015 and three in 2016, in areas claimed by the People's Republic of China.[83] After North Korea conducted a fourth nuclear test, claimed a satellite launch, and claimed an SLBM launch between January and April 2016, B-2As were deployed to RAAF base Tindal in Australia and the deployment of the THAAD missile system to South Korea was announced.[84] B-1Bs replaced the B-52s at Guam for CBP and they flew a profile over South Korea in response to a nuclear test in September 2016.

Carter authorized significant signaling activities vis-à-vis Russia, specifically, Polar Growl with its obvious allusions to the Arctic and Russian bears. On 1 April 2015, two pairs of B-52 Hs, from Minot AFB and Barksdale AFB, lifted off on a designated Bomber Assurance and Deterrence mission. One pair headed into the Arctic to practice navigation skills,[85] while the other pair crossed the Atlantic and conducted "dissimilar aircraft" interception training with Canadian, British, and Dutch fighters over the North Sea. Of note, USAF public affairs leaned on the fact that Polar Growl was there to demonstrate "compliance with national and international protocols and due regard for the safety of all aircraft sharing the air space,"[86] a shot at Russia's bad behavior over the North Sea with its flights.

Critics of American nuclear policies, however, alleged that the B-52Hs in the Arctic conducted "a simulated strike exercise against Russia. The bombers proceeded all the way to their potential launch points for air-launched cruise missiles before they returned to the United States."[87] It is not exactly clear whether this was the case. The four Polar Growl B-52Hs were, however, used to exercise NORAD. RCAF CF-18s scrambled from CFB Cold Lake and CFB Bagotville and were supported by USAF KC-135 tankers launched from Fairchild AFB and Bangor Air National Guard Base. The Polar Growl bombers were intercepted at the "northern and eastern extremity of the Canadian Air Defence Identification Zone."[88]

Polar Growl was separate from the other bomber deployments in 2015. In June 2015, three B-52Hs from Minot AFB deployed to RAF Fairford. These were assumed to be Bomber Assurance and Deterrence deployments publicly aligned with NATO exercises BALTOPS 2015 and Saber Strike 15. BALTOPS 2015 was conducted in the Baltic.[89] There was no nuclear play in either exercise, which makes the deployment of Minot AFB's B-52Hs odd. The proximity of two B-52H ALCM "Pez dispensers" in the Baltic Sea in close proximity to Russian Iskanders in Kaliningrad was clearly not coincidental. Concurrent with the B-52H deployment was the separate deployment of two B-2As to Fairford.[90] What exactly they were doing was unstated in the public domain, but USSTRATCOM specifically noted that "elevating Fairford to active status is a key component to this deployment due to the critical capabilities of this base."[91] This implied that bombers would be forward-based in the United Kingdom, just as they had back in the late 1940s during the Berlin Blockade and throughout the Cold War. And if the Russians leafed back into their Soviet-era intelligence reports, they would find that USAF bomber aircraft stationed in the United Kingdom had, as one of their target sets, the destruction of Moscow.[92]

In February 2016, three B-52Hs flew to Spain to exercise Cold War–era forward operating locations. The aircraft participated in exercise Cold Response, held in Norway, where they supported ground forces with conventional "play" flying extended missions from Spain. That said, USAF press releases emphasized the dual-capability of the B-52H force.[93] Like RAF Fairford, Morón Air Base in Spain played a significant role in the 1950s hosting B-47s targeting the Soviet Union, but it was also evident that the operation was designed to ensure that the Putin regime understood that Norway was under the nuclear umbrella as well as the conventional one. Indeed, several B-52s also deployed to RAF Fairford in June 2016 where they once again participated in BALTOPS 16 and Saber Strike 16.[94]

The follow-on to Polar Growl was Polar Roar, held on 3 August 2016. Polar Roar was a designated Bomber Assurance and Deterrence mission that involved the launching of five bombers, three B-52Hs and two B-2As. Supported by fifteen KC-135s and ten KC-10 Extender tankers, two Minot AFB–based B-52Hs "flew over the North Pole," came back over Alaska, were "intercepted by F-22s," and then dropped "inert weapons" in one of the areas of the Joint Pacific Alaska Range Complex (JPARC). At the same time, two B-2As departed Whiteman AFB and flew a profile over the Pacific Ocean, to the Aleutians where they were intercepted by F-15s and then conducted "an inert weapons drop" at JPARC. The last B-52H (there were supposed to be two, but one developed a fault) departed Nellis AFB, flew to the North Sea, and then over the Baltic to Estonia, where dissimilar interception training was conducted with the NATO Air Policing Mission.[95]

As with Polar Growl, critics of American nuclear policy displayed hypothetical maps depicting B-52s closely paralleling the north Arctic coast of Russia and B-2As flying to the end of the Aleutian Island chain, all in an attempt to portray Polar Growl as provocative activity.[96] What path, exactly, that the B-52Hs flew in the Arctic is not available in the public domain. In a media interview, Major General Richard Clark was asked about these extrapolations and would not be drawn out. Interviewer: "The Russians would look at that and see it as a dry run for an attack on targets inside Russia." Clark: "I guess they can draw the conclusions that they need to draw. . . . This was a significant exercise for us. We're training the way we might have to fight."[97] What is clear is that Polar Growl and Polar Roar demonstrated that the B-52H force could conduct Cold War–style transpolar operations. Two aircraft could release forty ALCMs, the equivalent combat potential of twenty B-52s from the 1950s carrying gravity bombs, or twelve B-52s from the 1960s carrying standoff missiles and bombs.

Bomber Assurance and Deterrence missions continued in late 2016. In August, there was a B-52H deployment to Fairford, joined by a pair of B-1Bs from Dyess AFB. These aircraft were employed in exercise Ample Strike, a conventional

exercise held in the Czech Republic.[98] The annual Global Thunder exercise was
held in November, with British participation highlighted by USAF media.[99] Global
Thunder generated "aircraft from normal day-to-day operations to nuclear capable
alert status" and concluded with a mass launch of twenty B-52Hs.[100]

Compared with previous activities prior to 2014, the 2015–16 period clearly
demonstrated a new and higher level of signaling for American nuclear forces.
Indeed, the deliberate emphasis on forward deployment and revitalization of
facilities in the United Kingdom and in Spain harkened back to the 1958–64
period of Reflex Alert deployments, where B-47 bombers were rotated to forward
bases in those countries. The three forward deployments in 2016 alone and the
activities of the aircraft while deployed clearly indicated that NATO was firmly
under the American strategic nuclear umbrella, with an emphasis on the Baltic
states and Norway. This coincided with the establishment of the NATO enhanced
Forward Presence (eFP) after the NATO Warsaw summit in July 2016 and thus
backstopped these moves. Polar Growl and Polar Roar was the next level of
backing for this new, emergent deterrent system, with connection made between
NATO and STRATCOM with the flight of CONUS-based B-52Hs over the Baltic.
As Secretary of Defense Ashton Carter noted, "We developed formal war plans
for a European conflict for the first time since the Cold War."[101]

Controlled Chaos, 2017–2019

The accession of Donald Trump to the presidency brought with it a unique foreign
policy style in which events in Europe were by no means front and center on
the agenda. The kaleidoscopic whirlwind of personal diplomacy coupled with
instantaneous pronouncements on social media generated a situation whereby
perpetual crises over China, North Korea, Iran, Syria, the war against ISIS, and
Afghanistan muddied the waters when it came to signaling to support coherent
American policy. At least, that is the impression that existed if one was caught
up in the polarized nature of American political rhetoric. Yes, issues regarding
Central Command tended to dominate when James Mattis was secretary of
defense, but Mattis' eventual replacement, Mark Esper, understood and spoke
the language of signaling.[102]

The days of STRATCOM conducting Polar-series exercises were gone for the
time being thanks to the 2017 "honeymoon" with Russia, but Trump tripled the

number of FONOPS directed at the People's Republic of China in 2017 and 2018.[103]
When it came to the July–September 2017 showdown with Kim Jong Un, B-1B
bombers escorted by F-15s flew near the Korea coast to underscore the American
position.[104] Despite the heated rhetoric over North Korea, there appears to have
been no overt movements of strategic nuclear forces for signaling purpose.

However, three B-1Bs and three B-52s deployed to Fairford in May as part
of a self-identified BAAD. They were also used concurrently to support NATO
Exercise Arctic Challenge in Norway and, later, BALTOPS 17 and Saber Strike.
Two B-2As joined the rest of the force on 9 June but did not participate in the
NATO exercises, which implies they were conducting other unspecified activities.[105] During the course of BALTOPS 17, an official statement by Russia claimed
an Su-27 from Kaliningrad intercepted and escorted a B-52H that it claimed was
"flying over neutral waters parallel to the Russian border."[106] All of this activity
occurred at the same time NATO's eFP forces deployed permanently to the
three Baltic states and Poland. The BAAD deployment ended on 17 June, with the
B-52Hs conducting an intercept exercise with RCAF CF-18s based at Iceland.[107]
There was now an increase in numbers of aircraft involved, but their activities
were consistent with previous operations over the past three years.

In the latter half of 2017, B-52Hs were deployed on the Red Flag Alaska exercise,
something that was relatively new. Previously, Red Flag Alaska focused on fighter
operations. Historically B-47s had been deployed in the 1958–64 period on Reflex
Action alert with nuclear weapons at bases in Alaska to strike targets in the
Soviet Far East but were withdrawn by 1965. The presence of B-52Hs in Alaska
and their demonstrated ability to conduct operations there was affirmed by Red
Flag Alaska.[108] Two B-1Bs and a B-52H revisited RAF Fairford in August. These
aircraft, in addition to their other activities, deployed to Poland and Slovakia to
visit air shows, clearly sending an assurance message to those Eastern European
NATO members.[109]

Finally, Global Thunder 18 was executed in October–November 2017, with a
strong public emphasis on B-2A operations.[110] Russian information operations
outlets sought to portray the exercise as prelude to "put its strategic bombers on
constant 24-hour alert . . . for the first time since the Soviet Union collapsed in
1991."[111] This was denied by USAF public affairs and there was no indication that
this was the case.[112]

In 2018, U.S. bomber deployments were renamed from Bomber Assurance and Deterrence missions to Bomber Task Force (BTF) missions. The first BTF mission consisted of the deployment of four B-52Hs from Minot AFB to Fairford in January to exercise "forward operating locations for strategic bombers" to assure and deter.[113] Operations included supporting Lithuanian Special Forces with employing precision-guided munitions, which involved B-52Hs operating over the Baltic states themselves.[114] Given the fact the B-52Hs from Minot were nuclear-capable B-52s and that their combat potential could include the use of cruise missiles from a forward launch point that close to Russia would not have been lost on the Putin regime.

A significant event in April 2018 was the tripartite strike against Syria. The employment of sarin nerve agent on two occasions followed by a devastating chlorine attack against an anti-regime suburb of Damascus produced an American–British–French response, which involved more than seventy aircraft (including two B-1B bombers), nine ships, and a submarine. Collectively these platforms fired approximately sixty Tomahawk and thirty-six other cruise missiles.[115] The Russian public response was limited to fulminations from Putin, Lavrov, and Antonov, the Russian ambassador to the United States, with vague threats about "consequences."[116] Interestingly, there were media reports of Russian *Kilo*-class submarines supported by a surface task group and ASW aircraft playing cat and mouse with a Royal Navy *Astute*-class attack submarine equipped with Tomahawks in the eastern Mediterranean. These reports asserted that the Russian task force prevented the British submarine from firing as part of the operation.[117]

A BTF deployment of B-52Hs to Fairford occurred in September. B-52Hs, supported by tanker aircraft, conducted a flight "up the Norwegian coast, over the Barents Sea, and Arctic Ocean" and back to RAF Fairford.[118] Another account said a B-52H turned "east in the Barents Sea and flew north along the west coast of Novaya Zemlya before turning west again north of Franz Josef Land. The plane then flew south along the west coast of Spitsbergen (Svalbard)."[119] This flight was clearly more provocative than the Polar series of flights as the aircraft was right in the Russian backyard and in close proximity to the Northern Fleet bases around Murmansk, the nuclear test site at Novaya Zemlya, and the disputed Svalbard Island. At the same time, a Bomber Task Force of B-2As and an E-6B Mercury deployed to Hawaiʻi and conducted operations from there. Wide coverage of the B-2A operations was given by the USAF.[120] The ambiguity of the Hawai'i deployment was notable, as much as the flight paths of the aircraft. Additional signaling may also have involved a resumption of FONOPS in Peter the Great Bay, off Vladivostok, using the guided missile destroyer USS *McCampbell* "to challenge Russia's excessive maritime claims" but also not coincidentally opposite the Russian Pacific Fleet headquarters.[121] The destroyer, like her sister ships, carried a ninety-six-cell vertical launch system capable of firing Tomahawk.

The diplomatic backdrop at this point was the U.S. withdrawal from the INF Treaty in response to Russian violations of it by deploying Iskander missile systems to Kaliningrad and the St. Petersburg area. According to National Security Advisor John Bolton, "the Russians were agitating, playing on European's fears we were abandoning them, leaving them defenseless."[122] This occurred during the annual Global Thunder exercise. There is nothing in the public domain that indicates that Global Thunder 19 was any different from any previous iteration, but USAF public affairs ensured that there were copious color close-ups of eight B-52Hs at Minot AFB situated on the "Christmas Tree" alert facility, all fully loaded with ALCMs, and another sixteen B-52Hs in an increased ready state.[123]

From January to July 2019, U.S. signaling activity dramatically accelerated. In January, the B-2A BTF deployment to Hawai'i was repeated with three aircraft.[124] Russia claimed five B-52s were operating over the Norwegian Sea in late February, practicing to attack northern Russia.[125] In March, a BTF consisting of six B-52Hs, four of which were nuclear capable, arrived at RAF Fairford.[126] There was no scheduled NATO exercise at this time, but some of the operations included training with Latvian, Estonian, and Polish forces. One account asserts that the BTF departed its bases in the United States, but then instead of heading for RAF Fairford, two of the aircraft headed straight for the Russian air identification zone off Kaliningrad and made a 90- or 180-degree turn before reaching it and then recovered to Fairford. At one point, two Su-27 Flankers intercepted and escorted one of the bombers.[127] Russian information operations outlets exaggerated the affair and claimed one B-52 was driven out of Russian airspace. Another claimed that the aircraft was practicing an attack against the Baltic Fleet base at Baltiysk. Photographs of the B-52H released by Russian outlets clearly showed it was carrying a pair of practice (blue) Joint Direct Attack Munitions, despite the fact that they emphasized the nuclear capabilities of the aircraft in their propaganda output.[128] That said, Russian conduits discussed with some alarm the combat potential of B-52Hs operating in Eastern Europe, pointing out that the aircraft "could easily strike Russia with 140 cruise missiles" with nuclear warheads.[129]

Six more B-52Hs rotated in April. This BTF flew sorties "over Norway, Iceland, Hungary, Latvia, and the Baltic Sea." Major General James Dawkins, Eighth Air Force commander, emphasized, "Make no mistake, Russia is certainly watching what we're doing, but most importantly our allies are watching."[130] On 17 June, three B-52Hs departed Minot AFB, crossed the Atlantic, and split up. One B-52H flew over the Baltic, where it was intercepted by Russian Su-27s, and then it returned to Minot AFB. The other two headed straight for occupied Crimea, which prompted a scramble by Russian interceptors. The two B-52s turned away with one diverting to the United Kingdom with an apparent engine fire.[131] On 11 July a B-52H departed Barksdale AFB and mounted a Baltic Sea flight off Kaliningrad before returning to base.[132] At this point, a Russian information operations conduit expressed concern about the potential threat posed by B-52H flights over the Norwegian Sea and the relationship between cruise missile ranges and specific Russian targets.[133]

Of interest here is the demonstration of global capability as well as the ability to monitor the Russian response both physically and electromagnetically while making the point that Russian anti-access/area denial capabilities established in both Kaliningrad and occupied Crimea could easily be put at risk if the Russians attempted to use those capabilities to isolate Eastern Europe. Concurrent with these exercises, two B-52Hs stationed on Guam departed on a mission profile that took them to the Russian-occupied Kurile Islands, an activity that had not taken place since the Cold War.[134] The positioning of the B-52Hs demonstrated that Russian Pacific Fleet facilities in Vladivostok as well as the critical ballistic missile submarine base at Petropavlovsk could simultaneously be put at risk from there.

The only secretary of defense that has been open and explicit about the use of signaling in the memoir literature is Mark Esper. He tried to orient Trump generally on the issue of signaling and deterrence after Iran shot down an MQ-4 Global Hawk in international airspace. Esper believed Trump wished to escalate the situation to force the Iranians to back down.[135] He also argued that deterrence would be restored through a demonstration of resolve, but Trump did not share his views on the issue.[136] Regarding Europe, in 2020 Esper ordered destroyers to the

Barents Sea for the first time in nearly three decades, directly challenging the Russian Northern Fleet. U.S. bombers also flew throughout Europe, including over Ukrainian airspace, and ground forces increased training with NATO allies.[137]

For the last half of 2019, there was the appearance of steady-state general deterrence and signaling, but in fact signaling was ramping up. A BTF to RAF Fairford consisting of a pair of B-2As was conducted right after the Putin regime declared that the INF Treaty no longer existed on 5 August.[138] This was followed by a four ship B-52H BTF to RAF Fairford in October. The four aircraft proceeded across the Atlantic, where one landed at RAF Fairford, while the other three flew directly to the Baltic Sea area off Kaliningrad, turned around, and recovered to RAF Fairford.[139] This BTF also included a twelve-hour mission over the Black Sea, region unspecified,[140] but from Russian accounts this mission involved a run at Putin's dachas in Sochi before peeling away, flying to Georgia, turning around over the Black Sea, and returning to base.[141]

Of critical importance was a significant mission undertaken in the Barents Sea. Three B-52Hs, escorted by Norwegian F-16s, flew to the northern tip of Norway and split up "into three different directions; one stayed in the western Barents Sea, one flew northeast to the coast of Novaya Zemlya and one flew along the coast of the Kola peninsula towards Kap Kanin at the entrance to the White Sea." A USAF RC-135 Rivet Joint and a U.S. Navy P-8 flew over the Barents Sea collecting intelligence during the operation.[142] The European BTF overlapped with Global Thunder 20, which was routine but did involve B-2A flights to the European area supported by KC-135 tankers stationed at RAF Mildenhall.[143] Russian sources claimed two B-52Hs were operating over the Norwegian Sea practicing strikes against bases in the Kola Peninsula.[144]

One important decision was to draw down the CBP on Guam. Esper favored Dynamic Force Employment (DFE), developing agile forces to deploy quickly and remain unpredictable while reducing wear upon the force.[145] The last CBP mission ended on Guam on 17 April 2020. Four B-1Bs replaced the B-52s on the first DFE deployment on 11 May.[146] The replacement of nuclear-capable B-52Hs with conventional B-1Bs suggests there was a shift in planning focus from the Pacific region to confronting Russia and that more B-52Hs were required for this and related tasks. That said, two B-1Bs on a flight from Guam to Alaska "detoured" over the Sea of Okhotsk on 21 May. This move would have been deemed by the Russians as incredibly provocative given the sensitive nature of the Japanese Sakhalin Island claims, the presence of the Pacific Fleet's ballistic missile submarine base at Petropavlovsk, and the fact they view the Sea of Okhotsk as a Russian lake.[147]

Stepping It Up in 2020

Global uncertainty about the COVID-19 pandemic generated serious concerns over readiness, especially strategic nuclear forces. The situation did not significantly attenuate signaling operations. There were essentially four geographic clusters of activity during the year: Norway and the United Kingdom; the NORAD area; in and around Ukraine, including and especially the Black Sea; and the Baltic.

The year kicked off with a flight by two Tu-60 Blackjacks from the Kola Peninsula over the North Pole and into and through the CADIZ "days after a senior military officer warned that North America's early-warning system is outdated."[148] The response was unclear, but the flight was likely undertaken to probe NORAD and test responses under the conditions of the COVID-19 pandemic.

A cluster of intense activity manifested itself off Norway and the United Kingdom, most of it in March. Two days before a B-2A Bomber Task Force deployment to the Azores on 9 March,[149] two Russian Tu-142s, escorted by MiG-31 Foxhounds part of the way, were tracked and intercepted by Royal Norwegian Air Force F-35s and RAF Typhoons as they transited down close to the Norwegian coast. These were no ordinary Bears: one was a Tu-142MR Bear J, designed to communicate with ballistic missile submarines using very low frequencies.[150] This was a highly unusual development. It could have been a communications check or an attempt to reconnect with a submarine having issues communicating something that was obviously problematic with a ballistic missile submarine. One view was that they were collecting on the upcoming NATO Cold Response exercise, but this does not explain the presence of the Bear J.[151] The mission could have involved intercepting NATO member submarine communications.

A second intercept on 11 March, this time of two Tu-142 maritime reconnaissance aircraft, was conducted by the RAF west of the Shetland Islands, and then near Ireland, and to the Bay of Biscay, where French interceptors escorted them. The third intercept occurred on 12 March, with two Tu-160 Blackjack bombers following the same route as the Tu-142s, exploiting the information collected the previous day.[152] It is likely this was in part a demonstration of Russian capability in the face of COVID-19 as well as a test of the West's intercept capability under the new conditions. The Russian aircraft were not using transponders.

A pair of Tu-142s entered the Alaska and Canadian ADIZs over the Beaufort Sea also on 9 March, where they were promptly intercepted by USAF F-22s and RCAF CF-18s vectored on by an E-3 Sentry AWACS and supported by KC-135 tanker aircraft. Photos of only one of the Tu-142s were released, which suggests the other may also have been a Bear J communications aircraft. In any event, this was a ferret flight, possibly combined with a communications test or exercise, with intentions similar to the concurrent actions over the Norwegian Sea.[153] One account noted that the Tu-142s overflew the area where ICEX 20 was being conducted.[154]

In May 2020, the United States undertook a FONOP, for the first time since the 1980s, into the Barents Sea. The operation included a Royal Navy frigate and three U.S. Navy destroyers, plus a support ship. There was significant speculation as to the reason behind the implementation of FONOPS in the Arctic, but it was "to send a clear message of resolve to Moscow" in conjunction with other signaling activities.[155] The task force sailed off Norway and the Kola Peninsula and the Russian cruiser *Marshal Ustinov* sortied out to observe.[156] P-8A Poseidon and RC-135 Rivet Joint intelligence collectors tagged along.[157]

The Barents FONOPS coincided with a Bomber Task Force deployment of two B-2As and four B-52s to RAF Fairford. One purpose of this deployment was to "demonstrate to potential adversaries that, despite the coronavirus pandemic and ensuing complications, the Air Force has maintained its ability to carry out complicated, simultaneous missions."[158] Additionally on 29 May, a pair of B-1Bs flew on a round-trip mission from Ellsworth AFB, over the United Kingdom, and then over NATO's newest member, North Macedonia. They proceeded on to the Aegean where they refueled from Turkish KC-135s, then flew to the Black Sea, where they practiced long-range anti-ship missile operations, then to Ukraine, and finally back to North Dakota.[159] The B-1Bs were intercepted by Russian Su-27P and Su-30M fighters out of occupied Crimea, who filmed the B-1Bs. This footage was used by TASS as part of their ongoing information activities.[160] This may have been a response to an incident on 19 May when Russia ran a flight of

Tu-22M3 nuclear strike aircraft against NATO-member air defenses over the Black Sea, forcing Turkish F-16s, Bulgarian MiG-29s, and Romanian MiG-21s to intercept them.[161]

Four B-52Hs departed Minot AFB on 3 June, flew over the North Pole region, and linked up with Norwegian F-16 and F-35 fighters covering KC-135 tankers out of RAF Mildenhall in the UK, before proceeding on a route that took the aircraft to the Laptev Sea opposite the Russian LRA air base at Tiksi, and then back to Alaska where they were refueled again. This was an impressively demonstrative Arctic "round robin flight," reminiscent of 1950s and 1960s Cold War operations, and one that appears to be unprecedented during the Cool War, barring the release of information to the contrary.[162]

On 10 June, Russian LRA mounted a complex operation against the Alaskan ADIZ that occurred over the Chukchi Sea, Bering Sea, and the Sea of Okhotsk, probably in response to the 3 June operation. Two Tu-95MS bombers escorted by Su-35s and a A-50 AWACS aircraft were part of the first wave, and a second pair of Tu-95MS, also supported by the Russian AWACS aircraft conducted their flight. Interceptions were carried out by USAF F-22s supported by KC-135 tankers and an E-3C Sentry AWACS plane.[163] Four days after, two B-52H bombers from Minot AFB deployed directly to Eastern Europe to support BALTOPS.[164] State-controlled Russian media claimed that the B-52s "practised strikes against the western and northwestern parts of the Russian Federation, approaching St. Petersburg at a distance of 180 kilometers, without encountering any opposition from [the] Russian Aerospace Forces."[165] Essentially the B-52s operated over the Narva region of Estonia, which put all of western Russia in ALCM range.

At some point a decision was made to mount a Bomber Task Force operation to Alaska, and in short order three B-52s landed at Eielson AFB on 17 June. From there, they proceeded to the Beaufort Sea for interception exercises with RCAF CF-18s and USAF F-22s and other activities related to "strategic unpredictability, and operational unpredictability."[166] They arrived the same day the by-now-familiar Russian "package" of two Tu-95MS, two Su-35s, and an A-50 AWACS probed the Alaskan ADIZ.[167] Twenty-four hours later, Facebook and Twitter posts with footage of B-2As refueling off northern Norway appeared.[168]

The action continued into the summer. On Independence Day, three to five Tu-95MSs and Il-98 tankers escorted by fighters "headed for the American borders

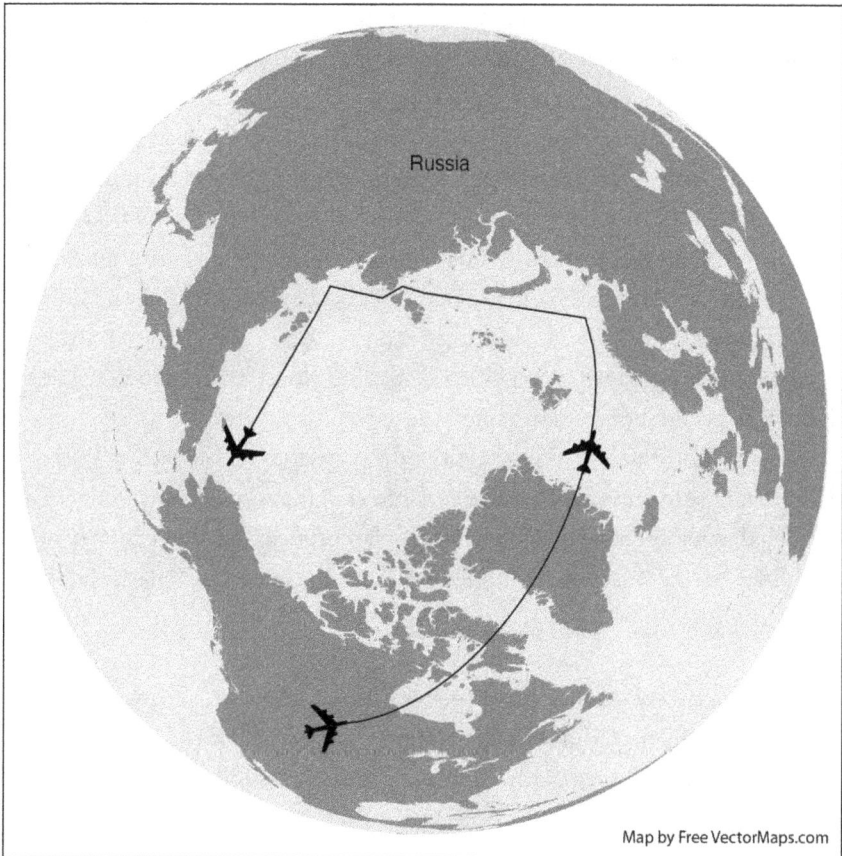

MAP 1 ◆ B-52 flights in the Arctic Basin, 3 June 2020 *Created by the author*

to practise combat missions" as "an enchanting congratulations [for] the Americans."[169] The same day, two B-52Hs launched from Minot, headed over Alaska for the Arctic, made a sharp turn, and paralleled the Kamchatka Peninsula and the Kurile Islands to Japan.[170] In mid-August, Long Range Aviation deployed a number of Tu-160 bombers—probably to Anadyr—where they joined Tu-95MS and Tu-22M3 bombers. This weeklong exercise involved "the tasks of relocating aviation equipment to operational airfields, flights with aerial refueling, the combat use of aviation weapons and other tasks for their intended purpose will be worked out." Russian sources noted that new jamming systems would be used against NORAD air defenses.[171] During the exercise, two Tu-95MS bombers "flew near the Japanese borders after Tokyo's provocation near the Kurile islands to work out strikes, especially after Tokyo's rhetoric about the need to take four islands [in the Kuriles] by force."[172]

These operations were accompanied by the Russian Pacific Fleet guided missile submarine *Omsk* which "surfaced relatively close to St Matthew Island" midway between Alaska and Siberia.[173] The *Omsk* is an *Oscar*-class SSGN that carries twenty-four P-700 Granit cruise missiles, each with either a conventional or a 500 kt yield nuclear weapon out to a range of 625 km. This placed the air defense system in western Alaska at risk. In effect, the *Omsk* demonstrated that they could take down radar sites and forward operating locations for the F-22 fighters in order to permit bombers to exploit the gaps.

Exercise Allied Sky overlapped with other activities in August, but the backdrop to it is convoluted and requires some explanation. USAF public affairs insisted that it was a scheduled exercise, but it morphed into a collection of other signaling activities that lasted into September. At the same time Belarus exploded with unrest on 9 August after a rigged presidential election that retained Alexander Lukashenko in power.[174] The situation quickly escalated over the next two weeks, prompting a violent crackdown by security forces, which in turn resulted in EU sanctions on Belarus. Lukashenko claimed that NATO members Lithuania, Latvia, and Poland were behind the unrest. This was followed by an offer of military assistance of 200,000 troops by the Putin regime.[175]

At the same time, Russian forces were gearing up for exercise Kavkaz 2020 in September. There was a snap exercise held in the Western and Eastern Military Districts totaling around 150,000 personnel and 400 aircraft.[176] The purpose of the snap exercise was unclear as the summer unfolded, but there was concern at high levels that the Putin regime could use Kavkaz 2020 as well as the gargantuan snap inspection exercise as a pretext to launch another assault in Ukraine.[177]

On 21–22 August, six B-52Hs departed the continental United States, flew up to the North Pole, to northern Norway, and recovered at RAF Fairford.[178] Six days later U.S. European Command announced that four of the Fairford B-52Hs would fly a tour over all thirty NATO nations in Europe in one day. Another pair of B-52Hs departed Minot AFB and flew over Canada for Europe with the stated purpose of "[sending] a clear message to potential adversaries about our readiness to meet any global challenge."[179] One B-52H pair deployed to the Black

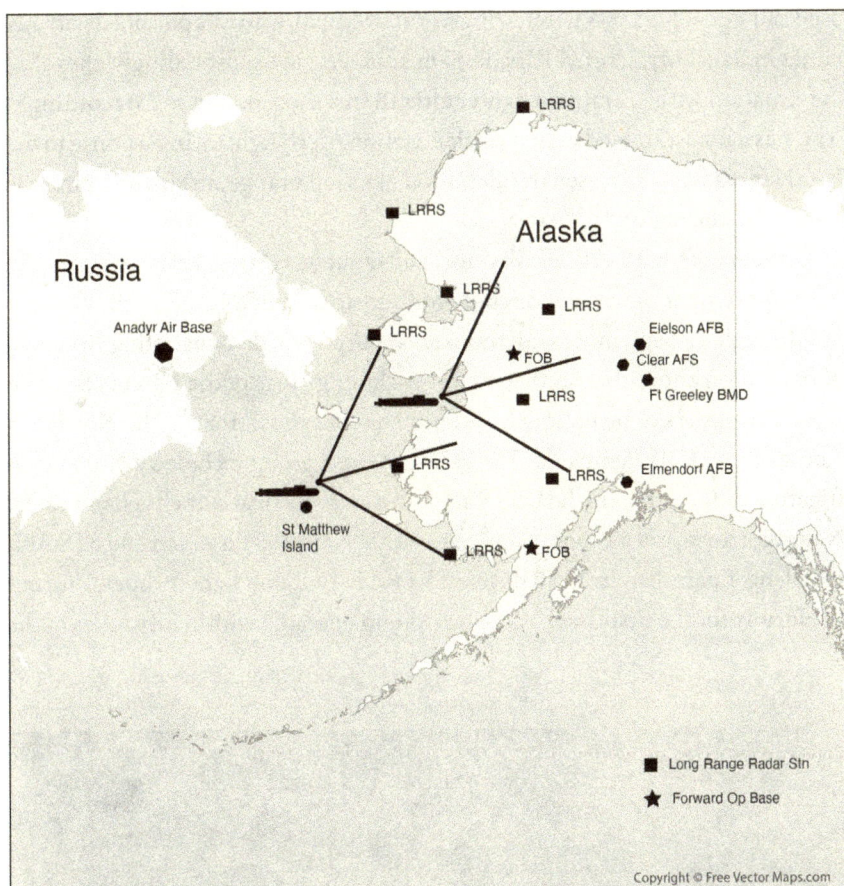

The map contains the following labels:

Russia

Alaska

Anadyr Air Base

LRRS (multiple locations)

Eielson AFB

Clear AFS

FOB

Ft Greeley BMD

Elmendorf AFB

St Matthew Island

Long Range Radar Stn

Forward Op Base

Copyright © Free Vector Maps.com

MAP 2 ◆ Position of the guided-missile submarine *Omsk* in relation to potential air defense targets in western Alaska. A second submarine operating in Norton Sound equipped with cruise missiles puts the second line of Long Range Radar Stations at risk. *Created by the author*

Sea, where they were intercepted by Su-27s and subjected to harassment that included "causing turbulence and restricting the B-52s ability to maneuver."[180]

Six days later on 4 September, two or three B-52Hs departed RAF Fairford and headed for Ukraine, where they flew a racetrack orbit on the coast of the Sea of Azov. Several U.S. and British ISR assets operated in the same area at the time.[181] Another account noted the call signs of the aircraft involved, and call signs themselves became part of signaling as the Cool War progressed. The B-52s used call signs "Julia 51–53," referring to Yulia Skripal, who was poisoned with her father in Salisbury, United Kingdom, in 2018. Perhaps most tellingly, they flew in formation with Ukrainian fighter aircraft in a show of unity.[182] Accordingly, "the Russians scrambled with [Quick Reaction Alert] fighters in response to the B-52H incursion and raised the alert level of its long-range surface to air missile systems in the region."[183]

Concurrent with Allied Sky and subsequent related activities, several NATO-member navies conducted another surface FONOPS in the Barents Sea throughout September with the express purpose of demonstrating "freedom of navigation above the Arctic Circle—including areas of the Barents Sea that Russia considers to be nationally-administered waters," that is, the North Sea Route off the Kola Peninsula. The FONOPS task group included a Norwegian frigate, a U.S. destroyer (the USS *Ross* with ninety vertical launch cells), a Royal Navy destroyer and a support ship, plus a U.S. Navy P-8 Poseidon and a Danish Challenger patrol aircraft, all covered by RAF Typhoon fighters operating out of Norway for the first time.[184] The task group operated within fifty miles of the

PHOTO 2 ◆ Armed Russian Su-27 Flanker harassing a B-52H in international airspace at close range off Ukraine, 30 August 2020 *U.S. Air Force*

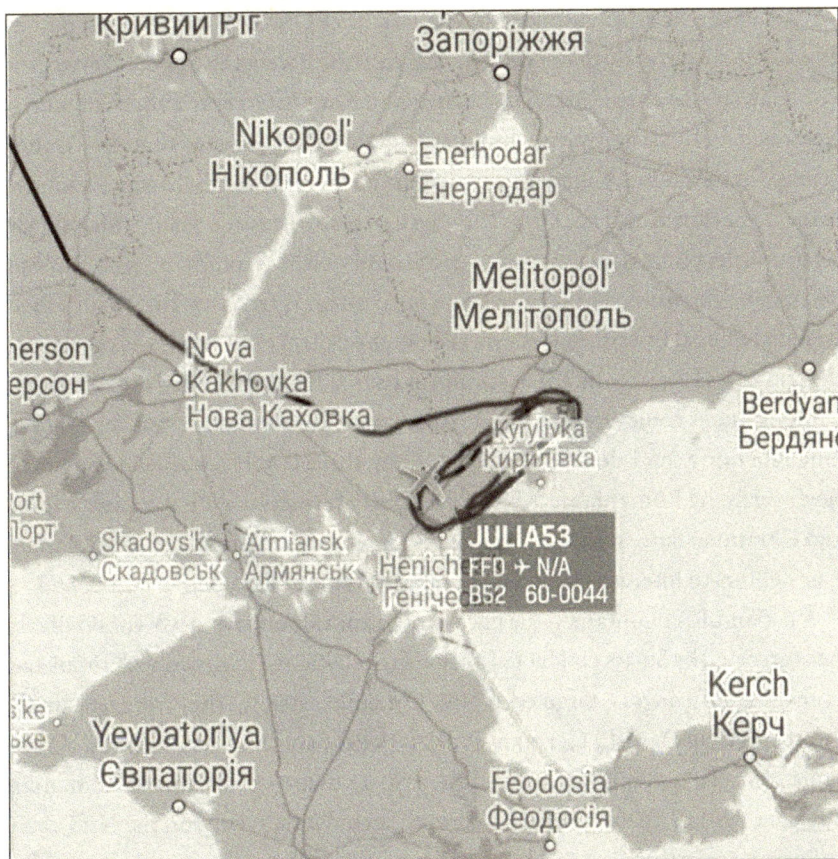

PHOTO 3 ◆ B-52 flights over Ukraine, 4 September 2020. The B-52 Bomber Task Force used Julia call signs, in reference to Yulia Skripal, who was poisoned by Russian agents in Salisbury in 2018. BTF B-52 flights this far forward should be contrasted with the subsequent administration's replacement of them with manned and then unmanned ISR platforms from 2021 to 2022. *flightradar24*

Russian coast.[185] This particular FONOPS likely had multiple objectives—the ongoing issue of territorial claims was one: "FONOPS in the Arctic may not force Russia to abandon their claims, it would demonstrate their impotence to enforce them, essentially rendering them moot in the eyes of our Arctic allies."[186] The other was to contribute to the other deterrence and assurance operations directed at Russia because of the deteriorating international situation.

Part of the Russian response to the Barents FONOPS was to run a pair of Tu-160 Blackjacks escorted by MiG-31BMs over the Barents Sea, to the Norwegian Sea, and then over the Atlantic. Norwegian F-16s intercepted them off north Norway.[187] A Russian information operations conduit asserted that the Tu-160s headed for the United Kingdom and that the purpose for the flight was a response to the "appearance near the airspace of the Crimean Peninsula of a British military electronic reconnaissance aircraft, which could well have interfered with Russian S-200 air defense systems." Russian sources noted that the Tu-160s "rushed straight to Great Britain, having worked out the delivery of nuclear strikes against the country's largest military bases and cities. . . . The British Air Force under no circumstances could prevent the Russian [strategic bombers] from turning the kingdom into a semi-desert."[188] The same day, three B-52Hs proceeded on a mission overflying Romania and Ukraine, flying in formation with Romanian F-16s and Ukrainian Su-27s. Russian air defense forces in occupied Crimea scrambled four fighters to intercept the three bombers over the Black Sea.[189]

Russian LRA maintained the pace over the next four days. Two Tu-160 Blackjacks escorted by Su-35s conducted a tour of the Baltic Sea that involved "simulated cruise missile firings . . . targeted at NATO member states." They were repeatedly intercepted by Danish, German, Italian, Polish, and Swedish fighters. At the same time, Tu-22M3 Backfires and their Su-27 escorts took runs at Romanian airspace, while a Tu-142 with its escorts collected intelligence on the NATO air defense forces response.[190]

Overlapping with the Barents FONOPS and European air operations was an unannounced NORAD "dynamic force employment operation" held on 20–23 September.[191] This operation employed AWACS aircraft and northern forward operating locations to support interception operations in all northern NORAD regions.[192] It also involved the alerting and forward deployment of an American Air National Guard F-16 unit from Colorado to Goose Bay, Labrador, and the exercising of forward operating locations.[193] What the NORAD press release neglected to mention was that on 19 September, two Tu-160 Blackjacks conducted a twenty-five-hour operation involving three aerial refuelings from six Il-78M MIDAS tankers. These aircraft flew from Engels, to the Barents and into the Arctic, swung east to the Chutkotka Peninsula, then down to within fifty miles of Nunavik Island off Alaska where they simulated cruise missile launches before headed home.[194]

Then on 21 September, two Tu-160 Blackjacks escorted by several Su-35 fighters repeatedly entered the Alaskan ADIZ on three occasions and operated fifty miles from Nunivak Island near Alaska. F-22s were vectored on to the Russian force with E-3C AWACS. A Russian information operations conduit noted that "the flight was a demonstration of strength and determination of Russia, since the nuclear missiles in service with the Tu-160 are enough to successfully hit targets from a distance of several thousand kilometers."[195] From the position off Nunivak Island, Tu-160s could saturate all U.S. military installations in Alaska. At the same time as the last day of the NORAD operation, RCAF CF-18s that were deployed to Romania as part of the NATO Enhanced Air Policing mission intercepted Russian Su-27 Flankers off the Romanian coast.[196] Simultaneously, two Tu-160 Blackjacks from Engels air base flew into Belarus, paralleled the Ukrainian border, took a run at the Polish border, then flew north to Lithuania and then paralleled the Latvian border before withdrawing. This was immediately followed by six Tu-22M3 Backfires that dropped live weapons on a range near Baranovici. This may have been part of a bilateral Belarusian-Russian exercise, but Russian information operations conduits asserted that "Russia has practised nuclear strikes against Eastern Europe, the Baltic states, and Ukraine."[197]

Immediately after there was a three-day operation, dubbed Astral Knight, in which the latest B-52H BTF deployment to RAF Fairford operated over the Baltic to exercise bomber interception and air defense system tactics with Poland and the Baltic states.[198] This included a B-52H with the call sign LeMay35 operating adjacent to Kaliningrad, a clear refence to General Curtis LeMay and SAC.[199] Russian information operation conduits interpreted Astral Knight as something designed to "work out nuclear strikes on Belarus and Kaliningrad." One Russian observer noted that Belarus does not have long-range air defense systems and is vulnerable to an American airstrike.[200] This is a perfect example of how the concept of combat potential not only underpins Russian interpretations of NATO signaling activity, but it also underpins their own interpretation of their own activities.

This particular BTF deployment came to an end after the Kavkaz 2020 exercise ended. The finale was a mass launch of eight KC-135s in what appeared to be a minimum interval takeoff launch reminiscent of the Cold War–era SAC alert scrambles on 29 September.[201]

Activity on both sides reduced significantly but not completely in October and November 2020. The deployment of eight Tu-22M3 Backfires to Belarus, which was reported in Russian information operations conduits on 4 October, included the statement that they each had three Kh-22 stand-off missiles with 1 MT yield warheads, "which makes it possible to turn half of eastern and central Europe into a desert" and "Europe will not be able to prevent this in any way." The conduits asserted that this was a response to the earlier B-52H deployment over Ukraine.[202] It is unclear whether nuclear versions of the Kh-22 were in fact deployed to Belarus, whether attached to the aircraft while airborne, or where the aircraft were being stored. Again, the opaqueness of a dual-capable system added to the ambiguity underpinning the threat.

As LRA activities tapered off in October, three Tu-160 Blackjacks conducted what looked like a repeat of the Barents–northern Norway–GIUK gap run, but the third aircraft uncharacteristically stayed behind and orbited the Barents Sea. There was speculation that it was acting as a radio relay station for the other two.[203] And after a hiatus, the Tu-95MS Bears were back. Escorted by Su-35s and MiG-31s and supported by an A-50 AWACS, the Russian aircraft "loitered inside the Alaskan Air Defense Identification Zone . . . for about 90 minutes and came within 30 nautical miles of Alaskan shores." They were intercepted by F-22s supported by tankers at several stages of their flight.[204] Only one LRA operation was run in November, with RAF Typhoons intercepting Tu-95MS Bears that flew to and around the North Sea.[205] American activities also tapered off with a BTF of two B-52s deployed to RAF Fairford on 15 October.[206]

Joseph Biden was elected president of the United States on 3 November 2020 and a new era in U.S.–Russian relations began. Russia ran the Grom 2020 nuclear exercise and ended it with a bang: on 12 December four Bulava SLBMs were volley fired by the *Borei*-class ballistic missile submarine *Vladimir Monomakh* from the Sea of Okhotsk; the missiles landed on the Chinzeh Test Range. This probably fell into the general deterrence category. Such a volley launch had only been done once since the Cold War ended—back in 2018—and any significant overshoot of the reentry vehicles, depending on the location of the submarine and its angle to Chinzeh, could hit Western Europe. It is unclear whether the launches involved MIRVs or not.[207]

Conclusions

Of crucial importance is the evolution of 2000s-era U.S. nuclear war planning away from late–Cold War concepts to include signaling activities in the panoply of the new plans. These were deterrence mechanisms, not merely assured destruction "wargasms" from the McNamara era. These changes reflected the post-911 period's requirements for more sophisticated deterrence systems in the face of multiple threats of varying quality and quantity. This was the context for U.S. strategic nuclear signaling activities during the emergent Cool War. At this point, NATO nuclear deterrent—let alone conventional deterrent activities—were almost in caretaker status. U.S. and NATO forces conducted general deterrence activities from the 1990s up to 2014. For the most part, U.S. signaling was relegated to the Asia-Pacific theater and involved conventional as well as nuclear forces mostly directed at North Korea but increasingly shifting toward the People's Republic of China.

Russia's assault on Ukraine in 2014 forced a rapid reconsideration of deterrence and signaling. Both conventional and nuclear signaling tools were activated as the NATO alliance struggled to understand exactly what was happening and its implications. It is clear that the application of Russian hybrid methodologies operated in a bandwidth where there was no clear deterrence system to counter them, as the Cold War–era deterrence system no longer existed and the current system was dependent on Russia's goodwill. The divided response by the Obama administration to the Ukraine invasion coupled with its confused response to aggression in Syria ensured that there would be no strong response to Russian movements. The United States and its allies were on the defensive, and the Putin regime, for the most part, retained the initiative throughout this period. There were exceptions, to be sure. Secretary of Defense Ash Carter's authorization of B-52 operations in the Arctic was one. Reinstituting Global Strike Command operations in the United Kingdom and then Spain while increasing the number of bomber assurance and deterrence activities over NATO countries was another. Increased information operations regarding the specifics of bomber operations to ensure credibility was yet another.

The Trump administration's approach to all of this was initially uneven, and it was only after the use of cruise missiles against Syria in response to chemical weapons use were the lines drawn between the Putin regime and the United States. From there, deterrence signaling ramped up, with the evolution to Bomber Task Force operations to replace Bomber Assurance and Deterrence missions all over the globe but particularly in Western Europe and the Arctic. Freedom of Navigation operations conducted by the U.S. Navy off Russian shores increased, as did B-52 flights directed at Kaliningrad. Steady Global Strike Command information operations underpinned all of this activity. Secretary of Defense Mark Esper ensured that these operations were meaningful signals, specifically B-52 flights in the Barents Sea basin off Murmansk, Novaya Zemlya, and Svalbard; runs against Kaliningrad; North Pole runs; and operating B-52s over Mariupol, Ukraine, and the Sea of Azov, as well as against occupied Crimea. Despite the uneven messaging by the Trump administration at the political and diplomatic level, military-level messaging was strong. We know that the Putin regime initiated planning for the further invasion of Ukraine prior to the accession of the Trump administration. It may even have considered making a move until the erratic Trump administration proved to be problematic. It is likely it decided to wait and see what would come in the 2020 U.S. presidential election before moving forward.

CHAPTER 4

◆ ◆ ◆

Buildup

2021

The tumultuous transition of power from the Trump to the Biden administration in January 2021 was swiftly followed by a round of Russian strategic nuclear signaling that in turn backstopped unscheduled large ground forces exercises in March. The first phases of the Russian plan to invade Ukraine again were underway. The Putin regime employed a variety of means to keep the world guessing as to its intentions during the year, while keeping its options open. The deteriorating situation between Belarus and its NATO neighbors—another conflict that had been simmering since 2020—assumed new dimensions in the context of the Ukraine situation. At every step of the way, strategic nuclear forces were exercised in support of these activities. At the same time, an evolved Russian information ecosystem was in constant play, ready to amplify signaling or generate confusion as required. That ecosystem played a central role in Russian efforts well into 2022 and 2023.

Russian Information Warfare

Russian information warfare is used to govern what the Putin regime wants us to see and believe while it pursues its objectives. The objective is to signal and manipulate Western policymakers and those who influence them in support of these goals. Its concurrent objective is to influence the Russian domestic environment to protect its power and maintain public support. In some cases, Russian information operations are structured to do both. There is nothing comparable in Western nations in terms of scale, tempo, and intent. Russian information warfare includes general information operation lines; specific, targeted information lines; and opportunist information operations lines deployed through conduits. These lines can be grouped into information campaigns of short and long duration. There is some debate, however, as to how this is all coordinated or initiated. Generally, there is a top-down view where all aspects are initiated and centralized, and a bottom-up view where concepts are generated at lower levels and approved by higher levels.[1] The reality is somewhere in between, depending on the nature and requirements of the particular information operations line and its objective.

Putin and the *siloviki* will have a dominant role. There is a central coordination mechanism, though the specifics of where it is positioned and how it works are difficult to determine. Such a body can act as a clearinghouse for bottom-up concepts and as a dispatch mechanism for top-down concepts. This mechanism likely resides in the Presidential Administration, possibly as a subcommittee of the Security Council with connectivity to other presidential administration bodies.[2] There is likely some form of intelligence feedback structure to provide measures of effectiveness for it. There is also probably some organization to deconflict deliberate disinformation campaigns from direct messaging.

The Russian information warfare system appears to be able to select and use combinations of information operations channels or conduits in support of its objectives. These combinations can illuminate how important the messaging is as well as its audience and intended effects. The target audiences include Western social media; Western mainstream media; those who provide advice to Western policymakers; and Western policymakers themselves.

The first cluster includes the television outlets Rossiya1 and RT. These are direct extensions of the state through which information operations can be employed directly. The second includes RIA Novosti, TASS, and Interfax, which use internet channels to disseminate information. Again, these permit the Putin regime to make direct and immediate statements. Overlapping both clusters is Putin himself, using direct statements and "interviews." The third cluster includes the Russian Ministry of Foreign Affairs and Russian embassies abroad.

MoD "media" | Ru Milbloggers
GRU Disinformation
Ministry of Defence
Military Movements:
-exercises
-ATC data
MoD-affiliated Military Pages
PUTIN "central coordination mechanism"
Western Policymakers
Western Advisors
Western Think Tanks
Mainstream Media
Social Media
Western Russian Sympathizers
Putin statements and "interviews" | Ministry of Foreign Affairs Statements
Intelligence Channels | Cyber
-Sputnik -Rossiya 1 -RT
Russian Online and Print "Media"
-RIA Novosti -TASS -InterFax

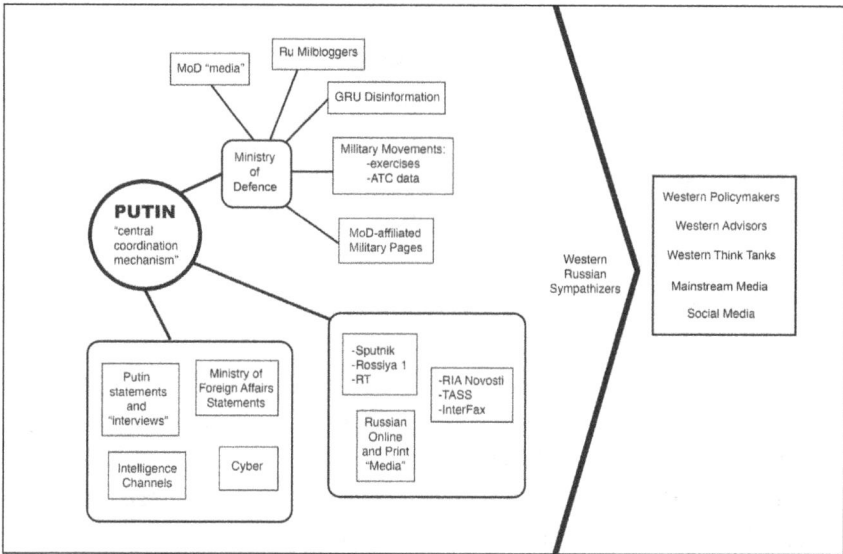

CHART 1 ◆ Russian Information Operations

The fourth are Russian online and print "media" and social media. These include Russian intelligence agency–linked news organizations; news organizations that are associated with the Kremlin; media outlets belonging to Russian oligarchs (particularly Yevgeny Prigozhin and Konstantin Malofeev); Kremlin "amplifiers," Telegram (social media platform) analytical "sources," and finally, Russian overt media.[3] In addition to these are the original "free" media outlets from the 1990s that were slowly hijacked and co-opted by the Putin regime by 2006.[4] In general terms, controlled "sources" on Telegram could feed overt and Kremlin-associated media, which could then be proliferated by the "amplifiers," some of which were overseas, and then fed back into overt media to support official positions.

On the social media front, Yandex was in play in the late 1990s, and VKontakte, the Russian Facebook equivalent, started in January 2007 (though Twitter was in use in 2006). By 2014 VKontakte could be considered a regime-influenced resource. Blogging was popular throughout the late 1990s and early 2000s, but it was infiltrated by "the Kremlin school of bloggers" by 2009. In terms of crossover capability, RT figured out how to use YouTube for its purposes around 2011. By

2013 the Internet Research Agency was in operation, as was Telegram. Social media could be used to generate information as well as to amplify it.

The fifth cluster surrounds the Russian Ministry of Defence (MoD). The MoD has its own "media" that makes direct statements. Military movements, including exercises and nuclear force movements, are amplified through Russian MoD media. Other information operations conduits include the plethora of what can be called chauvinistic military outlets online that are influenced by the Russian MoD and have primarily domestic audiences. These amplify Russian MoD statements and military force movements. Overlapping all of this are what are euphemistically called "milbloggers." Seemingly independent commentators in the world of social media, these are not directly controlled but are strongly influenced by the Russian MoD.

Part of this fifth cluster are disinformation operations conducted by or coordinated by the GRU. The Internet Research Agency is one mechanism. In direct support of military operations, there are geographically focused military units that conduct "battlefield information isolation; Information protection of troops, disinformation, and counter-propaganda; Civil-military interactions; and information support."[5] The Russian strategic nuclear forces, RVSN, LRA, and naval forces, likely have their own equivalents. It is clear that the Russian intelligence agencies—the FSB, SVR, and GRU—have their own organizations for special tasks.[6]

Of importance to this study is the relatively new information operations line associated with air traffic control (ATC) information. Most aircraft have some means of illuminating their presence to air traffic control systems, and by 2020 computer applications to capture and present this information were available for home computers and then mobile phones. This information was available globally, including flight movements inside Russia. Generally, military aircraft only "squawk" (reveal their location by transponder) under specific conditions. And in some cases, certain ATC information modes can be "spoofed" to conceal the identity and type of aircraft. (Indeed, even if one type of aircraft is pretending to be another on ATC data, the signal is still sent to its audience.) Russian nuclear command and control aircraft are registered as civil aircraft, as they are also used for VIP transport outside of Russia. It became evident by 2021 that Russia used "squawking" nuclear command and control aircraft as a signaling tool in concert with other information operations channels. As these aircraft belong to the presidential administration, they can serve as direct messaging tools.

These and other non-publicly visible military movements constitute a separate intelligence channel. The audience for these movements clearly includes not only Western military and intelligence staffs but also serves as amplification of the information via aficionados to social and mainstream Western media. Indeed, it remains possible that some social media amplifiers for this information are extensions of the Russian intelligence part of the ecosystem, though for the most part ATC hobbyists, like their amateur radio cousins, are non-malevolent enthusiasts. Similarly, U.S. nuclear command and control aircraft, some reconnaissance aircraft, and tankers are visible to the public on ATC systems. These have a specific signaling role in the context of the early deterrent phases of American nuclear war planning. We will see this play out in detail in 2021 and 2022.

Unrest in Belarus and Pressuring Ukraine: January–April 2021

The international event that received the most focus at the time was the 6 January takeover of the U.S. Capitol building by a coalition of elements who mistakenly believed the U.S. presidential election had been stolen. The effect of this situation on the American deterrent remains opaque. Two prominent journalists asserted that China, Russia, Iran, and North Korea were on "high alert," but this term is not defined or explored in any detail in their work.[7] One result of the uncertainty surrounding the insurrection was the postponement of U.S. operations in the Pacific region to signal China that there was positive control over American military forces.[8] On the other hand, publicly raising questions on the state of nuclear command and control and making comparisons to Nixon in the 1970s in public by Speaker of the House Nancy Pelosi could have been problematic in that answering those questions in detail danced close to revealing the conditions and procedures under which the deterrent system operated under conditions like this. Similarly, American journalists making incorrect, not to mention inappropriate, comparisons to alleged "mad man theory" likely generated even more paranoia in Moscow, Beijing, Tehran, and Pyongyang.[9]

The Russians appear to have been taken by surprise by the events of 6 January. Most Russian news outlets crowed that this was payback for U.S.-sponsored "color revolutions." Others, including Dmitry Medvedev, asserted it was a signal of American decline as a world power. Still others viewed the events as the precursor to another American civil war. It was even compared with the Maidan events of 2014. A joke circulated: How can this be a color revolution if there was no American embassy in Washington, D.C.?[10]

The only publicly observable nuclear force movements on 6 January 2021 were a scheduled Bomber Task Force deployment of three B-52Hs to RAF Fairford.[11] This was more along the lines of general deterrence signaling, assuming the movement of bombers was limited to three. These aircraft later participated in flights over the Black Sea off Crimea on 25 January, where they were intercepted by Russian aircraft. Russian media deemed this to be a "provocation" as usual.[12] This was four days after the Biden inauguration, so the messaging was likely that the deterrence situation was back to the nominal level established before the election. Indeed, a second Bomber Task Force deployed to the Middle East as tensions with Iran increased.[13] The Putin regime continued to make hay over B-52Hs conducting exercises with Romanian F-16s and were especially concerned about the aircraft "conducted training scenarios using air-to-ground missile (JASSM) tactics," implying they could be used against Russian forces in Crimea.[14]

The backdrop for all of this was ongoing unrest in Belarus. The rigged election in the summer of 2020 was met with escalating protests which were met with lethal force by the security services. As usual, this was believed by the Putin regime and especially by Belarusian leader Alexander Lukashenko to be an attempt by "dark Western forces" at a "color revolution." This was nonsense, as Belarusian street humor indicated: A citizen was pulled over by the police for speeding in the vicinity of a riot. They proceeded to drag him out of the car and beat him. The driver protested: "But I voted for Lukashenko!" To which the cops replied: "Cut the crap. *Nobody* voted for Lukashenko." Sanction after sanction and vocal condemnation was applied to Belarus by the international community into the fall of 2020 as the unrest spread, particularly after a crackdown on independent journalism was carried out by the regime and significant instances of mass detainment and torture were revealed. This state of affairs escalated throughout early 2021, which gave the Putin regime concern within their "color revolutions" strategic framework. Lukashenko was even seen to be carrying personal weapons in public.[15]

The situation in Ukraine shifted perceptibly in February as the Ukrainian government implemented actions to reduce the Putin regime's ability to influence Ukrainian politics. Specifically, Putin's friend Viktor Medvedchuk as "the leading voice for Russian interests in Ukraine" wielded numerous soft power tools, including a television propaganda concern, involvement with a major pro-Russian political party, and involvement with the gas pipeline issue. The television stations were shut down, Medvedchuk's assets were seized, and his considerable influence attenuated.[16]

A flurry of Russian maneuvering took place throughout February in response to what was happening in Belarus and Ukraine. It is also likely that the Putin regime wanted to test the newly elected Biden administration. This started with a 4 February announcement that Tu-22M3 Backfire bombers would continue to demonstrate Russian airpower and drive NATO ships in the Black Sea away from Crimea as part of a large-scale exercise. The announcement noted this had successfully occurred three times already.[17] Imagery of Tu-22M3s each equipped with three Kh-22 cruise missiles was released and that the aircraft "[checked] the readiness of a domestic bomber to strike at enemy forces" by live launches at a test range with the suggestion that "the Russian Aerospace forces could use their bombers to simulate an attack on American missile destroyers," a clear reference to concerns about Tomahawk-equipped destroyers operating in the Black Sea constituting a strategic threat.[18] The next day, Tu-22M3s supported by an A-50 AWACS aircraft tracked and carried out a simulated strike against the USS *Donald Cook* and USS *Porter* as "a worthy response to US provocations." According to Russian sources, the American ships reacted in some fashion, probably by using their fire control radars.[19]

The same day, Tu-160 Blackjacks were launched on a "simulated nuclear strike" in the Arctic. Two Blackjacks, with a third acting as a communications relay, conducted a twelve-hour mission that notionally "attacked" a base in northern Norway that was about to host B-1Bs on a Bomber Task Force deployment. The aircraft then flew to a point off Iceland, simulated an attack on Keflavík, Iceland, then flew to a point off Greenland and simulated a nuclear strike on Thule Air Base. The bombers were supported by MiG-31 interceptors stationed at Novaya Zemlya.[20]

These events should not be seen in isolation. A curt Russian MoD announcement on 21 February stated that three thousand members of air assault forces were deploying to the Ukrainian border "for large scale exercises . . . to seize enemy structures until the arrival of the main force."[21] Some seven thousand more personnel were activated or deployed in support of this snap exercise. In retrospect, this was probably the first phase of the Russian deployment for the February 2022 assault, but it could also have been preparations for an intervention in Belarus. Or it could have been both. The West's response was predictable but low key. Four B-1Bs did deploy to Ørland air base in Norway on a scheduled BTF deployment "to send a clear message to any adversary," even though they were conventional aircraft.[22] The British-led Joint Expeditionary Force mounted a five-vessel patrol in the Baltic Sea for a "joint demonstration" deployment.[23]

In early March, Russian-backed forces in Donetsk and Luhansk generated an abnormal number of ceasefire violations on the line of control. As a result, the Ukrainian forces moved military equipment forward to eastern Ukraine in case this problem expanded. This gave the Russians an excuse to conduct a series of SCRIs to "deploy sizeable elements belonging to the 41st Combined Arms Army to the Pogonovo Training Range south of Voronezh." This was followed by VDV movements and artillery assets from the 58th Combined Arms Army to Crimea. Consequently, "by mid-April, Russia had elements of four [Combined Arms Armies] with additional airborne units in Crimea . . . [which included] elements of the 90th Tank Division . . . and a surface-to-surface missile battalion from the Iskander equipped 119th Missile Brigade." Analysis suggested that "the Russian force posture that emerged in late March and early April thus indicated preparation for multi-front combat operations against Ukraine."[24]

It was later learned that the original plan for the latest phase of the Ukraine invasion evolved in March 2021 and that these early buildups pressured Western governments to continue negotiations, prepositioned equipment near Ukraine, and allowed a Russian assessment of Ukraine's international supporters.[25]

These SCRIs were accompanied by LRA deployments, which themselves may have been LRA SCRIs. The first consisted of a pair of Tu-96MSs escorted by Su-35s, on a nine-hour flight "practising strategic strikes against Japan." JASDF F-15s conducted the intercept.[26] Another involved the deployment of at least three Tu-95MSs from Engels Air Base to Olenya on the Kola Peninsula.[27] It is unclear what American activity was conducted in response to Russian activity. STRATCOM, SPACECOM, EUCOM, the U.S. Army, and the Canadian

Armed Forces conducted Global Lightning from 8 to 12 March, though this was a scheduled command post exercise "designed to train joint and combined forces to assess operational readiness in creating conditions for effective deterrence."[28] A Bomber Task Force consisting of two B-1Bs and two B-2As deployed to the Arctic on 18 March. The B-2As flew to the Azores, refueled, flew up to the GIUK gap and rendezvoused with the B-1Bs, which departed Norway, to a point north of Iceland. These B-1Bs "flew a wide loop over the Norwegian Sea" to the rendezvous, suggesting their mission was more than meeting up in the air.[29] Media noted that the B-1Bs conducted "warm-pit refuel" operations at Bodø Air Force Station in Norway, demonstrating flexibility in the extreme northern environment.[30] Operations included "deployment in and around the Barents Sea."[31] The geographic location of Ørland, incidentally, permitted the B-1B force to intervene and support the Baltic states and Poland rapidly with conventional munitions if required, with the implications that Russian Iskander in Kaliningrad could be targeted as well. The B-2As lurked somewhere in the background, demonstrating that nuclear force backed up the conventional force.

The Putin regime, however, turned up the heat on 26 March with exercise Umka 2021. The Russian MoD released high-quality footage of three ballistic missile submarines breaking through the polar ice in the Arctic near Franz Josef Land. The three submarines were identified as two Delta IV SSBNs and one Borei-B–class SSBN. This was, in the context of all previous Russian signaling, a massive show of force. Each Delta IV carried 16 R-29RMU Sineva SLBMs each with 4 reentry vehicles (RVs) of 500 kt yield or 10 RVs of 100 kt yield, for a potential total of 128 to 320 nuclear weapons.[32] The Borei-class carries 16 Bulava SLBMs each carrying 6 to 10 RVs of 100 or 150 kt for a total of 96 to 160 nuclear weapons.[33] Together, this force was capable of unleashing between 224 and 480 nuclear weapons. This was accompanied by a Russian state-controlled news blitz that found its way into Western media.[34] In essence, this information operation conveyed that Russian ballistic missile submarines could operate under the Arctic ice pack with impunity and could flood North America with SLBMs with little or no warning. Indeed, flight times were significantly reduced the closer the submarines were to North American targets.

What the Russians did not reveal through their means was that Umka 2021 actually involved more than three submarines, four of which surfaced through

PHOTO 4 ◆ Russian exercise Umka-21 on 26 March 2021 involved the very public simultaneous surfacing of three ballistic missile submarines carrying up to 480 nuclear weapons in the Arctic during the initial troop buildup opposite Ukraine in March–April 2021. High-resolution footage of the operation was distributed widely across all Russian information means. Subsequent imagery suggests that at least two other Russian submarines were involved. *Russian Ministry of Defence*

the ice, with the fifth being of an unidentified class. Image analysis from a Russian source using Maxar satellite imagery also suggested that a torpedo was used to blow a hole through the ice. Other Russian analysis suggested that the hole was made by an SLBM test launch, though there was no public confirmation of this from other sources.[35] Umka 2021 was immediately followed by a flight of two Tu-95MSs and two Tu-160 bombers that flew from the Kola Peninsula to the Norwegian Sea and then to the North Sea. They were continuously tracked by Norwegian and RAF fighters.[36] During this period, Russian Il-96–300PU nuclear command and control aircraft made runs from their base in Moscow to Novosibirsk and back, clearly testing communications.[37]

At the same time, Ukrainian military briefings to media noted with alarm that there were ten Russian brigades stationed behind the front lines in occupied Crimea and Donbas, with another ten moving up. An exercise held in Crimea "practised air and sea landings of a naval infantry brigade" that created a lodgment and five more battalions linked up with the landing force. The Russian buildup in the Rostov area consisted of four brigades that were on a one- to two-day readiness level, and that 19,000 tons of fuel had been delivered to the Donbas region.[38] The next day, U.S. European Command moved its alert level from "Possible Crisis" to "Potential Imminent Crisis."[39]

The fact that it would have taken weeks to ready and deploy such a large force to the Arctic suggests it was not in reaction to other moves that were taking place. The fact that it occurred right after the first tranche of Russian troops deployed as part of the SCRIs suggests that Umka 2021 could have been used to underscore what these troops could be used for. There were several possibilities. The first was that these were Russian preparations to intervene in Belarus against an ostensible color revolution and Umka 2021 was a message to the (nonexistent) "dark Western forces" to stop interfering and keep out if the "revolution" succeeded and Russia moved in to retake the country. The second is that Umka 2021 backstopped the Russian troop deployment and was a signal to Ukraine to stop the alleged buildup to liberate Donbas. The third is that Umka 2021 was designed to signal the Biden administration to back off on supporting Ukraine. The fourth is that Umka 2021 was a larger signal that Russia would go all in to seize all of Ukraine, and for the Biden administration to stay out of it. Of note, the protests in Belarus ceased by 25 March, one day before Umka 2021.[40]

Any suggestion that Umka 2021 was some form of limited signal must be put against Russian moves taken in April. The day of the Umka 2021 announcement, two Tu-142 Bear maritime patrol aircraft flew to the North Sea, followed by two Tu-160 Blackjacks, which were met by Norwegian, British, and Belgian interceptors.[41] The next day, footage of eight Russian helicopters, including attack helicopters, violating Ukrainian airspace near Sumy, was released on a Telegram channel "associated with Russian security forces."[42] But it was Shoigu's 13 April public statement that gave observers pause. He confirmed that the early March deployments had taken place, and he identified the formations. He implied this was "a defensive reaction" to NATO exercise Steadfast Defender.[43] The problem, however, was that exercise was not run until May 2021; it was essentially a brigade-sized force plus support, and it was conducted in Romania, Bulgaria, Hungary, and Portugal.[44] One analysis noted that the March deployment "in addition to the already deployed 87,000 troops was likely made intentionally as part of a deliberate campaign to make Russian forces deployed near Ukraine look more formidable, threatening, and persuasive."[45] One day before the Shoigu announcement, Putin visited Engels Air Base, home of the Tu-160 force and its associated nuclear weapons, accompanied by an Il-96–300PU presidential command post aircraft, a Tu-214PU command post aircraft, and a third unidentified command post aircraft.[46] And the day of the Shoigu announcement, LRA was in the middle of conducting what was probably a snap exercise or SCRI.[47]

The sum of Russian activity produced a phone call to Putin from Biden on 13 April, which was "part of Biden's effort to put Putin on notice to expect a chillier relationship than the one Putin had with Trump." In that conversation, the items that were discussed included the Navalny poisoning, Russian cyberattacks, interference with the 2020 American presidential election, and importantly, Biden "warned Putin not to start a new military incursion into Ukraine," something that Putin did not deny, and instead Putin focused on "the U.S. Government's treatment of Native Americans and its decision to drop atomic bombs on Japan."[48] The two leaders agreed to meet at some point in the future to discuss differences. At this point there was some debate in the U.S. intelligence community. Some remained convinced the movement of troops were only exercises, while others, like Director of National Intelligence Avril Haines recalled, "There was definitely a moment where I recognized that this is not just a force buildup for a diplomatic effort."[49]

Three days later, the Russian "mobility exercise" was deemed completed. Shoigu announced "a drawdown" of the 41st and 58th Combined Arms Armies and the deployed VDV formations. The troops returned to their bases deeper in Russia, but they left their equipment in place at the training areas. The excuse was that they would return and use it in September during the scheduled Zapad 2021 exercise.[50] Indeed, the estimate was that only 12,000 of the deployed troops actually returned to their bases.[51] If the Putin regime wanted Biden's attention, it got it. But flights of Tu-160 Blackjacks against Norwegian airspace continued, as did Tu-22M3 operations directed at NATO member naval bases throughout the Mediterranean all the way from Syria to Sardinia.[52]

When Russia mounted its initial troop buildup opposite Ukraine in March–April 2021, there were significant deviations from the baseline when it came to strategic nuclear command and control forces. The first visible departures consisted of flights by Il-96–300PU aircraft and later by Tu-214PU and Tu-214SR

PHOTO 5 ◆ The Il-96–300PU presidential command post aircraft (*top*) is shrouded in mystery; it is equipped with full-spectrum communications systems. The second aircraft is the rarely seen Tu-214SUS airborne communication station aircraft. This particular Tu-214SUS shadows President Putin when he travels around Russia to ensure continuity of communication in the vast stretches of Russian territory. *Author's collection*

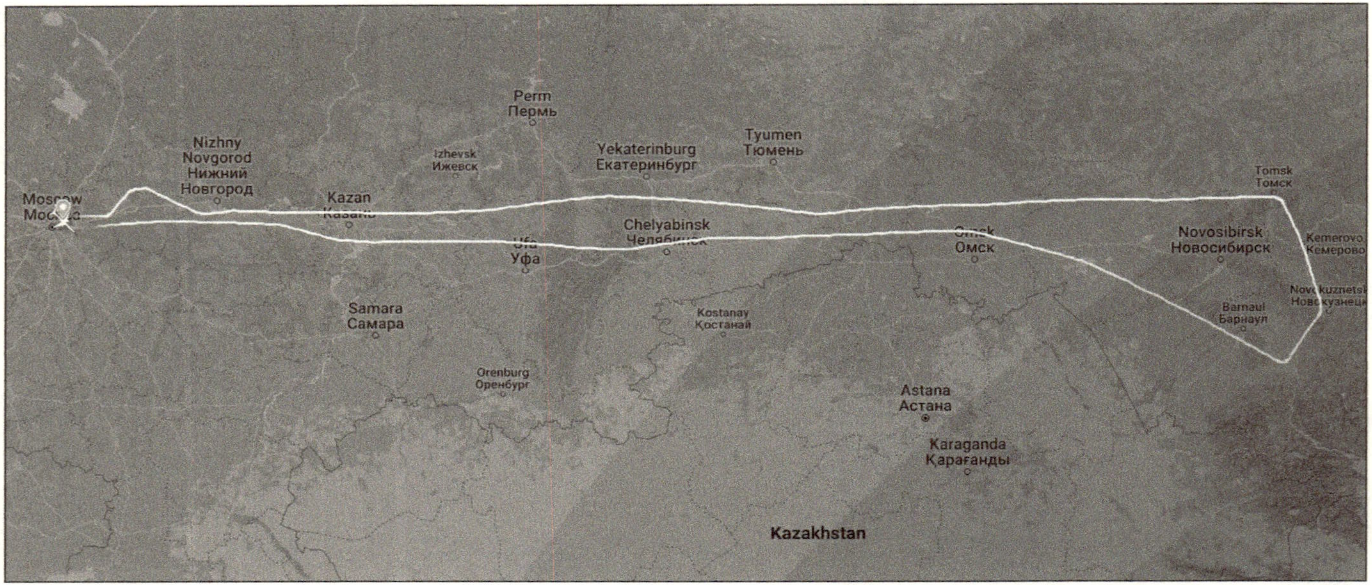

PHOTO 6 ◆ A typical Novosibirsk Run, checking communications between Moscow; Il-96–300PU, Tu-214PU, and Tu-214PU-SBUS aircraft; and the six ballistic missile divisions east of the Urals (the 29th Missile Division is near Lake Baikal). *flightradar24*

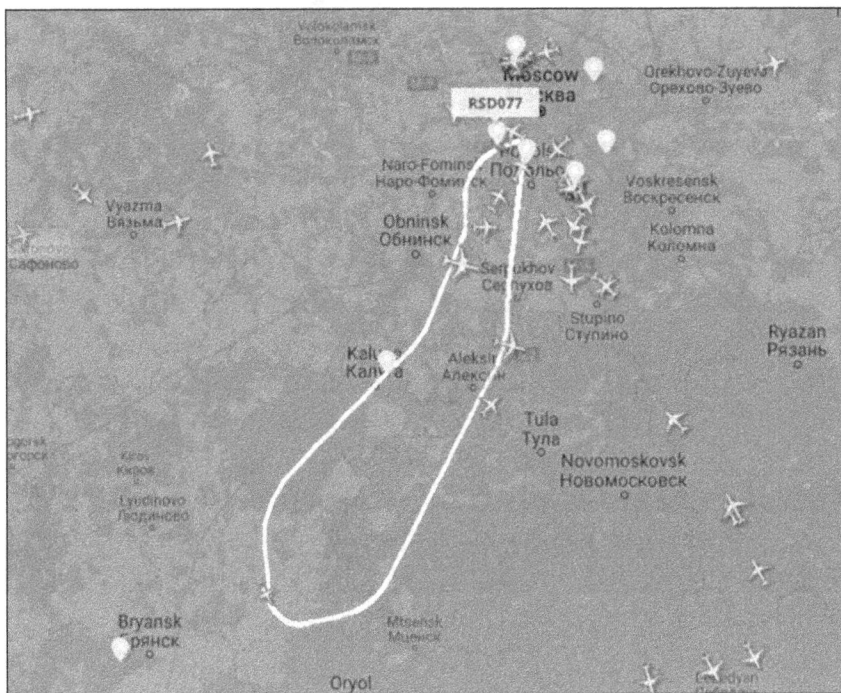

PHOTO 7 ◆ The Kozelsk Loop around the 28th Missile Division (MD) southwest of Moscow. The 28th MD was believed, until late 2022, to have had an Emergency Radio Communication System–like system based in some of its silos. Command and control aircraft regularly checked in with this system for the "studio audience" in the West as much as to ensure operational connectivity. *flightradar24*

aircraft. These are hardened presidential command post aircraft that carry communications means that cover the low-frequency to ultra-high-frequency communications bands. Some carry a battle staff. In February and March, this aircraft type was employed in what will hereafter be called the "Kozelsk Loop," and by the end of March they were deployed on what will be called the "Novosibirsk run."

The Kozelsk Loop is significant. The aircraft departs its base from a special holding compound at Vnukovo airport, flies down to the Kozelsk area southwest of Moscow, loops around a specific geographic point, and returns to Vnukovo. The flight takes about an hour. Kozelsk is the location of an ICBM field that dates back to the Cold War. It was later upgraded to house the 28th Guards Missile Division consisting of fourteen TOPOL-MR, then YARS missiles in silos.[53] The bulk of the Russian strategic rocket forces consists of mobile ICBMs situated in bases across the country all the way out to Irkutsk. Kozelsk is the westernmost anchor of these missile forces. The 28th Guards Missile Division is believed to have a capability similar to the U.S. Cold War–era Emergency Rocket Communications System (ERCS) in which a number of ICBMs were equipped with ultra-high-frequency transmitters designed to give "go code" orders to the entire missile force as the missile traveled over them.[54] The flight of the command and control aircraft on the Kozelsk Loop appears to be some form of communications check. They are likely equipped with a Russian counterpart of the American Airborne Launch Control System (ALCS), which permits the battle staff to communicate directly with the silos and order a launch if necessary. Importantly, such a communications system could theoretically have a range of two hundred miles or more, and thus the aircraft does not have to be this close to the missile field to communicate with it. This suggests that the flight had a concurrent crisis signaling function.

The Novosibirsk run has the command post aircraft depart its compound at Vnukovo and fly a mission out to Novosibirsk (home of the 39th Missile Division), where it orbits and then returns to the Moscow area: such a flight can take ten or more hours. The aircraft passes over and checks in with likely ground entry points and associated RVSN forces at Vladimir, Nizhny Novgorod, Perm, Tyumen, Novosibirsk, Yekaterinburg, Izhevsk, and Yoshkar Ola. Given the distances involved it is crucial for Russian strategic forces to be able to communicate with far eastern forces, and thus an aircraft is required to also relay communications because of the curvature of the earth. Such an arrangement presumes that Russian communications satellite constellations would be disrupted in the early stages of war, and thus the Putin regime is demonstrating that they are not as dependent on them as American forces are.

In terms of frequency, the Kozelsk Loop appears to start at the end of February 2021, though it probably existed before this, and was supplemented by the Novosibirsk run in the last half of February. In general terms, these flights were done once every two weeks, with the Kozelsk and Novosibirsk runs doubling up one week. At the height of the April war scare, however, the Kozelsk Loop was executed every week and supplemented by a run to Omsk (home of the 33rd Missile Army headquarters) in place of the Novosibirsk run.[55]

Consequently, the combination of command post aircraft activity, bomber activity, and submarine activity was significant during the spring 2021 war scare. After this phase of the crisis wound down, there were no visible flights on the Kozelsk Loop or Novosibirsk run until a Kozelsk Loop was conducted much later in July.

It is evident that Russian strategic nuclear forces were exercised in April 2021 and those movements deviated from the norm. It is equally clear that Russian authorities wanted these movements to be seen as they were visible on ATC systems, as well as on social media, and any cursory analysis of previous behavior would easily demonstrate this deviation.

The pattern of U.S. nuclear command and control aircraft movements during the March–April war scare significantly changed. From January to mid-March 2021, the average number of E-6B Mercury planes in the air any given day was one or two. Occasionally three were in the air simultaneously with a small number of trailing wire antenna deployments over water, but for the most part this amounted to a shell-game-like repositioning. On one occasion four were up along with an E-4B, which was likely an exercise.[56] By mid-March, E-6Bs started testing their ALCS over ICBM fields in Wyoming and Montana. In April there were additional days when four E-6Bs were aloft simultaneously. On 11–12 April, three E-6Bs conducted trailing wire antenna deployments over the Pacific, Atlantic, and the Gulf of Mexico. This was followed on 12 April with three sequential E-6Bs aloft, followed by an E-4B and a pair of E-6Bs, one of which conducted trailing wire antenna operations over the Gulf of Mexico, followed by four E-6Bs of which two deployed over the Atlantic.[57] The number of ALCS communications checks increased significantly to six in the last half of April. Cumulatively, this was a dramatic increase in U.S. nuclear command, control, communications, and intelligence (NC3I) activity compared to January and February.[58]

Steady State … or Not: May–June 2021

The decision was made by the Biden administration "to create an alternative path that would involve Russia de-escalating around Ukraine," an effort focused on diplomatic efforts.[59] May 2021 featured a scheduled and predictable B-52H Bomber Task Force to Morón Air Base in Spain and a subsequent Allied Sky tour of NATO Europe. One B-52H deployed to Norway to "integrate" with Norwegian air force operations around 17 May.[60] The next Russian move was to deploy two TU-160

Blackjacks to "practise missile strikes against NATO" on 19 May, apparently in response to a visit by an "American nuclear submarine" to Norway. The Tu-160s "worked out strikes against NATO facilities in Norway, including strikes against military air bases with F-35 fighters, air defense systems and radars, located in Northern Norway."[61] This was conducted simultaneously with the launch of two Tu-95MS from Engels escorted by Su-27 fighters which conducted a simulated strike on the Romanian port of Constanta, where ships from NATO countries were visiting.[62] The American response was to fly an undetected B-52H to Estonia and pop up for the benefit of the Russian air defense system. Russian media "intercepted" the B-52H after the fact and claimed it was "carrying nuclear weapons."[63]

Meanwhile, the situation in Belarus simmered. By late March the security forces had gripped the open unrest, but Lukashenko, taking a page from the Russian anti–color revolution manual cracked down on all independent media. By 21 May, there was no non-state-controlled media reporting on the subject. Two days later, Belarusian air force MiGs intercepted a British commercial aircraft and forced it to land in Belarus, where Belarusian journalist Raman Pratasevich was arrested. The EU and the United States immediately slapped sanctions on Belarus.[64] As a retaliatory measure, Lukashenko announced that "we used to stop drugs and migrants. Now you will have to catch them for yourselves."[65] In effect, "Belarusian authorities actively enable migrants from the Middle East to travel to Belarus by facilitating tourist visas, . . . allowing them to travel to the border area with Poland, Lithuania, and Latvia."[66] This situation escalated later in the summer and fall of 2021.

Russia ramped up TU-22M3 Backfire operations in the Mediterranean later in May. Stationed in Syria, they took a run at Israeli air defenses, followed by a missile training run by three Tu-22M3s at three U.S. Navy amphibious ships in the eastern Mediterranean the next day. On the 28th further mock attacks were made against NATO-member ships to "send an unambiguous signal to the North Atlantic Alliance" with Kh-22 missiles "equipped with a thermonuclear warhead with the power of up to 1 MT." Gibraltar was mentioned as a possible target.[67] This may have been in response to what Russian sources claimed was a "provocation" by a B-52H that was working with MQ-4 and RC-135 reconnaissance aircraft to collect information on occupied Crimea's air defense system.[68]

Putin and Biden met in Geneva on 16 June 2021. During that meeting Biden raised the issue of "strategic stability," though the details were not released. Biden's definition of "strategic stability" was not provided but was related to establishing "mechanisms" to deal with it, which included "bilateral dialogue" specifically on new weapons "that reduce the times of response, that raise the prospects of accidental war." Though unstated, this was a reference to hypersonic weapons.[69]

At this point there were two events taking place. The first was a large-scale NORAD Exercise Amalgam Dart, a nine-day exercise that started 10 June. This involved all three NORAD regions, including operations over the Beaufort Sea supported from forward operating locations in the Canadian Arctic and Thule Air Base, and the deployment of mobile air defense and radar systems by the United States to an undisclosed location.[70] The second was the deployment of B-52Hs to the Arctic during Amalgam Dart. Four B-52Hs launched from Morón, Spain, and flew north of Norway, over the Arctic Basin, and into the northern Pacific, then back to Barksdale AFB in Louisiana.[71] This twenty-seven-hour mission received substantial mainstream media coverage.[72] When they were over the Bering Sea, Russian forces scrambled MiG-31 and Su-35 fighters as the bombers paralleled the Kamchatka Peninsula. The B-52H force was preceded by a U.S. Navy EP-3E Aries electronic warfare aircraft, which was also intercepted and observed by Russian fighters.[73] During the exercise, Russian Tu-160 Blackjacks supported by MiG-31 and Su-35 fighters flew from Engels Air Base to the Baltic Sea, where they were intercepted by Swedish, Danish, and Italian fighters. Italian F-35s maintained a close escort of the Russian fighters.[74]

It should be noted here that the B-52H flights over Ukraine, specifically the ones over Mariupol and the Sea of Azov, had ceased with the advent of the new administration, as had the close flights over the Barents Sea off Murmansk. Operations over the Baltic countries continued. The Ukraine flights were replaced with unmanned aerial vehicle (UAV) operations conducted by unarmed MQ-4 Global Hawks operating opposite the Donbas.

Overlapping with these activities were the exploits of two Russian naval task groups and supporting air elements in the Pacific. One task group consisting of a cruiser, frigate, two corvettes, and a tanker "hunted" a hospital ship, a hydrographic vessel, a corvette, and a large space control ship (the *Marshal*

Krylov) acting as a U.S. aircraft carrier task group. Other participants included an unknown number of submarines, Il-38 maritime patrol aircraft, three Tu-142MA antisubmarine aircraft, and supporting Il-78 tankers. Eventually the Russian task force made its way to the Hawaiian Islands and operated around thirty miles from Pearl Harbor. Three *Burke*-class guided missile destroyers and a U.S. Coast Guard vessel shadowed the Russian ships, while F-22 Raptors conducted intercepts on Russian aircraft approaching the islands.[75]

There were likely multiple objectives underlying their activities. Laying off the iconic American Pacific base was a clear enough signal on its own, but there were discussions that the *Marshal Krylov* was present to monitor scheduled tests of the American BMD system, which were allegedly canceled because of the naval operation.[76] Another reason floated in the Russian information environment was that the task force worked out the problems of attacking the islands with nuclear weapons.[77] Practicing attacking an aircraft carrier was not new, but it had not been done in some time.[78] And there was no discussion of submarine activity, which had to have been intense as much as it was unseen. A further purpose raised in the Russian information environment was that these operations might be designed as an information counter or as a response to NATO-Ukrainian exercise Sea Breeze held in the Black Sea.[79] A Russian military website noted the U.S. Navy response was surprisingly not aggressive and exclaimed that "Biden is a weakling!" The website proceeded to allegedly quote from American social media: "Under Trump this would not have happened"; "Russians just know that Biden, a patient with dementia, is a weakling"; "It's not for nothing that this happens immediately after Putin's meeting with Joe"; and "I think retarded Joe didn't warn Putin not to mess with Hawaii."[80] This likely reflected the prevailing attitude in the Putin regime's leadership caste.

During the May–June period, U.S. NC3I aircraft deployments dropped slightly but spiked on 13–14 May. An E-4B flew near Camp David and the Ravenrock complex while three E-6Bs were aloft, with one over the Gulf of Mexico and one conducting an ALCS check over the Montana ICBM fields. This was followed the next day by four sequential flights of E-6Bs, followed by an E-4B orbiting Lake Superior and a simultaneous E-6B flight over the Pacific Ocean, and then a single E-6B during the course of the day. A similar series of flights took place 23 May.[81] In June, U.S. NC3I aircraft flights dropped to a level of activity similar to March, which was still significantly higher than in January–March, indicating an elevated interest in sustaining nuclear command and control operations.[82]

Multiple Crises: July–September

Several concurrent events suggest that it was at this point that the Russian plan existed, but parts of it continued to be exercised: these moves took place during the run-up to the Zapad 2021 exercise scheduled for September. These included "artillery units conduct[ing] strikes against command posts" involving Iskander missiles units, and airborne units trained at "conduct[ing] landings behind enemy lines." Bridging exercises were conducted at four locations. Rosgvardia units were also exercised in the summer of 2021, including "a scenario [that] assumed that the military authorities had taken over a civilian port and commandeered civilian and military cargo vessels." These lower-level exercises "involved border guards, customs officers, and personnel from the Ministry of Emergency Situations and Ministry of Internal Affairs." Zaslon 2021 was an eighteen-day Rosgvardia exercise to test neutralization of "threats to state and public security" involving rear area security operations. It involved four thousand personnel operating near Rostov. Notably, there was Zashchita 2021, an exercise where a joint emergency group including Rosatom units responded "to a simulated radiation hazard." Another exercise had the 27th NBC Brigade "deployed to Kurchatov to deal with a simulated accident at the Kursk nuclear power plant."[83] Notably, Kursk is identical to the power station at Chernobyl.[84] Other training included "the testing of a new financial system to support military units during wartime when IT systems may not function."[85] Importantly, "in July 2021, the 9th Section of the 5th Service of Russia's Federal Security Service (FSB) was enlarged into a directorate tasked with planning the occupation of Ukraine."[86]

It was at this time that the Belarus-EU border crisis, not uncoincidentally, was reinvigorated. Belarusian leader Lukashenko threatened to permit unregulated migration and narcotic smuggling from Belarus to adjacent EU countries as retaliation for sanctions related to the April crackdown. Then very little happened until late July.[87] Was the crisis with the EU states deliberately aggravated by the Lukashenko regime at the behest of the Putin regime to secure Belarus and distract NATO and EU member states as part of its overall plan to continue with its invasion of Ukraine?

There was a protracted and sophisticated campaign conducted by Russia and Belarus. The basic modus operandi was the recruitment or luring of vulnerable people from the Middle East to Belarus; facilitating their travel to Belarus using visas and the Belarusian state carrier, Belavia; and using Belarusian security forces to coerce people to illegally cross the borders of Lithuania first, then Poland, and finally Latvia. Analysts struggled to identify the purpose of this activity.[88] The most obvious was that they constituted payback for the sanctions on the Lukashenko regime. Connected to this was that border area disruption made it more difficult for anti-Lukashenko elements to operate and communicate with the outside world. There is a variant of this in which Lukashenko used the situation to crack down on the Polish minority in the southwest of the country. Another was to extort money from the EU member states to "solve the problem" and at the same time "redistribute" elements of the Belarusian economy to benefit those connected to him.[89] The next most obvious objective was "to destabilize the European Union."[90] Another argument was that the activity was intended to "intimidate Europe."[91]

Russian officials, such as Sergy Lavrov and others, "expressed support for Belarus in its confrontation with Poland and have used the crisis to promote anti-Polish and anti-EU propaganda." This line was supported by Putin.[92] Here was a deeper game in that "any new wave of migrants on Europe's eastern borders may worsen anti-immigration sentiment and strengthen the arguments of parties and politicians critical of increased immigration,"[93] that is, the resurgent far right in Europe, the same anti-EU and anti-NATO parties and politicians assiduously cultivated by the Putin regime.[94]

Yet another layer was the employment of information warfare narratives by the Putin regime to support its long-term objectives. Lithuanian specialists identified four axes of attack: social, humanitarian, political, and conspiratorial. Social narratives were designed to spread fear of migrants in the target countries and that money will be spent on them and not citizens. Humanitarian narratives employed manufactured information accusing Lithuanian, Polish, and Latvian security forces of brutality against migrants. Political narratives focused on who was responsible for the migrant crisis whereby the "real" cause is "the military actions conducted by NATO and its members in the Middle East region." The conspiratorial narrative was that the "migration crisis is a pretext to implement measures used to control the people of Lithuania" and that the "migration crisis is a part of a plan to exterminate the white race," essentially a form of the so-called replacement theory.[95] Russian minority populations in the Baltic countries could easily become the functional equivalent of Sudetenland Germans at a moment's notice under these conditions in the same way pro-Russian Ukrainians in parts of eastern Ukraine functioned in 2014.

Analysts at the time interpreted Belarusian moves as part of a larger scheme of hybrid warfare. This Western lens was flawed: these moves actually constituted Gerasimovian hybrid methods in support of a larger Russian geopolitical policy and its strategic manifestations that philosophically dated back to the 1990s and was operationalized after 2005. They were not merely a limited regional operation.

U.S. NC3I aircraft activity in July consisted of a steady state of four to five flights sequentially during most days, with associated trailing wire deployment flights over water and ALCS checks. After mid-July this increased to six or seven flights. On 26 July the Beijing regime tested a hypersonic glide vehicle in a Fractional Orbital Bombardment profile. This produced the scramble of an E-6B on alert at the Patuxent River Naval Air Station, followed by two pairs of an E-6Bs and the deployment of an RC-135S Cobra Ball aircraft over North Dakota. Four E-6Bs then rotated with the aloft aircraft, with two conducting trailing wire antenna deployments over the Pacific and Atlantic. The numbers dropped to two E-6Bs aloft over the United States after the test.[96]

Before that point, however, the collapse of the Afghan government in August and its effects enters our narrative. By July 2021 Afghanistan's second city was overrun by the externally supported insurgency, and by August the opposition forces were on the outskirts of Kabul. A poorly executed withdrawal operation for NATO member nationals and Afghan personnel who supported ISAF during the twenty-year war turned into a fiasco, with TV imagery of Afghans clinging to the sides of C-17 transports and comparisons to Saigon in 1975 omnipresent in the global information environment.

To what extent did the failure of the Afghanistan operation in 2021 and the perceived failure of U.S. leadership play a role in convincing the Putin regime it could carry out more aggressive Ukraine operations without American interference? To answer that question, we need to examine how the Russian information warfare ecosystem handled the events and to what extent its output reflected internal thinking on the matter.

There were several informational dynamics in play. The use of the words "abandoned" and its synonyms peppered the discourse in pro-Kremlin news outlets, with the implication that the United States would abandon allies when the going got tough. It is not a great leap to see that information line directed at NATO allies and Ukraine.[97] Another more important dynamic was the idea, expressed in official Russian media, that Biden himself was problematic when it came to crisis decision-making. State-controlled outlets recycled selective quotes from the American political information ecosystem critical of the Biden administration, mostly derogatory interview quotes from Trump as well as Fox News and *New York Times* content.[98] The track record of the Biden administration in the eyes of the Putin regime was not one that inspired deterrence.

As the Afghanistan affair wound down and the destabilization of the EU continued, Russia mounted Zapad 2021. Preparations in the form of SCRIs and parallel exercises had been underway since July, and the first Russian troops also arrived in Belarus in July. The exercise geography simulating Lithuania, Poland, and part of Belarus in the northwest was used to simulate a Polish minority region. Zapad 2021's narrative involved Polish and Lithuanian "provocations" that expanded into a color revolution inside Belarus that was to be used as a pretext for Poland and Lithuania to invade the country, an invasion that would be countered with large-scale Russian and Belarusian combat operations. When the so-called provocations failed, Poland and Lithuania deployed forces to the border "disguised as exercises."[99] This was followed by air and missile strikes against Belarus and an advance on the ground to a depth of 150 km. This was repelled by the Russian 1st Tank Army and 41st Combined Arms Army, backed up with the 20th Combined Arms Army, which prepared the eventually successful counterattack.

There were several parallel exercises, including one involving the 61st Naval Infantry Brigade near Murmansk which was deployed to Kaliningrad to work with the naval infantry brigade stationed there. In late August, "missiles and artillery rounds were distributed from one of the arsenals located in the [Southern Military District] in conditions as close to combat as possible." Some 10,000 logistics personnel were involved. Additionally, "250 main battle tanks . . . were withdrawn from a storage and maintenance facility . . . presumably to form new tank units." Air operations involved all forms of airpower, but significantly Tu-22M3 Backfire bombers were employed in the exercises in massed air strikes.[100]

Further examination of the "activities of the [Southern Military District] in August and September thus amounted to nothing less than another ZAPAD-like exercise." Ultimately, "in September the Russian armed forces were conducting at least three large-scale exercises linked to ZAPAD and in support of operations in the western theatre of military operations." Significantly, there were command and control activities conducted at a high level to coordinate the activities of three military districts.[101]

A parallel exercise was also conducted by the Northern Fleet at this time. It involved some fifty ships "operating in the Murmansk region, Barents, Kara, and Laptev Seas, and along the Franz-Josef Land and New Siberian Islands." The ostensible objective was to control the northern sea lines of communications and included the destruction of an "enemy" surface action group. In addition, a separate antisubmarine warfare exercise was conducted with fifteen ships in the Baltic. As for nuclear operations, Zapad 2021 "did not appear to include a nuclear element. It featured only one Tu-95, which was used to test air defenses."[102] The possibility that Zapad 2021 might have been part of a larger trajectory of preparations for an attack on Ukraine was not mentioned in Western analysis. One conclusion was that "Russia practised going to war with NATO." And collectively, these exercises involved 200,000 Russian personnel.[103]

There were no visible activities of U.S. strategic deterrence in July, though STRATCOM held Global Storm in mid-August. This involved interoperability movements in the western Atlantic involving two B-52H, the SSBN USS *West Virginia*, and an E-6B Mercury command and control aircraft. The likely objective was to test the E-6B communications systems with the bombers and submarines. Global Storm also tested the ability to use helicopters to replenish an SSBN while it was at sea to extend deterrence patrols.[104] Three B-2A Spirit bombers deployed to Iceland and possibly to the United Kingdom on a Bomber Task Force tasking on 25 August.[105] None of this amounted to out-of-the-ordinary general deterrence or assurance signaling activities.

Events were punctuated by two People's Republic of China hypersonic glide vehicle tests, one of which entered a Fractional Orbital Bombardment configuration. The 12–13 August test produced considerable U.S. NC3I aircraft movement. Two aloft E-6Bs were replaced with another conducting a trailing wire antenna deployment over the Atlantic. It was reprised with another E-6B, but then an E-4B lifted off along with two more E-6Bs. An RC-135S Cobra Ball then came on station over North Dakota, and then another E-4B and three E-6Bs replaced the previous aloft aircraft. A second Cobra Ball flight also took place. A similar deployment took place on 14 August with E-6Bs over the Pacific as well as the Atlantic.[106]

In the run-up to Zapad 2021, visible Russian strategic nuclear force activity consisted of Kozelsk Loop communications checks. The fact this was done on two consecutive days was unusual, but it cannot be correlated with any other visible Russian activity.[107] During Zapad 2021, however, a Tu-214SR communications relay aircraft approached the Estonian border on 11 September. Though the purpose of this flight is opaque, it could have been a communications test with Iskander missile units located south of St. Petersburg. This was followed by a Tu-214SR orbiting east of Moscow, a Tu-214SR conducting a flight over the 27th Missile Army's area of operations east of Moscow, and a Tu-214PU-SBUS Ministry of Defence command plane which deployed to Nizhny Novgorod on 13 September. These significant movements were likely to ensure that Putin, who was visiting the Mulino Range which was involved in Zapad 2021 activities, could demonstrate that he was in contact with the strategic nuclear forces while outside of Moscow. The Tu-214PU-SBUS suggests that either Sergy Shoigu or Valery Gerasimov accompanied him.[108] This command and control package also allowed the Russian national command authorities to maintain communications with a pair of Tu-160 Blackjacks and their MiG escorts that conducted a run over the Baltic Sea, which Russian sources claim "led to a panic in the West" and resulted in intercepts by Danish, Swedish, Finnish, and French fighters.[109]

American activity during this time amounted to significant deployments of strategic reconnaissance aircraft, both for intelligence collection and for signaling. The basic deployment run at this time was a U.S. Air Force MQ-4 Global Hawk UAV over eastern Ukraine and a U.S. Navy P-8A Poseidon over the Baltic to keep an eye on Kaliningrad.[110] This changed in the days before Zapad 2021. On 7 September, a P-8A Poseidon orbited off the Kerch Strait, observing occupied eastern Crimea, the straits themselves, and Novorossiysk. The Biden administration had not run B-52Hs over Mariupol or the Sea of Azov as the previous administration had, so objectively the P-8A and the MQ-4 were a step down when it came to signaling: observant forward interest as opposed to an armed, in-your-face potentially nuclear umbrella.

The same day, however, an RC-135U Combat Sent flew a ring around Kaliningrad. There are only two Combat Sent aircraft, also known as "the U-boats," and in official terms, they provide "strategic electronic reconnaissance information to the president, secretary of defense, Department of Defense leaders, and theatre commanders." Combat Sents locate and identify "military land, naval, and airborne radar signals."[111] The significance of this mission is that Combat Sent is tasked from the highest level of the U.S. government and is generally used prior to an operation to update threat radar databases. The fact that the aircraft is tasked from a very high level itself can be a signal of interest. This deployment occurred right after Poland declared a state of emergency regarding the Belarus border situation and right before Zapad 2021.

There was overt, stepped-up surveillance throughout Zapad 2021. Generally this involved a combination of MQ-4 on the borders of the Baltic countries with Russia and Belarus; E-8 JSTARS aircraft with ground movement target indicator systems watching vehicle movements in Belarus; P-8A Poseidons over the Baltic off Kaliningrad; and the U.S. Army Artemis airborne battlefield surveillance system in the same areas. RC-135 Rivet Joint communications interception aircraft joined the Combat Sent exercise in running rings around Kaliningrad.[112] In the Black Sea, a P-8A Poseidon scooped up information by paralleling the coast of occupied Crimea. A follow-on mission had an RC-135 Rivet Joint fly this route but also parallel the Russian coast down to Sochi, down to Georgia, and then back along the track. This route was replicated with an MQ-4 soon after, and on 24 September Combat Sent flew this track with an RC-135 Rivet Joint.[113] Two days after, an RC-135 Rivet Joint departed the United Kingdom and proceeded to the Barents Sea, where it flew over to Novaya Zemlya, then off Murmansk, and back to RAF Mildenhall.[114]

Given the geographic locations, it is clear this activity was conducted to monitor Zapad 2021, likely to determine to what extent it constituted a threat or could be used as cover for an assault. The specific types of aircraft and their properties strongly suggest there was messaging involved; that is, the United States demonstrated it was dedicated to closely monitoring Russian activities that could pose an immediate threat to NATO and Ukrainian interests. But there were no visible strategic nuclear force movements to reinforce this messaging or to match Russian strategic nuclear force movements during Zapad 2021.[115]

The only visible anomaly on that front was the mass launch of seventeen KC-135 tankers from Fairchild AFB, probably a nuclear operational readiness inspection (ORI), on 29 September, ten days after Zapad 2021 completed its field phase.[116] It was at this point that Chairman of the JCS General Mark Milley determined that Zapad 2021 and ancillary Russian activities were "odd; it was much bigger in scale and scope than the previous year's exercise."[117]

While attention was focused on Eastern Europe, North Korea tested what appears to have been a hypersonic glide vehicle on 13 and 28 September. The U.S. NC3I aircraft response closely followed those conducted in August; twelve E-6Bs were aloft over the course of the day, with four up simultaneously with a Cobra Ball aircraft.[118] For the most part, the baseline pattern of U.S. NC3I aircraft flights remained slightly elevated throughout September, averaging three aircraft per day.[119]

Signaling Intensifies: October–December

In the fall of 2021, Secretary of State Antony Blinken first developed "an understanding of what the Russian leadership was actually thinking and planning for those forces," and General Paul Nakasone became "convinced the Russians are going to invade Ukraine." This produced a series of high-level diplomatic efforts by the United States to influence the Putin regime. Though direct American intervention was out of the question, National Security Advisor Jake Sullivan wanted to "put ourselves and the Western world and Ukraine in the best possible position to deal with it, and to Russia in the worst possible position to succeed." During those efforts, Director of Central Intelligence Bill Burns, flew to Moscow and came back convinced that those surrounding Putin had not been fully read-in to the Russian plan. Burns thought "the invasion of Crimea made [Putin] believe that we would do nothing. . . . He thought the Europeans were distracted"—which they were, attempting to deal with the Belarus border crisis.[120]

For the first half of October, there were visible signaling moves by both American and Russian strategic forces. The Russians continued to run Kozelsk Loops but not on the scale seen earlier in the year.[121] Combat Sent continued to collect off occupied Crimea and the Russian coast in the Black Sea. RC-135 Rivet Joint collection aircraft were deployed: one flew over all three Baltic countries; another conducted operations in the same area as the Combat Sent over the

Black Sea; yet another operated off Murmansk.[122] During one Rivet Joint mission off the Kerch Straits, Russian sources claim that Russia mounted an electronic attack on the American aircraft, disrupting its navigation system and forcing it to return to base.[123]

On 13 October Russia announced it was exercising the 35th Missile Division around Barnaul.[124] This was a significant move. The 35th Missile Division consists of thirty-six mobile Topol-M ICBMs. Each missile is mounted on a TEL and these vehicles are located in central bases. While conducting exercises or on alert, the TELs deploy from their bases with their security forces into the forests nearby in order to disperse them and make them harder to target. The Topol-M missile carries three or four 300–500 kt yield reentry vehicles or four to six with 150 kt yield reentry vehicles. From these locations the Topol-M can reach any target in Western Europe.[125]

The geographic location of Barnaul relative to American strategic nuclear forces is significant. These systems are likely targeted against Minuteman III silos in the northern United States. This exercise was a singular event and does not appear to have been coordinated with other Russian strategic nuclear forces, which should have happened in October.

The question of why Russian strategic nuclear forces exercise Grom 2021 was not run in the fall after Zapad 2021 is a curious one. In June 2021, there were official announcements that an exercise named Grom 2021 would be run in Armenia in September, but this was a small-scale conventional special forces exercise as opposed to a strategic nuclear forces exercise. It suggests that a decision to postpone or cancel Grom 2021 (the nuclear exercise) had been made by at least summer 2021, probably earlier.[126] The Russian Ministry of Defence announced only on 21 December 2021 that Grom would proceed during early 2022.[127] The higher-level staff and Russian leadership that are usually present for the exercise were busy with other activities, namely, finalizing planning and gaming the next phase of the invasion of Ukraine. The Putin regime attempted to time Grom 2022 as a signaling tool during the February 2022 offensive. This suggests that the February 2022 invasion timing had been determined at least by June 2021 and probably earlier.

It is clear that the Biden administration became increasingly concerned about Russian military activities in the vicinity of Ukraine at this time. This led to a

PHOTO 8 ◆ An example of the Topol-M MIRV mobile ICBM transporter erector launcher (*top*) and a satellite image of a pre-deployment base containing an estimated twenty-eight TELs. Note the TELs that are involved in camouflage training. On alert, the TELs deploy to forested areas to reduce vulnerability. *Russian Ministry of Defence and Google Earth screenshot by author*

visit by Secretary of Defense Lloyd Austin to Ukraine on 19 October to discuss "Russia's destabilizing actions in eastern Ukraine and in occupied Crimea."[128] This visit was underscored by a flight of two B-1B bombers and two KC-135 tankers that conducted an operation over the Black Sea off occupied Crimea and the Russian coast. These aircraft were intercepted by Russian Su-30 fighters.[129] The fact that conventional B-1Bs were employed instead of nuclear-capable B-52Hs is significant in that the messaging was a step down from the previous administration which employed B-52s over the Sea of Azov and Mariupol in 2020.

Austin was asked during the visit about Ukraine joining NATO, to which he asserted it was up to the Ukrainian people to determine their own foreign policy. Putin gave his response in a speech to the Valdai Club. In it he resorted to the standard propaganda lines about NATO expansion, that there was a repressed minority in Ukraine that needed a voice, that BMD launchers could be loaded with Tomahawks, and that it was only a matter of time before NATO had missiles that could strike Russia from "Kharkov."[130]

It is possible that Russian strategic forces were active at this time to underscore Putin's Valdai speech or that there were preparations for a missile test, though the specifics are unclear. The U.S. Air Force deployed an RC-135S Cobra Ball aircraft right off the Russian coast opposite Anadyr.[131] Cobra Ball is a JCS-tasked specialized aircraft equipped with electro-optical and other sensors designed to collect information of the characteristics of reentry vehicles as they reenter the earth's atmosphere or skim along it.[132] The aircraft has the ability to act as a portable early warning platform to supplement existing ground- and space-based systems, particularly if the latter are not functioning. The aircraft continued its operations in the Alaska area throughout late October.[133]

The suggestion that there was expanded Russian activity is supported with the launch of two Tu-95MS Bear missile carriers escorted by Su-35 fighters from Tiksi air base in the Russian Arctic. The flight path took them to locations opposite Alaska and then down the Kamchatka Peninsula. Russian sources claimed this was done in response to "numerous provocations from Washington," and the mission was designed as a practice strike to "work out the option of destroying the largest U.S. Air Force base in Alaska, Elmendorf-Richardson."[134] Russian nuclear command and control was also exercised during this period with a Tu-214SR communications relay aircraft operating near Kazan and then with a Kozelsk Loop.[135]

LRA then mounted a complex operation that included two Tu-160 Blackjacks, Su-24 escorts, and a A-50 Mainstay AWACS aircraft to probe Norwegian defenses. Bodø-based fighters handled the intercept.[136] The Russian aircraft did not use transponders. The same day, Tu-22M3 Backfires escorted by Su-27 fighters "practised strikes against a potential enemy." Russian media outlets linked these operations to "the unsuccessful statement of the German Ministry of Defence" who apparently "made an unfortunate statement about the need for military deterrence of Russia and the provision of appropriate military pressure on Russia."[137]

USSTRATCOM mounted its scheduled Global Thunder 22 on 1 November, exercising all three elements of the strategic deterrent forces, "with a specific focus on nuclear readiness." Global Thunder featured a very public alert launch of eleven B-52Hs from both Barksdale and Minot AFBs, in addition to validating "the nuclear operation process."[138] There is every indication this exercise remained confined to North America.

Global Thunder 22 included the deployment of an E-4B Nightwatch over Lake Superior on two occasions and the Gulf of Mexico on one occasion.[139] The significance of this lies in the location as well as frequency. During the Cold War, Strategic Air Command EC-135 Looking Glass nuclear command and control aircraft tested their five-mile-long trailing wire antennas over the lake as a safety measure. In case of emergency the antenna could be jettisoned without causing damage on land.[140] These antennas are designed to operate in the low-frequency and very low–frequency portions of the electromagnetic spectrum. Their function is to communicate with the U.S. SSBN and ICBM forces. On a day-to-day basis, E-4B trailing wire antenna operations are relatively rare.[141] As the crisis unfolded, however, E-4B operations like this took on signaling overtones when combined with other aloft assets.

The idea that there was a new crisis emerging from events regarding Ukraine, however, made its way into Western media at this time. Amateur open-source intelligence hobbyists and other interested analytical parties in Western social media turned their attention to the possibility that Russian forces had not redeployed after Zapad 2021 and what the implications of this were. This discussion moved into the mainstream media on 1 November, when *Politico* used Maxar satellite imagery and *Janes'* analysis to explain how Russian mechanized forces

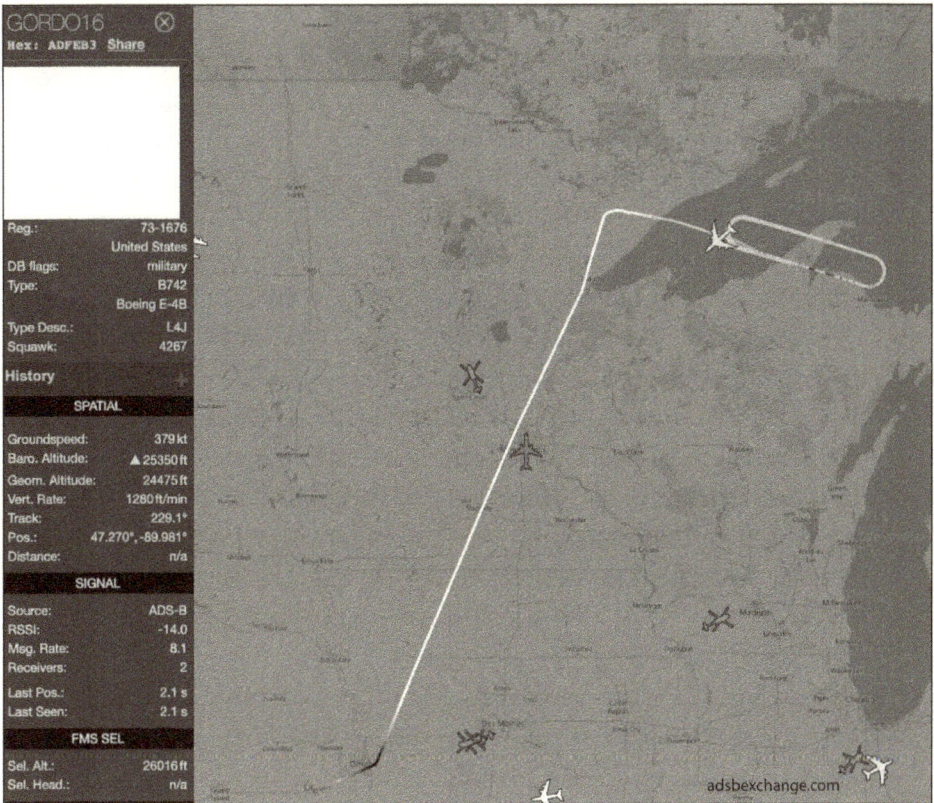

PHOTO 9 ♦ An E-4B Nightwatch NAOC orbiting Lake Superior during a communications check with USSTRATCOM forces. Michigan's Upper Peninsula has been home to a variety of strategic communications facilities dating back to the Cold War. *ADS-B Exchange*

were building up near Yelnya on the Russian-Belarusian border as well as in Bryansk and Kursk.[142]

The Belarus "coercive engineered migration" continued to generate problems well into November to the point where Lithuania declared another state of emergency.[143] Poland increased the number of military personnel on the border to 17,000 by this point, and the situation continued to be viewed by Western officials and observers through the lens of hybrid warfare.[144] It remains unclear what was going on in the Putin regime at this point. It had generated a situation whereby the Polish army was now on the Belarusian border, and the Russian forces retained 100,000 troops on the Ukrainian border, both in Belarus and in Russia.[145] On one level, "NATO" forces needed to be "deterred" to permit the "consolidation" of the situation in Belarus before operations against Ukraine could commence. There is no way that Lukashenko and Putin could have possibly believed that a single Polish division that was not deployed for combat operations could have posed a true threat to Belarus on the scale of the enemy force in the Zapad 2021 exercise. The tension level on the diplomatic front between Warsaw and Moscow, however, was increasing and a new round of sanctions was in the offing. Consequently, two Tu-22M3 Backfires made a run at Polish airspace from inside Belarus to underscore Russia's discontent with the situation.[146] This was further underscored by the flight of Tu-160 Blackjacks to the North Sea, where they were intercepted by Belgian F-16s.[147]

Visible activities by American forces from 1–18 November included three P-8A Poseidon and two RC-135 Rivet Joint flights to the eastern end of the Black Sea and the Kerch Strait and, interestingly, an E-8C JSTARS run off the Russian coast near Novorossiysk, which collected Ground Moving Target Indicator information on the Russian buildup in that area. Globally RC-135 Rivet Joints and an RC-135S Cobra Ball were active over the Chukchi Sea off Alaska. Surveillance aircraft also monitored the situation on the Belarusian border, and especially activities in Kaliningrad, using E-8C JSTARSs and RC-135 Rivet Joints.[148] The USS *New Mexico*, a *Virginia*-class SSN with twelve Tomahawk launch tubes, arrived at Tromsø, Norway in response to aggressive verbiage by Sergy Lavrov directed at the Norwegian government.[149] There was no visible bomber activity overseas during this period aside from a pair of B-1B Lancers operating in the Arctic. There was nothing visible backstopping, say, NATO secretary general Jens Stoltenberg's expression of concern over "large and unusual" Russian buildup and his warning to Putin "against taking aggressive actions."[150]

Indeed a generally overlooked event was the Russian destruction of a dormant Tselina-D electronic intelligence satellite by a direct ascent Pl-19 Nudol anti-satellite missile fired from Pletsetsk Cosmodrome on 15 November. The destruction of the Soviet-era satellite polluted its orbit with debris and threatened the International Space Station. The likelihood that this operation constituted signaling is high. The Pl-19 Nudol, which came in fixed as well as mobile versions, had been tested eleven times since 2014 but never against a live target. None of the other tests correspond to crisis maneuvering as this one did. The selection of an ELINT satellite as a target could easily be interpreted as a message to the United States while it was undoubtedly collecting on the Russian buildup using similar systems. In a larger sense, the dependency of American forces on satellites is well known and a space denial test would have emphasized this point.[151] The visible U.S. response was a sustained NC3I aircraft posture, with three E-6Bs launched, rotated with two, then those rotated with two more, as well as fifty-one tankers aloft, something that had not been done throughout 2021. The posture was reduced later in the day to one E-6B, then an E-4B, and finally a pair of E-6Bs.[152]

The visit by the Ukrainian minister of defense to meet with the U.S. secretary of defense on 18 November was met with a Russian probe of the air defense identification zones around Alaska when two Tu-95MSs and two Su-35Ss "practised a nuclear strike on U.S. territory"[153] but were intercepted by RCAF CF-18s supported by a USAF KC-135 tanker.[154] This probably took place over the Beaufort Sea. In an unusual move, two E-6B Mercuries conducted trailing wire antenna deployment flights over the Pacific Ocean and the Atlantic Ocean, while a third conducted an ALCS check over Wyoming, Montana, and North Dakota.[155] A pair of Tu-95MSs also probed NATO air defenses in the Baltic Sea right afterward but were intercepted by Portuguese F-16s operating from Lithuania.[156]

The Belarusian-Polish border issue simmered in late November, and a UN team confirmed that Belarusian security forces were complicit in the abuse of migrants. Western analysts concluded that the border activity was hybrid warfare but struggled to determine a motive underpinning it, making no connection to what was going on in and around Ukraine.[157] RC-135 Rivet Joint aircraft continued with Black Sea surveillance but started to focus on Transnistria as well, which suggests there was increased intelligence community concern about the Russian-backed "frozen conflict."[158]

In terms of Russian strategic nuclear force movements, the number of Kozelsk Loops increased to one a week, involving both Il-96–300PU and Tu-214PU-SUS specialized command aircraft. This coincided with Putin's complaints throughout the last week of November about how "U.S. nuclear-capable bombers flying over the Black Sea posed a threat to Moscow," as did "the U.S. missile defense systems in Poland and Romania," and he "expressed concern about NATO deploying missiles on Ukrainian territory that could have a flight time of seven to ten minutes to Moscow."[159] Since there were no more B-52Hs to intercept over the Black Sea, Russian fighters had to be satisfied with an RC-135 Rivet Joint and a Cl-600 Artemis instead. The Rivet Joint's collection track was very close to Sochi.[160]

There were announcements in Russian media in early December that "hypersonic weapons" would be used in the delayed Grom exercise, now dubbed Grom 2022.[161] The Putin regime would flourish hypersonic systems at several points in the war, so it is important to ground our discussion of these systems on the facts. Hypersonic missiles remain an evolving technology and are generally defined as any system that travels at Mach 5 or more to deliver its effects. Confusingly all reentry vehicles and some monoblock warheads achieve Mach 5 as they plunge to the target. For our purposes, hypersonic weapons include boost glide vehicles, hypersonic rockets, and hypersonic cruise missiles, or hybrid systems. The primary properties of hypersonic weapons that are of concern are their insanely high speeds, approaching Mach 10 for the cruise missiles and past Mach 25 for HGV, which generate difficulties in intercepting them and most important, produce significantly reduced reaction time.[162]

At this point Russia had three hypersonic systems deployed in small numbers: the Avangard hypersonic boost glide vehicle launched from ICBM silos, the Kh-47 Kinzhal air-launched ballistic missile, and the Tsirkon cruise missile. All three are nuclear-capable systems; the last two are dual-capable.[163] Russian interest in the systems from an information warfare perspective had many facets. The first was that it gave the Putin regime a card to play against American BMD in the ongoing general deterrence game: "Your BMD is useless, why bother deploying it?" The secondary intent was to intimidate the Europeans by demonstrating the NATO-based deterrent, in the form of fighter-bombers and B-61s, was vulnerable, an argument that could be leveraged by Russian sympathizers in the West into putting pressure on governments to get rid of them by arguing they were obsolete. The third objective was to intimidate those supporting Ukraine, as there was no apparent defense against these systems. Official Russian media had pumped up hypersonic *wunderwaffen* capabilities for some time. Izvestia, for example, had no fewer than 248 articles on the subject from January 2020 to November 2021 (by comparison, RIA Novosti lagged behind with eight articles).[164]

The irony of playing the hypersonic card after "expressing concern" about NATO deploying missiles to Ukraine with a short flight time to Moscow, knowing full well that there was no intent to do so and that the future American intermediate nuclear force-like system that was in the works *was itself* a response to Russia's breaking the INF Treaty with Iskander, was astonishingly cynical. Russia deployed both hypersonics *and* intermediate nuclear forces in 2021. NATO and the United States had built neither.

The Russian ground force buildup inexorably continued. The entire 41st Combined Arms Army was now situated on the Belarusian border with Russia. The 1st Guards Tank Army and 6th Combined Arms Army were encamped in the Voronezh area, while the 49th and 58th Combined Arms Armies were now in place on Ukraine's southeastern border. Units from the Southern Military District continued to flow into occupied Crimea.[165]

In early December, the Biden administration gave a background intelligence briefing on the situation to the *Washington Post,* which was a signal to the Putin regime ahead of a virtual meeting between Biden and Putin on 7 December.[166] The briefing asserted that it would be a "multifront offensive as soon as early next year" and showed a map depicting Russian troop dispositions. Blinken, on a trip to Latvia, gave a more detailed briefing to his European colleagues, and told them that they had evidence of a plan and that "Putin could quickly order an invasion of Ukraine." It was not known "whether Putin has made the decision to invade" but that Putin was putting forces in place to do so. This was all supported by "Russian influence proxies and media outlets [which] have started to increase content denigrating Ukraine and NATO, in part to pin the blame for a potential Russian military escalation in Ukraine" that was detected as starting in early November. Blinken urged Ukraine to "exercise restraint because, again, the Russian playbook is to claim provocation for something that they were planning to do all along."[167]

Intelligence collection with E-8C JSTARS over Poland continued as before. Six KC-135s were scrambled out of Fairchild AFB in a scheduled operational readiness inspection, and a pair of B-52Hs departed Barksdale AFB for a mission to the Sea of Japan.[168] Russian movements, however, were more extensive. A Tu-214SR communications relay aircraft flew an uncharacteristic flight path. The aircraft departed Moscow, flew to a point between the Estonian border and St. Petersburg, flew to the Caucasus, and then crossed the Black Sea and orbited over occupied

Crimea.[169] Russia also ramped up the Novosibirsk run with the same Il-96–300PU presidential command post aircraft on three occasions, two within forty-eight hours of each other.[170] These suggest a communications check with strategic missile forces, with the Tu-214SR flight appearing to have been designed to let observers known that the Putin regime was linking its strategic nuclear forces with the situation in Ukraine. The most important Russian signaling took place on 16 December. An Il-96–300PU presidential command post aircraft departed Moscow and flew to a point on the Russian-Ukrainian border opposite Kharkiv. This flight had not been conducted previously, nor has it since.[171] The deployment of such a scarce high-value asset within the envelope of the Ukrainian air defense system was not only incredibly provocative; it was another unmistakable signal linking Russia's strategic nuclear forces to the situation in Ukraine.

The visible American response to this consisted of a flight of three B-52Hs over southern Manitoba, Canada and the manipulation of the E-6B Mercury command and control aircraft force. B-52Hs rarely "squawked" while in Canadian airspace. U.S. NC3I aircraft and their activities, however, shifted. On 21 December an E-4B Nightwatch departed Lincoln, Nebraska and proceeded to a point near the Wyoming ICBM fields, then flew to Lake Superior, where it visibly orbited for several hours. Two E-6Bs were aloft at this time. Then two pairs of E-6Bs rotated aloft, conducting training wire antenna operations over the Pacific and Gulf of Mexico.[172] Global Thunder was long over, and this should be considered a significant American signal in the context of the situation as U.S. NC3I aircraft activity dropped off for the rest of December, averaging two to three aircraft per day.

It appears as though the message was received, and the Putin regime upped the ante. Two Tu-22M3 Backfires flew into Belarusian airspace and "practised strikes against Kyiv."[173] Starting on Christmas Eve (Western calendar) the Putin regime launched four Kozelsk Loop flights around the Kozelsk ICBM field and the Chekhov and Sharapovo command bunker complexes southwest of Moscow until 29 December. This involved a mix of Il-96–300PU and Tu-214PU-SUS aircraft and even included two flights in a single day, which again was unprecedented given the baseline behavior of the Russian NC3I aircraft fleet over the past year.[174] Then on 26 December, state-controlled Russian media announced that ballistic missile submarines of the Pacific Fleet based on the Kamchatka Peninsula "were urgently sent to sea."[175]

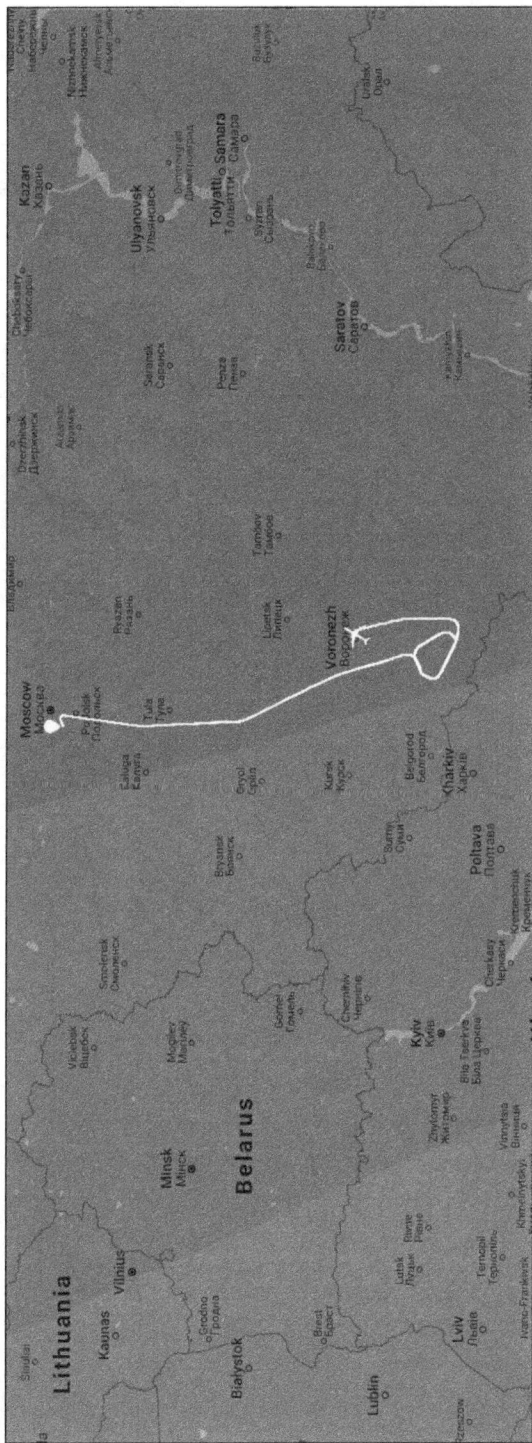

PHOTO 10 ◆ Russian presidential nuclear command and control aircraft RA-96021 on a 16 December flight toward the Ukrainian border. This flight had never been undertaken in the past, nor was it ever repeated. In the context of fraught diplomatic negotiations at the time, it was a clear Russian signal of the importance they placed on those negotiations and coercion directed at Ukraine. *flightradar24*

PHOTO 11 ◆ The U.S. Navy E-6 Mercury TACAMO aircraft (*top*) has the ability to communicate with ballistic missile submarines with a trailing wire antenna (TWA) and also possesses the Airborne Launch Control System that permits it to communicate directly with Minuteman III ICBM sites (U.S. Navy). The ADS-B Exchange screenshot depicts the deployment of the TWA by an E-6B over the Pacific Ocean. *ADS-B*

Putin met with Lukashenko in St. Petersburg on 29 December, at Lukash-enko's insistence. The Belarusian leader claimed there were 30,000 NATO troops building up in Poland, Ukraine was mobilizing 10,000 troops on its border with Belarus, and thus there needed to be a military exercise to act as a deterrent to "aggression."[176] Russian information operations ensured at this point that the request for this exercise was seen to be a Belarusian initiative.[177] The timing of the subsequent exercise, Unified or Allied Resolve (depending on the translation), was allegedly agreed to on 17 January 2022, with start and end dates of 10–20 February 2022. The latest phase of the invasion of Ukraine was actually scheduled to be launched on 20 February, the day after the delayed Grom strategic nuclear forces exercise was scheduled to be held and the day of the end of the Unified Resolve exercise. In effect, the pretext of NATO "aggression" against Belarus, established earlier in 2021, was employed to justify a Russian troop buildup in Belarus which in turn was really part of the larger plan to take Ukraine. The cover story established by Lukashenko and Putin was misdirection combined with establishing the legal basis for the Russian troop deployment.

The building tension led to another Biden-Putin conversation on 30 December. In blunt terms, Biden offered two concepts, one focused on diplomacy and de-escalation, the other pursuing deterrence and increasing the costs to Russia if it continued its aggression in Ukraine.[178] While the call was in progress, the alert E-4B Nightwatch, call sign Spice99, lifted off from Lincoln, Nebraska. It was followed by three E-6Bs, one flying over Wyoming near the ICBM fields to conduct an ALCS check and another over the Gulf of Mexico testing its trailing wire antenna.[179]

The seriousness of the international situation was not apparent to the voting public in the Western countries that were not adjacent to Russia and its Belarusian satrap. On New Year's Day, a British Royal Navy *Vanguard*-class ballistic missile submarine deployed from its base in Faslane, Scotland and was photographed doing so. These photos proliferated on social media. The significance of this is multifaceted. First, the *Vanguard* class can carry sixteen Trident II D5 submarine-launched ballistic missiles. Each missile has twelve slots for reentry vehicles. The British reentry vehicles are derived from the U.S. Navy Mk 4 reentry body and carry a British derivative of the W76 warhead.[180] With its load-out maximized,

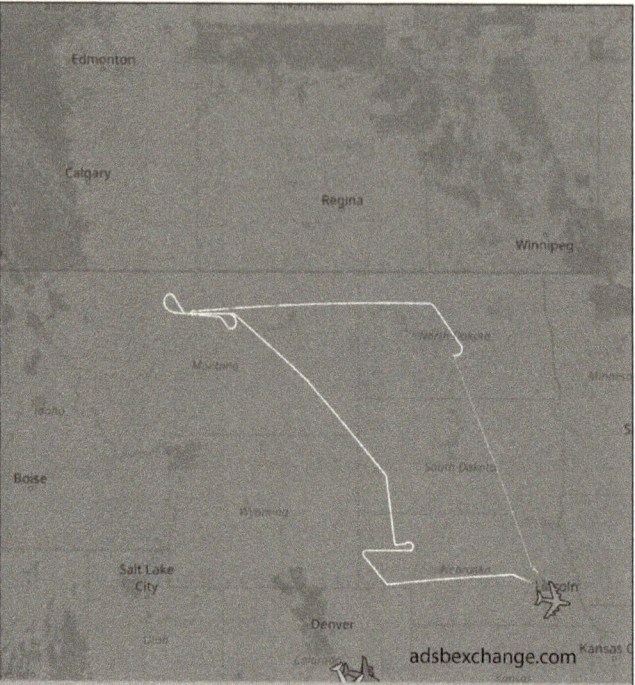

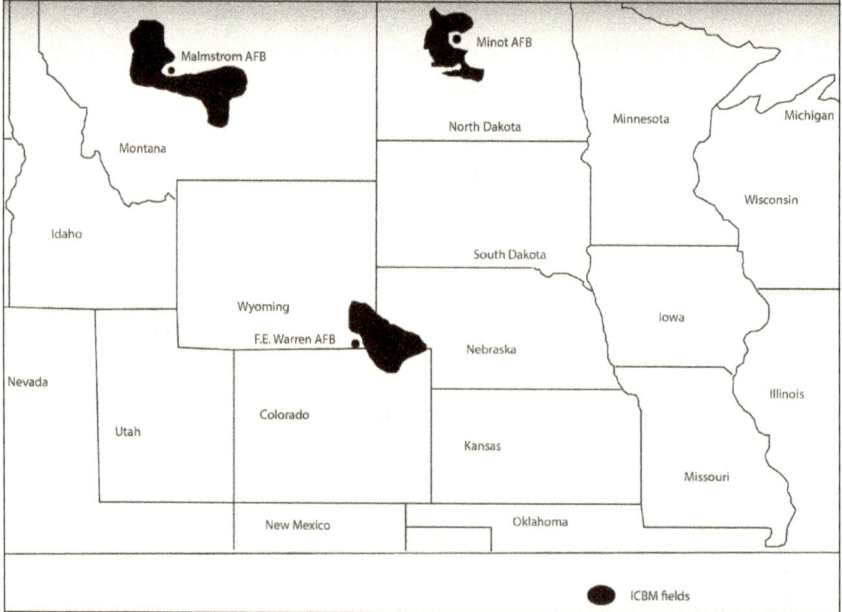

MAP 3 ◆ E-6B Mercury conducting a test of its Air Launch Control System over the ICBM fields in the northern United States on 21 November 2021 *ADS-B Exchange*

this submarine can carry 192 nuclear weapons.[181] Second, during the Cold War, deployments like this involved elaborate surface, subsurface, and aerial operations to ensure that Soviet nuclear attack submarines did not tail and track SSBNs. Ballistic missile submarine deployments were kept secret and not advertised in any way.[182] Third, the idea that any civilian government worker would work on New Year's except under exceptional circumstances is absurd. And finally New Year's replaced Christmas as the seasonal "super holiday" in the Soviet Union, and this tradition was retained after the collapse of the Soviet Union. As such, it is a day of great significance in Moscow.[183]

Conclusions

Despite multiple international events, the Belarusian border situation stands out as the most dominant one. There is little doubt that the Putin regime used the Lukashenko regime to mount a campaign of hybrid methods against Poland, Lithuania, and Latvia and, by extension, the EU. If the goal was to destabilize and embarrass the EU members, the campaign failed. If the goal was to confuse, mask, and distract from the moves to prepare the way for a campaign in Ukraine and secure Belarus as a base of maneuver for that, the campaign was a success.

The Putin regime's view that the United States under the new administration was weak, divided, and indecisive was progressively strengthened throughout 2021. First there was the events of 6 January, which fit the Russian view that this was a U.S. "color revolution." The second was the apparent shift from forward deterrence signaling over Ukraine with B-52s to a forward observation and presence posture using ISR aircraft. This occurred after the Umka 2021 exercise in the Arctic, a significant signaling event for which there was no visible response. The third was the lackluster response to the progressive Russian land forces buildup opposite Ukraine. In retrospect, diplomatic efforts throughout the first two-thirds of the year were insufficient and may again have contributed to the Putin regime's outlook on the new administration. The fourth was the tepid response to the Russian naval operation off Pearl Harbor. Finally the bungled American and allied response to the collapse of the U.S.-backed government in Afghanistan would only have reinforced this further. Calling out Putin without corresponding signaling appears to have been counterproductive.

It is not clear exactly when the Putin regime decided to move on Ukraine. Certainly the decision had to have been made in 2020 to move piece after piece into place throughout 2021 and conduct separate exercises that ultimately contributed to the whole in 2022. Russia maintained steady strategic nuclear signaling to backstop its buildup throughout the year. The decision to move to the next step was likely made in the fall of 2021. By this point, the Biden administration had come off the Afghanistan fiasco and directed more attention to the situation in Eastern Europe. Yet there was not a lot of strong signaling until December. And it was clear by the end of the year that the situation was far more serious than originally understood and that traditional diplomatic methods were of increasingly questionable value. It was time for the Western powers to up their game.

CHAPTER 5

◆ ◆ ◆

Triumph of Disbelief

January–February 2022

The Russian plan to take over Ukraine was multifaceted and intricate. The conventional buildup throughout 2021 was designed to keep the Putin regime's options open and secure Belarus while at the same time applying as much pressure as possible to intimidate and disrupt NATO and the EU. If Ukraine did not simply give in, there were other options in the Russian plan, which was honed throughout January 2022.[1] The objective of the plan was the removal of the Zelensky government and its replacement with a pro-Russian one covertly waiting in the wings: the objective was not necessarily to mount a full-blown conventional ground invasion to take over the country and then install a government. Those forces were there to backstop the removal process, whichever option presented itself, and essentially then be invited in by the new pro-Russian government as occupation forces, fighting against uncoordinated local opposition as required. The explicit and implicit threat of Russian nuclear weapons use accompanied by the manipulation of strategic nuclear forces and supporting information operations to reinforce those threats were integral to this plan.

The trigger event for the Russian plan was to have been a traditional Soviet-era "false flag" attack conducted by a composite unit of covert operators against civilian or military targets inside Russia or Russian-occupied Donetsk. There were options such as a phony attack on civilians, shelling occupied Donbas from

Russia, destroying chemical facilities to harm civilians, or using a fake Ukrainian unit to employ chemical weapons.[2] Collateral information suggests that child casualties would increase the emotional shock value of the event.[3] All variants were to employ coordinated information warfare techniques to enhance the experience and deepen the justification as required. In March, British intelligence confirmed that the Russian assault was supposed to have been led with a chemical weapon false flag operation.[4]

Volodomyr Zelensky, the cabinet, and senior military personnel were scheduled for assassination by teams inserted into Kyiv before the operation commenced.[5] The Ukrainian leaders' bodies were to be destroyed, possibly with rented mobile crematoria,[6] to generate confusion as to whether there still was a Ukrainian government. A coup de main using airborne and special forces to seize an airhead near Kyiv would facilitate the deployment of special operations teams from Rosgvardia SOBR units,[7] who would rapidly move into Kyiv, occupy key government buildings, and arrest the next tier of government leaders so they could be "tried" in "court" in order to intimidate the rest of the government apparatus. Rosgvardia population control troops would follow. The plan relied on speed and psychological shock to succeed.

That was the end state. To get there involved weeks of shaping operations to ensure that there was no outside intervention and to forestall the extension of American, British, or French nuclear deterrence over Ukraine. In essence, the shaping operations focused on NATO's northern flank, the Baltic and eastern flank, Western Europe in depth, the Mediterranean, and Canada. All of this activity was designed to distract from events surrounding Ukraine and keep NATO off balance. Other Russian activities to improve Russia's strategic advantage on the strategic nuclear plane were also conducted using the shaping activities themselves as cover. Ultimately the Russian plan failed and was countered with an unprecedented coordinated American and allied response that included signaling with strategic nuclear forces.

Distractions and Intimidation: January 2022

While the world was holding its breath over the crises in Eastern Europe, unrest broke out in Kazakhstan. Protests over cost-of-living issues erupted in Zhanaozen on 2 January, followed by sympathetic demonstrations in Almaty the next day.

Clashes with law enforcement escalated, and on 5 January Almaty and Aktau airports were seized by protesters, statues were torn down, and some police units sided with the protesters.[8] Pro-Russian Telegram channels encouraged ethnic Russians to form self-defense cells in northern Kazakhstan.[9] That day, President Kassym-Jomart Tokayev declared a national emergency, dismissed the cabinet, and replaced his political rival, Nursultan Nazarbayev, who chaired the Kazakh National Security Council.[10] Tokayev went further: he used social media to declare that 20,000 insurgents not speaking Kazakh were invading, declared there was an external terrorist threat to Kazakhstan, and invoked the CSTO defense relationship for assistance.[11]

Kazakhstan's importance to the Putin regime cannot be understated. First there was Baikonur Cosmodrome, the Russian space launch complex, and the Sary Shagan ABM test site. Second was the geographic expanse of Kazakhstan in terms of border length: it was essentially Russia's soft underbelly. Third there were the resource extraction and economic relationships with Russia. Fourth there was a significant Russian minority population. Social and political chaos in Kazakhstan would have generated significant complexities in Russian internal and external policy.

Russian information operation conduits immediately labeled events in Kazakhstan as a "color revolution." Thematically the approach varied from assertions that it was a Soros conspiracy using nongovernmental organizations (NGOs), to a novel combination of "color revolution mixed with Islamic Jihad." A retired Canadian diplomat was trotted out by a Russian information conduit to "confirm" some or all of this. Eventually Dmitry Medvedev came out on 28 January after it was over and declared the fraternal peoples of both countries staved off a "color revolution."[12] The immediacy of this labeling should not surprise us, given the predilections of the Russian establishment to view any unrest in "post-Soviet space" as being externally generated.

Russian military transport aircraft regrouped for an intervention, including the movement of seven Il-76 transports from Belarus to Yekaterinburg and the repositioning of large An-124 aircraft from the 224th Flight Unit. On 7 January twelve Il-76s and An-124s departed Chkalovsky air base near Moscow loaded with elements of the 45th Spetsnaz Brigade and portions of the VDV formations from Ivanovo and Ulyanovsk, and a "peacekeeping company" from the Belarusian army. Russian IL-76s picked up Tajik "peacekeepers" from Gissar. In effect, Russian Spetsnaz landed and secured Almaty airport, and a follow-on force that initially numbered three thousand troops flowed in after them.[13] Nine days after the first riots, Tokayev declared the situation under control, and "CSTO forces" were withdrawn on 20 January.[14]

In essence, the unrest was a collection of locally based protests that were unconnected. Shutting down the internet across the country and employing crushing force, including 13,000 arrests, with the resultant speedy withdrawal of the CSTO forces strongly suggests there was no larger organization in the country coordinating resistance. One of the more compelling theories as to what happened and why revolves around the possibility that Tokayev detected a coup plot or some other shift of power involving Nazarbayev and used the situation to scotch it.[15] Russian motivations were multifaceted. Russia had been using soft power techniques to draw Kazakhstan back into the fold. At the same time, the intervention "allowed Russia to act out the role of great power." Tokayev obsequiously asserted that "we support Russia's position regarding the indivisibility of security in the Eurasian space,"[16] a statement that essentially was an expression of support for Russian intervention in Ukraine on the eve of the operation.

What observers overlooked at the time were Russian moves related to the Ukrainian operation while the Kazakhstan situation was in play. Significant military resources were employed for a short time in Kazakhstan but nothing that drew away from the main effort in Belarus and Ukraine. Indeed, Estonia requested reinforcements for NATO's eastern flank while the Russian airlift into Kazakhstan was in progress, likely concerned that the Kazakhstan operation was a distraction for operations against the Baltic countries or Ukraine.[17]

And then there was the Svalbard cable event on 7 January, the day Russian troops intervened in Kazakhstan.[18] Svalbard hosts the SvalSat satellite ground station with an antenna field consisting of a hundred antennas that "gives the station a unique position to provide all-orbit support to operators of polar-orbiting satellites."[19] It is one of two such arrays in the world and has a relationship to intelligence collection satellites, especially ones used to collect information on ship movements in the Barents Sea.[20] The Svalbard array is linked to Norway via undersea fiber-optic cables. These cables were found deliberately severed exactly at the point where the ocean drops steeply from an undersea shelf to almost three kilometers in depth, thus complicating and delaying any repair job.[21] The implications of the Kazakhstan intervention and the Svalbard event are intriguing. The severance occurred on 7 January, the same day as Russian forces intervened and two days after the request for CSTO troops was made, also the same day a Russian trawler passed over the cable in that area twenty times.[22] Within days the European Space Agency synthetic aperture radar satellite Sentinel-1B ceased operations, leading to an extensive investigation.[23]

The same day Sentinel-1B failed, North Korea launched a hypersonic glide vehicle. The U.S. intelligence apparatus knew the test was going to take place, and a Cobra Ball aircraft was dispatched to an orbit southwest of the abandoned ABM facilities in North Dakota two hours before the event in order to observe the earth's horizon to the west.[24] Two E-6B Mercuries were already in the air, presumably having taken off in the morning as part of their baseline routine, with one conducting a trailing wire antenna deployment. An E-4B Nightwatch was also up, but unusually it was orbiting NORAD headquarters at Colorado Springs. The North Korean launch took place at 1727 hours EST. Three minutes later the Federal Aviation Administration (FAA) issued a national ground stop order for the U.S. West Coast airports, including ones in Hawai'i, and instructed aircraft to land. This lasted five to seven minutes.[25]

Two hours after the ground stop incident, four E-6B Mercuries and another E-4B Nightwatch were up, and tankers were scrambling from the former SAC airbase at Dow in Maine. Two of the E-6Bs conducted simultaneous trailing wire antenna deployments, one over the Atlantic, the other over the Pacific. On 11 January, there were three of the four E-4B Nightwatches, one E-6B Mercury, and forty-three tankers visible. A P-8A Poseidon was also squawking off Murmansk. P-8As were also operating off King's Bay, Georgia, home of the Atlantic Fleet SSBN force.[26] Global Strike Command then mounted an Agile Combat Employment (ACE) mission and made it public. This involved the dispersal of Barksdale-based B-52s to the abandoned Eaker AFB, which now houses a Cold War museum dedicated to the SAC.[27] ACE missions are defined as "a proactive and reactive operational scheme of manoeuvre executed within threat timelines to increase survivability while generating combat power. When applied correctly, ACE complicates the enemy's targeting process, creates political and operational dilemmas for the enemy, and creates flexibility for friendly forces."[28]

The status of U.S. strategic command and control forces on 12 January included two E-6B Mercuries up with one deploying a trailing wire antenna over the Pacific, Cobra Ball over North Dakota, and unusually an E-4B Nightwatch over Lake Superior, flying the same trailing wire antenna track as the one flown in October 2021, all at the same time. There was also a simultaneous launch of E-6Bs from Tinker AFB and Lincoln airport later in the day. Three B-52s using Fear call signs squawked over Des Moines, while KC-135 tankers were dispersed over Manitoba and New Brunswick.[29]

The FAA ground stop was the result of the FAA acting "out of precaution prior to NORAD's final determination" on the threat. NORAD classifies adversary missile tests as threats or nonthreats. There apparently was a lack of consensus that generated some confusion.[30] The ground stop of aircraft on the West Coast suggested to some observers there was concern that the North Korean hypersonic vehicle could have been an electromagnetic pulse (EMP) generator, with the obvious implications of this during a time of crisis.[31] The U.S. State Department sanctioned Russian and North Korean entities engaged in hypersonic vehicle technology transfer the next day.[32] This does not explain the movements of the other aircraft, particularly those that moved after the North Korean test had taken place.

What did take place was an unscheduled Russian large-scale exercise that coincided with a meeting of the NATO-Russia Council in which Russian representatives demanded that NATO stop admitting new members and to withdraw forces from former Soviet-controlled areas in Eastern Europe.[33] The exercise involved several thousand troops conducting exercises, including live-fire exercises, right on the Ukrainian border.[34] The combination of the Svalbard disruption, Estonian concerns, the Sentinel satellite failure, the North Korean hypersonic test, plus the exercises on the Ukrainian border warranted the signaling that took place: testing nuclear command and control systems and processes, and demonstrating that the USAF could rapidly disperse and operate its bombers from the handful of vulnerable base hubs. Cobra Ball continued to demonstrate that the U.S. system was not totally dependent on satellite surveillance.

Visible Russian nuclear forces activity was limited but unusual. The Kozelsk Loop was flown by the Tu-214SUS special communications aircraft: that in itself was not strange. However, an aircraft squawking as an Il-96–300PU presidential command post landed at Domodedovo airport and taxied to a ramp area that was protected by fifteen mobile antiaircraft missile systems. Domodedovo is less than a twenty-minute drive from the command bunkers at Chekhov and Sharapovo. This posture was held for at least three days.[35] The other unusual movement was the abrupt U-turn over Tomsk taken by the Ministry of Internal Affairs Tu-204 used by Vladimir Kolokotsov and his rapid return to Moscow.[36] If the Russians were trying to use Kazakhstan as a distraction to improve their position relative to Ukraine or use their activities as a whole to bully NATO members, they failed. U.S. national security advisor Jake Sullivan expressed the American position in a press briefing: "We're ready either way, diplomacy or respond robustly to any naked aggression."[37] The U.S. NC3I deployment that day consisted of an E-6B over the Pacific deploying a trailing wire antenna, replaced with two E-4Bs and two E-6Bs, one of these deploying a trailing wire antenna over the Gulf of Mexico. This was followed by the launch of a third E-4B, with two more E-6Bs, one of which headed to the Gulf of Mexico to deploy a trailing wire antenna.[38]

The events of 14 January demonstrate that the Putin regime shifted to focus on disrupting the situation in the Baltic area. Three Russian amphibious landing ships that departed their base in Murmansk arrived in the Baltic and joined up with three more from Baltisk.[39] This prompted an alert in Sweden and the boosting of defenses on the island of Gotland.[40] The Swedish view was that they were a ripe target because they were neutral and "the island could become the 'new Crimea,' from where Russia would control access to the southern Baltic Sea region."[41] Reports of unidentified UAVs operating over Swedish nuclear plants appeared.[42] The Russians also established a NOTAM adjacent to Finland, implying they would exercise military forces there.[43]

On 17 January the U.S. State Department reported a planned Russian "false flag" operation designed to trigger the Ukraine operation. These revelations spread like wildfire throughout Western mainstream and social media.[44] It took the Putin regime three days to issue a denial, which suggests that they were caught off guard by the U.S. move.[45] It took the Putin regime's information warfare machine another fourteen days to respond with a counter-allegation through a Russian tabloid that Ukraine was going to stage a false flag chemical attack in the Donbas.[46] The best that the Putin regime could do in the short term was to revert to abuse on RT in the form of Anton Krasovsky, who opined, "We will invade Ukraine. We will take your Constitution and burn it on Maidan, and

we'll burn you too. Ukraine is our land!"[47] The Chief of Naval Operations and
U.S. Strategic Command both tweeted on 15 January that the *Ohio*-class ballistic
missile submarine USS *Nevada* had arrived at Apra Harbor, Guam.[48]

This was a significant nuclear flourish. Ballistic missile submarines rely on
stealth as part of their deterrent properties. To deliberately reveal where one

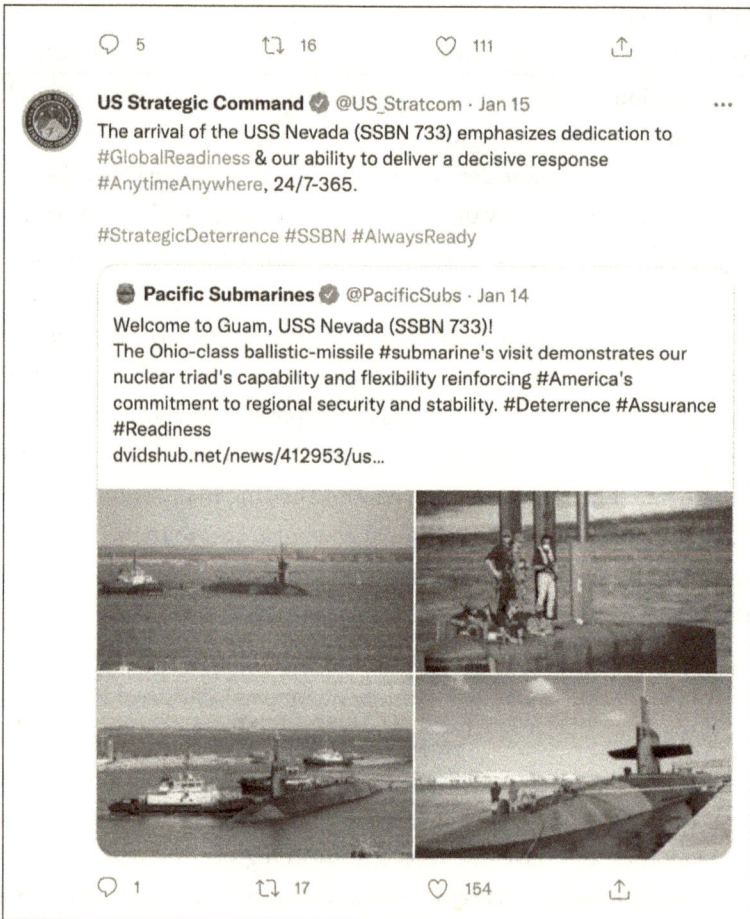

PHOTO 12 ◆ U.S. Strategic Command tweet on the ballistic missile submarine USS
Nevada deployment. Deliberate public confirmation of the locations of ballistic missile
submarines and especially using social media was a dramatic and significant break in
U.S. policy. *USSTRATCOM*

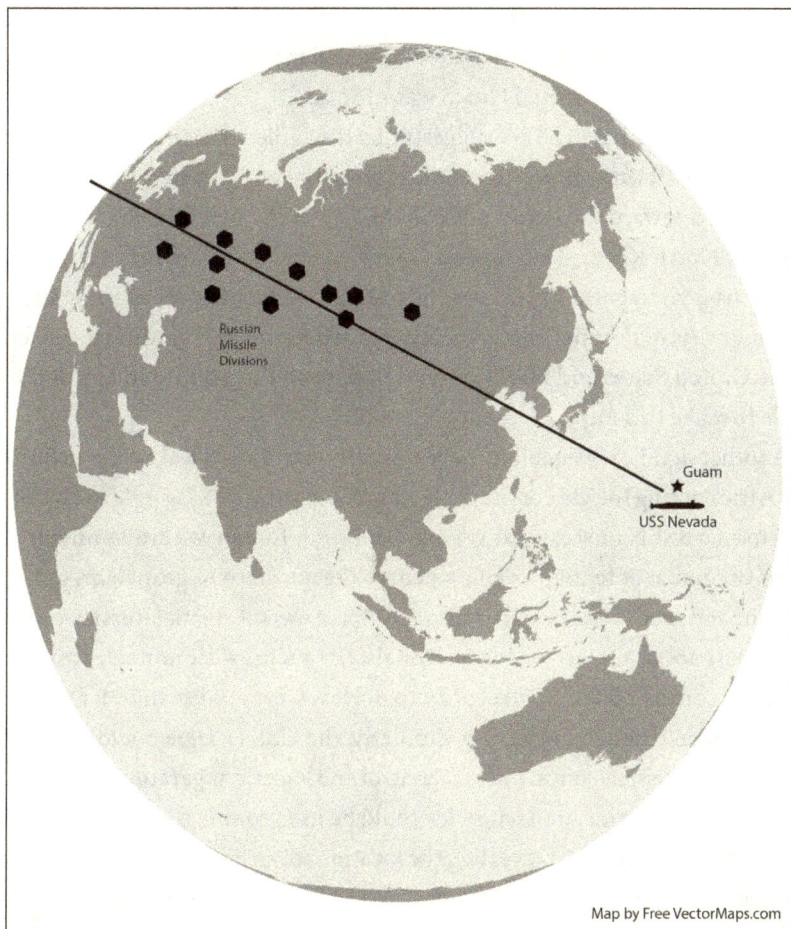

MAP 4 ◆ Range gates for the USS *Nevada*'s Trident missiles from Guam across Russia relative to Russian mobile ballistic missile bases *Created by the author*

was operating was a calculated decision to signal. The geographic position of the USS *Nevada* at Guam and the range of the Trident II D5 missiles permitted target coverage of all Russian RVSN forces, which were located in a more or less straight line across Russia.

In effect, SLBMs could hit all of the Russian mobile ICBM bases before those systems could deploy to their protective obscurity in the forests. To underline the point that this was not a routine unzipping of a U.S. capability, an E-6B Mercury nuclear command and control aircraft arrived at Anderson AFB a few hours earlier and its presence also circulated on social media.[49] Contextually, the last time an SSBN openly visited Guam was in 2014.[50]

Using a Pacific-based SSBN suggests that the audience included North Korea as well as the Putin regime. The combat potential of the system seems to have been geared toward the Russian RVSN force, but the potential flight path also was over North Korea. This suggests knowledge of collusion between the two parties vis-à-vis hypersonic vehicles and testing, and then between testing and the ongoing crisis in Europe. The larger aspects of deterrence regarding the defense of the United States and NATO would have been the priority in any decision made to make this move.

Another significant signaling move was a tweet by U.S. Naval Forces in Europe and Africa noting the visit of the *Ohio*-class submarine USS *Georgia* to Cyprus.[51] This former SSBN now carried approximately 154 Tomahawk cruise missiles. In terms of combat potential, a deployment to the northern Aegean Sea region put Ukraine and southwestern Russia in range. There was substantial Russian concern in the past about the use of conventional systems to take out nuclear systems and their command and control prior to nuclear weapons use in order to generate confusion through ambiguity. Similarly, the USS *Georgia* could effectively saturate the Russian command and control and logistic apparatus preparing for operations in Ukraine, or its missiles could be used to sink the entire Black Sea Fleet. The combination of revealing the locations of an SSBN and an SSGN from different cardinal points is significant. The SSGN could take out Russian nuclear command and control with conventional weapons, thus blurring the nature of an attack to confuse the remains of Russian nuclear command and control while the SSBN took out Russia's strategic nuclear forces before they could deploy.

Russian strategic nuclear force reaction to the submarine reveal was delayed by nearly forty-eight hours. Until they could mount a response, Dr. Yevgeny Fedorov from the Duma's Economic Committee publicly fulminated on the matter and recommended that Russia attack the Nevada Test Site with a nuclear weapon to demonstrate "that Russia is serious."[52] At the same time, Russia suddenly issued NOTAMS for areas off Novaya Zemlya, ominously implying the possibility of a nuclear weapons test.[53] Maintaining pressure on the Baltic front, a Russian "transport aircraft" came close to violating airspace over Gotland; a NOTAMS

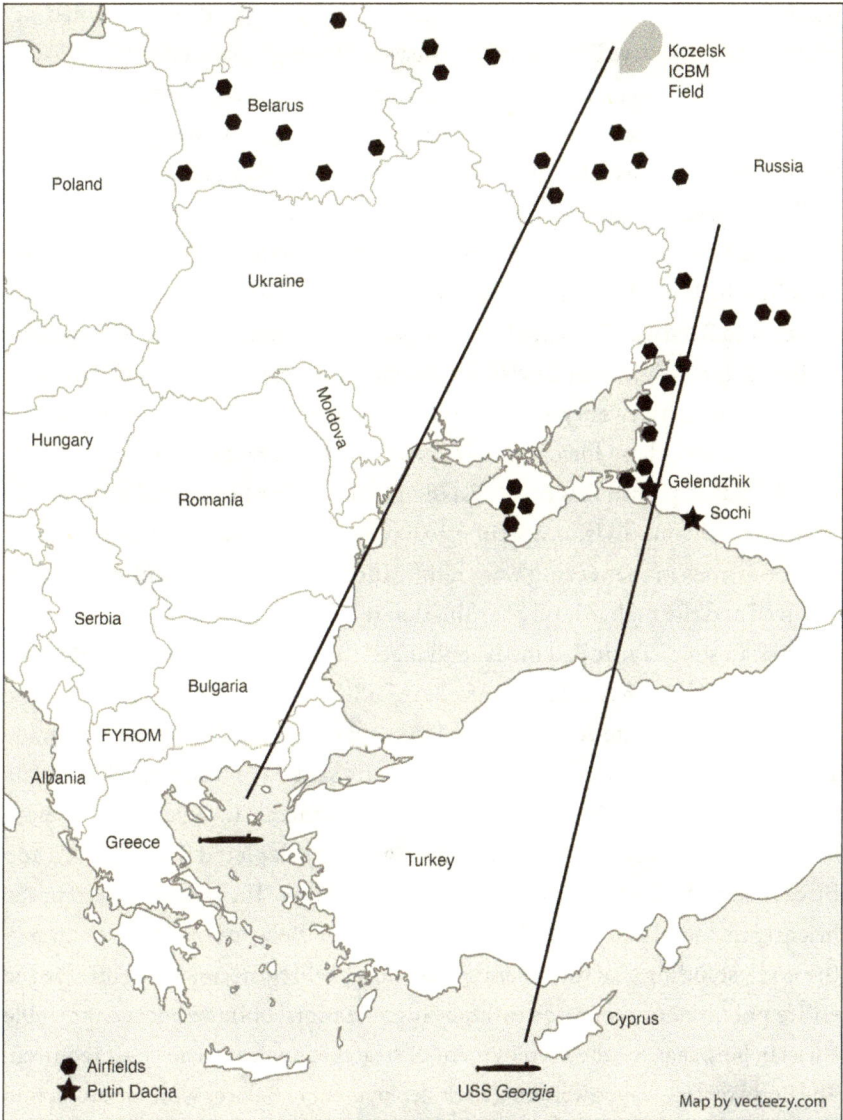

MAP 5 ◆ Range gates for the USS *Georgia*'s Tomahawk cruise missiles from the Aegean Sea relative to targets in Russia *Created by the author*

was issued for an area adjacent to Finland; and blatantly, a Russian-owned 747 overflew two sensitive Finnish military bases. This flight entered the NOTAMS area, proceeded over the border with Finland, flew directly for the Baltic, and proceeded to Leipzig.[54] This has the feel of the Putin regime lashing out at neutrals because they were frustrated in the wake of the USS *Nevada* and USS *Georgia* reveal.

The Russian response came on 17 January when the Belarusian chief of defense announced the start of the "combat readiness phase" of the Allied/United Resolve exercise which would run from 10 to 20 February. Russian troops ostentatiously moved into Belarus.[55] This was backstopped by maneuvering two RVSN missile divisions: the 35th Missile Division at Barnaul and the 14th Missile Division at Yoshkar Ola, with twenty-seven and thirty-six mobile ICBMs, respectively.[56] Geographically, these formations are positioned to target U.S. Minuteman III ICBM fields. Up until this point, Russian mobile ICBM units had only openly been exercised one division at a time, so having two divisions from two separate missile armies maneuvering was significant both from the number of forces involved and the higher-level coordination required.

Russian state-controlled media enhanced the move by hinting that missiles were deployed to "an Arctic Military Base," alluding to the old SS-5 missile base near Anadyr opposite Alaska,[57] and Putin's spokesman Dmitry Peskov made a strange, unprovoked remark that essentially said, "We won't discuss missile deployments to Kaliningrad." In the public joint Russian-Belarusian briefing for the Allied/United Resolve exercise, graphics depicted Topol-M, Tu-160 Blackjack, and a Delta IV ballistic missile submarine. The language used in the briefing included the phrase "Together with the Belarusian side, we came to the understanding that for general protection it will be necessary to involve the entire potential of the state's military organization" oblique but recognizable Russian language for the employment of strategic nuclear weapons if required. The Putin regime was now linking their deployment to Belarus with the protection afforded by their strategic nuclear systems and, by extension, their operation in Ukraine.[58] It was on 18 January that the Putin regime significantly ramped up information operations in state-controlled and affiliated media with a specific focus on nuclear war fear themes.[59]

High-value U.S. Air Force aircraft appeared once again to fly unique patterns near iconic Cold War or nuclear-associated facilities, patterns not seen previously in 2021 or later in 2022.[60] On 18 January, an E-4B Nightwatch operating from the former Lincoln AFB, took off, banked, and flew over the SAC Museum at Ashland, Nebraska. The command and control aircraft banked again, overflying the museum a second time in a pattern that looked like an infinity sign. Three E-6Bs were aloft, with two conducting trailing wire deployments over the Atlantic and Pacific sequentially.[61] Over in Europe, USAF F-16s assigned to the NATO nuclear dual capable aircraft (DCA) mission out of Aviano, Italy practiced delivery profiles over the Adriatic while squawking on ATC flight trackers.[62]

Russian LRA mounted a complex operation on 20 January. Four Tu-22M3 Backfires launched from a base in western Russia, aerial refueled over the Barents Sea, and conducted a simulated airstrike against facilities on Franz Josef Land.[63] At the same time, two Tu-160 Blackjacks launched from Engels, refueled over the White Sea, and probed NORAD air defense identification zones.[64] The operation may have been coordinated by a Tu-214PU command plane that staged from St. Petersburg. The combination of the two movements demonstrated that the Backfires could destroy air defense systems to open a corridor for the Blackjacks, which would move through the gap and fire their cruise missiles. NORAD responded with an unscheduled "synchronized multi-region, aircraft operation from four locations in Canada and the US" that demonstrated "counter-cruise missile capabilities."[65] This was the day CSTO forces were withdrawn from Kazakhstan, but more important, it was right before Blinken met with Lavrov in Geneva and warned him about interfering with NATO members.[66]

Curiously an aircraft appeared on Ukrainian ATC over the Dnipro River north of Kyiv, proceeded to overfly Bovary district, then over the government district downtown, and finally exited the city south over the river where it disappeared before it could be intercepted. This aircraft identified itself on ATC as an Su-57 fifth-generation fighter, of which only the Russian air force possesses, but the flight parameters suggest some form of UAV that was spoofing its identity.[67] It is possible this was a Russian ferret flight to probe Ukrainian defenses and to signal that Russian airpower could "go downtown" with impunity.

Additionally, trainspotters noted 2S7 self-propelled artillery being moved toward Ukraine. The 2S7 is a dual-capable system capable of firing the 3BV1 nuclear round which has an alterable yield between 1 and 10 kt. The 2S7 can also function as a siege gun using conventional rounds. There were fewer than sixty

2S7 in the Russian inventory and eight of them were seen on a train along with their support vehicles. This was probably not deliberate signaling using a nuclear system. The trainspotter footage was one of hundreds of such examples that were taken at this time of various Russian vehicles being deployed. It appears likely that the Putin regime was unable to control trainspotters and their posts and instead permitted the activity to enhance information operations measures generally.[68]

The next piece of the Russian information warfare puzzle was the Irish Sea flap that started after 20 January when the Russians announced upcoming global naval exercises. A surface action group of five ships led by the *Slava*-class cruiser *Ustinov* were tracked by a Norwegian P-3 two days later headed from the Barents into the Atlantic.[69] One of these exercises was a live gunnery exercise that was declared for a space southwest of Ireland in the Atlantic but might be used to cover submarine communications cables being severed. Other speculation noted the dual-capable nature of cruise missile systems embarked on Russian warships and the ability to threaten European capitals from "behind" in the way Tu-160 flights over the North Sea could.[70]

Public analyses missed that the planned exercise area was near the main operating base for the French *force de dissuasion* SSBNs. The presence of Russian ships in this area could force deploying French SSBNs to use other routes through shallower waters, making them more susceptible to detection. As with other Russian activities at the time, this announcement was yet another distraction while the Russian buildup took place.

The first of these occurred on 24 January, when the Putin regime embarked on a four-day period of strategic nuclear forces movements. Russian Ministry of Defence media released footage of four Tu-95MS aircraft operating from Engels air base near Saratov. The descriptor specifically pointed out they were the cruise missile variant, with footage focused specifically on one of the cruise missile pylons. The pylons in this footage were empty.[71] The following day, for a two-day period, two RVSN missile divisions were exercised: the 54th at Teykovo east of Moscow and the 7th northwest of Moscow. This was accompanied by Kozelsk Loops done by Tu-214PU command planes.[72] In terms of geographic position, these formations are likely targeted against European targets, and the 7th is the closest TOPOL-M formation to the NATO area. Indeed, a Russian

PHOTO 13 ◆ Signal shift: the first screenshot comes from a Russian Ministry of Defence film released on social media on 2 February 2022. The Tu-95MS Bear is shown with no cruise missiles mounted on the wing pylons. On the 21 February 2022 footage released by the Russian Ministry of Defence, the aircraft are clearly shown with the pylons loaded.

Russian Ministry of Defence

information conduit bragged that this mobile ICBM deployment was "held near the borders of NATO."[73] Within twenty-four hours, the 29th Missile Division deployed its Topol-Ms into the forests near Irkutsk.[74] A Tu-214-PU-SUS airborne communications station also conducted a Kozelsk Loop, suggesting there was higher-level coordination for the missile division exercises.[75] This was followed by a Tu-95MS exercise held off Alaska, using Il-98 tankers to demonstrate extended range operations.[76] Additionally, there were movements of Tu-22M3 Backfires from Belaya air base north of Irkutsk on 27 January. This was not advertised for public consumption.[77]

All of these movements were accompanied by an unannounced "combat readiness check" of the entire Western Military District from 25 to 29 January.[78] It is evident that these movements were linked: the strategic nuclear forces were maneuvered for public and governmental consumption, but the conventional forces were not: those movements were directed at the Western intelligence apparatus, with implied linkage between the two series of exercises. On 27 January, a TASS "interview" with Dmitry Medvedev expressed the regime's position that Russia had "the right to use strategic nuclear forces in response to an attack involving weapons of mass destruction against us or our allies and in response to any other threat to our countries existence."[79] This statement must be situated in the context of Russian ideology and what "existence" consists of, as discussed in Chapter 1. Note also that this statement sets the backdrop for Russian false flag operations involving chemical or biological weapons as an excuse to use nuclear weapons in Ukraine.

For comparative purposes, USAF F-16s from Aviano conducted training over the Adriatic on 24 January, as they had on 18 January, probably simulating nuclear delivery profiles.[80] Also around this time, the German signals intelligence service *Abteilung Technische Aufklärung* allegedly intercepted discussions in Russian quarters about using nuclear weapons against Berlin, Ramstein Air Base, and Büchel Air Base. This has been interpreted in German circles as a deliberate information operation to intimidate the German government.[81] The Luftwaffe did conduct an exercise on 26 January that employed Tornado aircraft from Büchel Air Base, a NATO DCA base with an American nuclear weapons custodial detachment. The Tornadoes practised eluding a Typhoon fighter defense to strike notional targets on the north coast of Germany.[82]

Not coincidentally, RT carried a story on 27 January titled "Russia warns of NATO nuclear threat: The military bloc is training its members to unleash atomic hellfire."[83] Previously, VPK asserted that *Arleigh Burke*–class destroyers were the spiritual descendants of the Cold War–era American SSBNs that used to operate from Rota, Spain. The underlying message of the piece was multifarious. First it reflected Cold War–era Soviet fear of cruise missile–equipped U.S. Navy ships, particularly in the Black Sea. Second it supported the argument that there was a larger ongoing threat to Russia from the southwest connected to Ukraine, thus justifying the invasion.[84]

In addition to crisis signaling, this collection of LRA and RVSN movements was yet another distraction so that Northern Fleet forces could move into place in the Barents Sea. Elements of the Northern Fleet and the Arctic Expeditionary Group began to deploy into the Barents Sea for unspecified exercises.[85] This had the appearance of securing this zone so that Russian SSBNs could deploy to the Arctic, or it could have been to exercise something similar to the Cold War–era bastion strategy.

Concurrently Russia commenced an information operation targeting potential operations in Latin America. RT published articles suggesting future naval Russian basing in throughout the region while K-politika discussed the potential of military bases in Venezuela. There was a clear attempt to connect the 2018 Tu-160 Blackjack deployment to Caracas to present events. Gazeta.ru mentioned Russian submarines armed with Zircon hypersonic cruise missiles were operating off the U.S. East Coast, while VPK compared the Cuban Missile Crisis with "the Vietnam syndrome." The underlying implication was that Russia was prepared to repeat the 1962 Cuban situation using Venezuela as a base, and if the United States meddled there to stop it, it would become "Vietnam."[86] An indicator of how serious the Biden administration took this scenario was the unprecedented open deployment of a USAF RC-135 Rivet Joint off Venezuela to collect information on the situation and to signal U.S. concern.[87] There were also P-8A Poseidon operations off the U.S. East Coast.[88]

The United States attempted to convene a meeting of the UN Security Council, making the announcement on 28 January. The Putin regime's response, in addition to deploying an Il-96–300PU on a Novosibirsk run to test RVSN communications,[89] was to use its information conduits to publicly highlight it had already conducted a "check of combat readiness" of the units and formations of the Western Military District from 25 to 29 January. These announcements included pictures of Topol-M mobile ICBMs.[90] Additional stories on Büchel Air Base formations and their defensive measures against UAV and saboteur attack followed, which in turn was accompanied by a barrage of articles on what amount to Russian nuclear *wunderwaffe*, including the R-36 (SS-18) ICBM, the Skiff containerized underwater ballistic missile, various "unstoppable" hypersonic glide vehicles, and the Poseidon unmanned underwater vehicle. The readiness of the SSBN *Knaz Oleg* was also emphasized. Various Russian military commentators were trotted out to comment on how "defenseless" the United States was in the face of these systems.[91] What went overlooked in most of the public discourse was the fact that Russia reestablished NOTAMS near Novaya Zemlya, which suggested preparations for a test of an unidentified weapons system.

At this point none of this information was picked up and retransmitted into Western mainstream and social media, though this changed later in 2022. The points of interest for Western analysts were the readiness of the Western Military District and the mobile ICBM force, so the other material likely was designed for a domestic Russian audience, in effect, "Russia has a formidable, modern, and unstoppable nuclear deterrent, so do not fear NATO nuclear attack when we go into Ukraine again."

Rivet Joint intelligence collection missions were flown right off Sochi, and there was the hint that collection missions were occurring over the Sea of Azov as well.[92] Throughout the course of the day, three pairs of E-6Bs rotated aloft, with one aircraft in each pair conducting trailing wire deployments over either the Pacific, Atlantic, or Gulf of Mexico.[93] Curiously, U.S. Fleet Forces put up a Twitter post with pictures of a surfaced submarine at sea, location unspecified, noting that the SSBN USS *Wyoming* conducted "an exchange of command and crews at sea" to demonstrate "the continuity and operational flexibility" of the SSBN force.[94] A crew change of a ballistic missile submarine at sea is rare, and this move was intended to demonstrate that the United States could sustain SSBNs on station rather than wasting time sailing back to Bangor or Kings Bay, which in turn implied that the United States wanted to maintain target coverage beyond what was considered the precrisis norm.

Whatever was going on near Novaya Zemlya, and possibly in other locations, clearly concerned the Biden administration. Plane spotters in the United Kingdom identified a WC-135W Constant Phoenix radioactive materials detection aircraft arriving at RAF Mildenhall on 29–30 January.[95] Constant Phoenix is equipped to detect and collect samples of radioactive materials in the atmosphere. The last time a WC-135W deployed to Europe was in August 2021 when radioactive material was detected off Bornholm when a Russian *Oscar* II–class cruise missile submarine surfaced and had to be towed back to a Russian naval base.[96] Not coincidentally, public restrictions on accessing RAF Fairford were established on 31 January. By this point U-2/TR-1 reconnaissance aircraft were operating from this location and possibly conducting high-level atmospheric sampling. These moves suggested anticipation that Russia would test a nuclear weapon or possibly a nuclear-powered system like the Burevestnik cruise missile, like the one that exploded at the Nyonoksa test range in August 2019.

The Putin regime maintained the pace of strategic nuclear force signaling throughout. The 54th Missile Division at Teykovo exercised on 31 January, deploying its mobile ICBMs.[97] Of significance, LRA units were briefed on the legality of "color revolutions," that is, whatever happened in the future, Russia's response was legal and moral in the defense of Russia.[98] While the exercise was underway, a Cobra Ball was aloft over North Dakota, and three E-6Bs rotated aloft, one conducting a trailing wire deployment over the Atlantic.[99]

Throughout January there was more Russian strategic nuclear force signaling taking place than at any point since 2007, both in quantity and quality. It clearly was designed to both backstop Russian efforts globally in conjunction with multiple concurrent distraction and intimidation information warfare activities.

The Lead-Up to Invasion: February 2022

In early January, there were a series of advisories warning that Russia would conduct cyberattacks including attacks against critical industrial control systems in Canada.[100] In late January the Canadian foreign ministry came under a multiday cyberattack which shut down Canada's diplomatic communications system.[101] While this was in progress, significant demonstrations started in Ottawa and at some ports of entry on the Canada-U.S. border. These demonstrations were initially organic coalitions of citizens frustrated with COVID-19 vaccination mandates and their effects on livelihoods and commerce. Although the anti-vaccination movement reflected legitimate concerns, there are studies that demonstrate that Russia had deliberately stimulated anti-vaccination sentiment several years before the Wuhan pandemic emerged, and the events of February 2022 in Canada were ultimately founded to some extent on long-term Russian information operations. Studies have also demonstrated that those who were anti-vaccination tended to not support measures taken against Russia in response to events in Ukraine.[102]

As the protests expanded in February the prime minister evacuated the capital. More ports of entry were interdicted with demonstrations, including the key economic link at Detroit-Windsor. Within days the effects were felt throughout industries in both countries that were still recovering from the Wuhan pandemic. It remains unclear as to what extent Russia directly or indirectly supported the demonstrations.[103] Additionally, the Canadian financial system itself came under some form of attack by an unknown entity attempting to undermine the entire Canadian economy.[104]

Russian information conduits mounted a coordinated operation intended to inflame the situation, question the viability of Canadian democratic institutions, and degrade Canada's international standing.[105] The implicit messaging was that Canada was in the midst of its own 6 January 2021 event and that democratic institutions no longer functioned. Every Canadian government action was used to support these themes to discredit the system with the intent of augmenting disorder.

What has been overlooked in the analysis of this crisis within a crisis was a major flourish of Russian strategic nuclear forces that took place from 2 to 7 February, three to four days after the protests started. The first cluster on 2 February was a flight of three Tu-160 Blackjacks, three Il-78 tankers, and an A-50U AWACS that flew over the Arctic Basin toward North America.[106] The second was the movement of two Tu-95MS missile carriers, two Tu-142 reconnaissance or submarine communications aircraft, several MiG-31 interceptors, and an A-50 AWACS. This formation took off from Engels and the Kola Peninsula, proceeded over the Barents Sea, where the Tu-96MSs and Tu-142s continued to the Norwegian Sea, and then to the North Sea. These were intercepted and tracked by Norwegian and British fighters.[107] Other movements included a Tu-95MS exercise in the Amur region.[108] Concurrent to the LRA operations, the 42nd Missile Division stationed around the Nizhnyaya Salda area conducted an exercise with its mobile ICBMs.[109] At the same time, Russian air defense forces in the St. Petersburg area conducted a snap air defense exercise using S-400 missiles.[110] During the course of these operations, a Tu-214PU conducted a Kozelsk Loop flight.[111] That day, the United States deployed three E-6Bs at the same time, simultaneously conducting

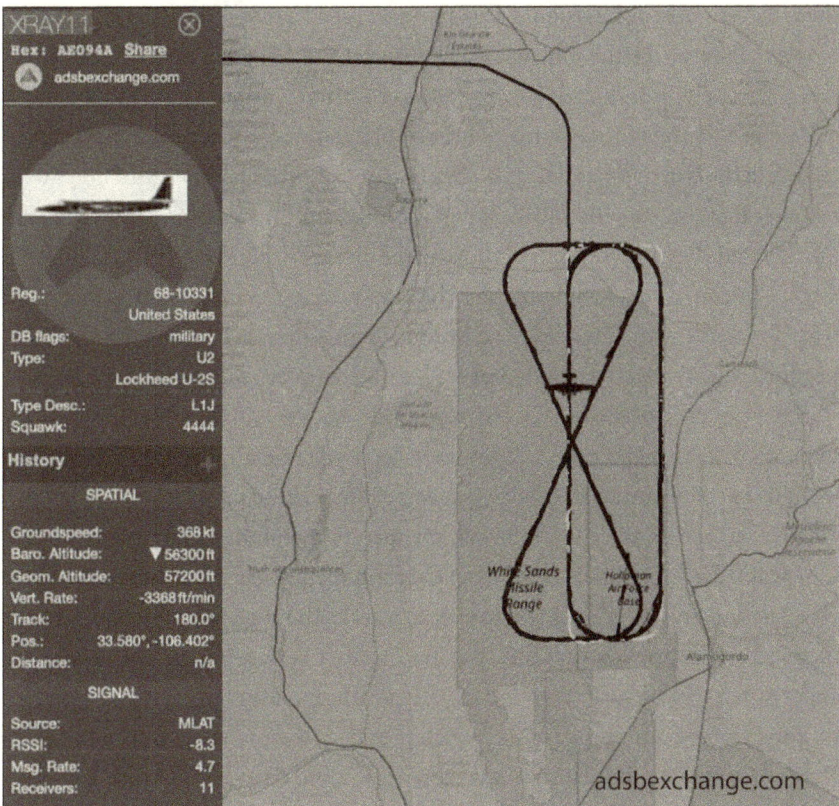

PHOTO 14 ♦ A U-2S orbiting over the Trinity site in New Mexico. As with many openly squawking strategic aircraft during this period, there was a pattern of them squawking near sites associated with historical nuclear weapons activities. This included an E-4B NAOC flight that drew an "x" directly over the SAC Museum at Ashland, Nebraska. This type of activity ceased by April 2024. *ADS-B Exchange*

trailing wire deployments over the Pacific, the Atlantic, and the Gulf of Mexico. A Cobra Ball aircraft was also aloft. A U-2S aircraft was seen conducting loops above the Trinity test site in New Mexico.[112]

These movements alone were qualitatively and qualitatively significant, but they were followed by intercepts of Tu-95MSs and Tu-142s over the Norwegian Sea the next day, with the Tu-95MS headed out over the Atlantic toward North America and the Tu-142s flying to the Irish Sea. And on 4 February, the 54th Missile Division at Ivanovo exercised its mobile ICBMs. Tu-22M3 Blackfires conducted operations over Belarus on 5 February (and again on 8 February), and then the 54th Missile Division at Teykovo exercised its mobile ballistic missiles the next day, followed by the 39th Missile Division at Novosibirsk.[113] A Tu-214PU-SUS and a Tu-214PU conducted Kozelsk Loop communications checks throughout this period.[114] The U.S. response consisted of deploying a Rivet Joint to Alaska to collect in the Russian Far East; a Cobra Ball over North Dakota with a northwest facing track; three E-6Bs with one conducting a trailing wire operation over the Pacific; and then the rotation of those three aircraft with three more later in the day.[115] The Russians also launched the Kosmos 2553 reconnaissance satellite from Pletsetsk, which would have produced an enhanced posture in NORAD headquarters if it were launched over North America.

Superimposed over this was a sub-operation whereby MiG-31Ks equipped with hypersonic Kh-47 Kinzhal missiles were deployed to Kaliningrad. This deployment was accompanied with its own specific information operations.[116] All of this activity was accompanied by information operations emphasizing several themes. One was the alleged "failure" of NATO to intercept Russian bombers. Another was Russian civil preparedness for nuclear war. These were accompanied with material on the prowess of the Tu-160 Blackjack force. References to the *Belgorod* super-submarine appeared yet again.[117] This sub-operation likely was connected to Putin's press conference with French president Emmanuel Macron in which Putin expressed glib "concern" that the war "could drag NATO into a nuclear war." Putin specifically stated that if NATO helped Ukraine take back Crimea, Europe "would be automatically drawn into a military conflict with Russia."[118]

It is evident these events were designed to intimidate, divert, and disrupt. The European powers were still coming to grips with the nature and extent of the Russian problem vis-à-vis the Baltic flank, the northern flank, and their

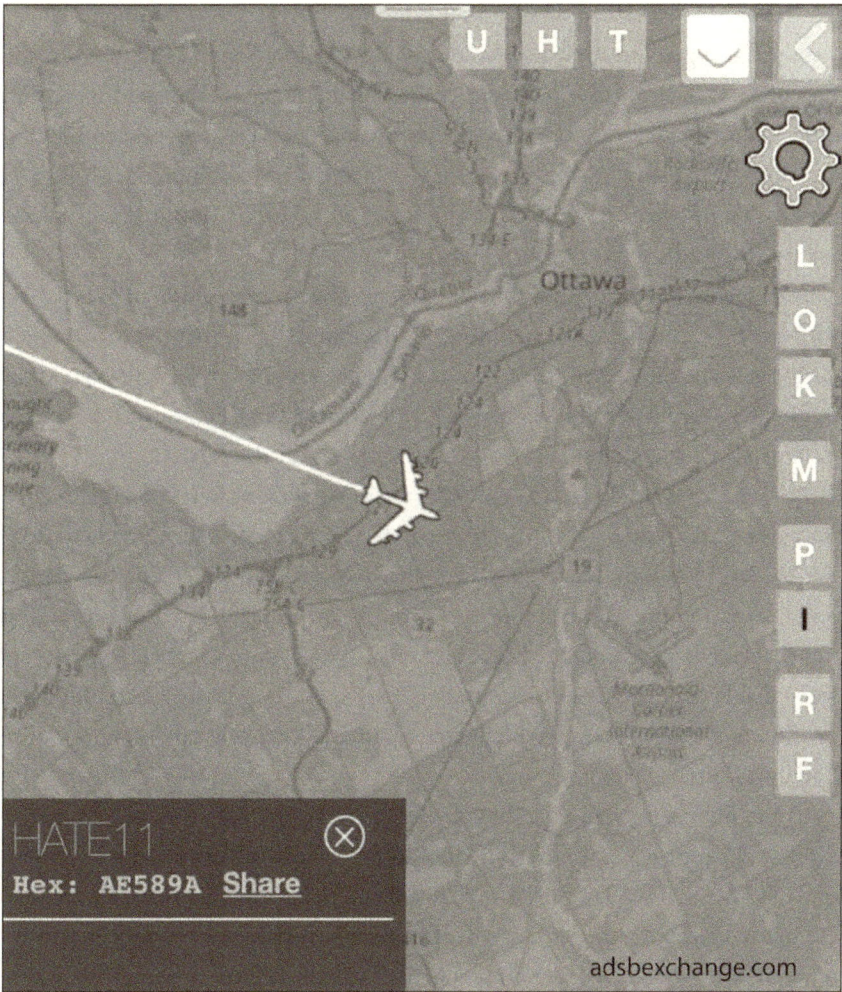

PHOTO 15 ◆ A B-52 flight on 9–10 February 2022 over Ottawa, Canada, during domestic unrest and repeated Russian Long Range Aviation operations in the Arctic involving Tu-95MS and Tu-160 nuclear bombers. This B-52 launched independently from the rest of a B-52 BTF mission to the UK, overflew Ottawa, continued to a point over the Atlantic where it refueled, and continued on to RAF Fairford. *ADS-B Exchange*

relationship to the Ukraine situation. The North American component, however, was designed to apply psychological pressure on Canada and to stress the
Canadian-U.S. relationship during a period of acute crisis. The fact this was all
done simultaneously demonstrates that this was not a series of disconnected
activities. It is likely that this flourish was also timed to coincide with Putin's
visit to Beijing during the Winter Olympics and that Communist China was
yet another audience.

Intelligence collection in and around Ukraine by American aircraft did change,
with an MQ-4 Global Hawk, an E-8C JSTARS, and an RC-135 Rivet Joint deployed
far forward on the edge of the occupied Donbas region. An RC-135 Rivet Joint
was even intercepted by Russian fighters off Crimea. On 8 February, F-16s out of
Aviano Air Base conducted visible delivery runs over the Adriatic.[119]

More important, a Bomber Task Force mission involving four B-52Hs launched
from Minot AFB on 9 February. Three of these proceeded over Canada to RAF
Fairford. The fourth departed Minot and overflew Ottawa, Canada's capital
before crossing the Atlantic and recovering at RAF Fairford.[120] The messaging
was that Canada was under the U.S. umbrella. That morning, forty USAF tankers
were visible over the United States; two E-6Bs, one conducting a trailing wire
operation over the Atlantic which were rotated with another pair; and a Cobra
Ball over North Dakota.[121]

On 10 February the joint Russian-Belarus exercise Allied Resolve with an
estimated 30,000 troops kicked off. This was followed by a Russian announcement
of NOTAMS in the Black Sea, which in effect blockaded Ukraine by air and sea.
Concern was expressed through diplomatic channels, and the JCS chairman
telephoned his Belarusian counterpart.[122] British defense minister Liz Truss
flew to Moscow to meet with Sergy Lavrov.[123] Canada chose to evacuate its
embassy from Kyiv to Lviv and to pull out its training teams working with the
Ukrainian army.

Russia asserted that the ASW cruiser *Marshal Shaposhnikov* had repelled an
American *Virginia*-class submarine operating in the Sea of Okhotsk, a claim
adamantly denied by the United States.[124] A Russian information conduit
announced that "NATO destroyers" that enter the Black Sea and "approach
Crimea will be immediately destroyed from the shore," highlighting the Bastion
anti-shipping missile system but also underscoring Russian concerns about

Tomahawk-equipped U.S. guided missile destroyers.[125] American ISR operations remained robust in and around Ukraine, including MQ-4 Global Hawks on the Belarusian border, the Russian border, occupied Donbas, the periphery of Crimea, and down to southern Russia off Novorossiysk. A P-8A Poseidon skirted occupied Crimea and flew to the Sea of Azov on 12 February. This was followed by Rivet Joint flights over the Black Sea on 14 February.[126]

It was at this juncture the U.S. government released more information on Russian false flag plans. Citing the discussion in an emergency meeting in the White House Situation Room and a press conference, American media quoted National Security Advisor Jake Sullivan, who asserted that the "U.S. is firmly convinced that Russia is looking hard at the creation of a false flag operation to justify an invasion," "something that they generate and try to blame on the Ukrainians as a trigger for military action." Sullivan said that any subsequent attack would likely begin with "aerial bombing and missile attacks" ahead of "the onslaught of a massive force."[127]

On the northern flank, two Tu-160 Blackjacks flew over the Barents and then down to the Norwegian Sea.[128] Russian NOTAMS went up again between Murmansk and Novaya Zemlya on 14 February, implying there would be some form of *wunderwaffen* or nuclear weapons test. Russian information conduits latched on to the presence of the WC-135 Constant Phoenix in the United Kingdom, connected it to the possibility of a Burevestnik missile test, and used it to highlight the alleged "fear" in the United States of this unique Russian system. The attention directed on the WC-135 had an echo effect with Western media, in that the presence of the aircraft appeared in Western social media, then Western mainstream media, and then it was used by Russian information operators for their domestic and international purposes to enhance apprehension over nuclear weapons systems.[129] TASS also announced at the same time that a regiment of silo-based UR-100N ICBMs equipped with Avangard hypersonic glide vehicles from the 13th Missile Division at Yasnyy had just taken up "combat duty."[130] Again the existence and deployment of an allegedly unstoppable nuclear delivery system was intended for both audiences and built on the web of informational relationships established to convey intimidation.

The American response was curious. The BTF was in the process of deploying to Europe, and three E-6B Mercury command planes were airborne. After the

Novaya Zemlya NOTAMS and the Avangard announcement, a B-52 flew over the Dugway Proving Ground, a chemical and biological weapons testing range, while squawking on ADS-B. The bomber flew in a pattern over Dugway, spelling out the word "BAD."[131]

The next day Russian NOTAMS throughout the Barents Sea continued, seriously interfering with Norwegian fishing.[132] In the Baltic, the Russian ambassador to Denmark produced documents from the 1940s, asserted that American troops could not be stationed on the island of Bornholm and ominously stated that "Russia will be forced to consider the implications it will have on the relationship between Russia and Denmark."[133] Russian concern over Bornholm apparently related to the security of missile forces in Kaliningrad: they believed that American HIMARS systems, if deployed to Bornholm, would threaten the missile systems situated in Kaliningrad and thus undermine Russia's ability to coerce NATO.[134]

That day, two B-52Hs departed RAF Fairford, flew over Stavanger, Norway, up the center of the country, and exited over the Norwegian Sea about one-third of the way up where they orbited for some time.[135] This was the location whereby British V-bombers during the Cold War would loiter awaiting their go-codes before proceeding to their targets in the Soviet Union. From here, these B-52Hs had the combat potential to put Russian targets at risk west of the Urals. Russia overtly withdrew the MiG-31K/Kh-47 systems from Kaliningrad to Soltsy Air Base, ostensibly to reduce noise complaints from locals.[136]

Concurrent with these activities, the Ukrainian financial system, military systems, and all government websites came under massive cyberattack.[137] Russia also established NOTAMS over part of the Pacific off the Kamchatka Peninsula, which forced commercial aircraft in transit from North America to Asia to alter flight plans.[138] Russian placed two nuclear command and control aircraft up: one in the Moscow area, a Tu-214VPU, and a Tu-214PU command post that flew to Ulyanovsk and then Tyumen, and back. At this time, Defense Minister Sergy Shoigu was en route to Syria in his Tu-214PU-SBUS command plane to inspect Russian forces and underscore Russian regional capabilities. He followed a gaggle of Tu-22M3 Backfires and MiG-31Ks with their Kh-47 hypersonic missiles headed to Syria. This was accompanied by a blast from Russian information operation conduits extolling the virtues of the allegedly unstoppable MiG-31K/Kh-47 combination as a response to "U.S. backed extremism" in the Middle East.[139] While Shoigu was in Syria, three Russian Su-35 fighters deliberately harassed a U.S. Navy P-8A Poseidon in an "unsafe and unprofessional" manner while it was going about its activities over the Mediterranean.[140]

Accounting for this burst of Russian activity is relatively straightforward. The Russian Duma adopted legislation recognizing the so-called Donetsk People's Republic and Luhansk People's Republic as independent countries, and it is clear the Putin regime wanted to underscore this while continuing its efforts to distract elsewhere. This was, in fact, the first stage of annexation of these occupied Ukrainian areas into Russia proper. It also set the "legal" (Russian law) basis for invading adjacent parts of Ukraine.[141]

The B-52Hs from RAF Fairford returned to their orbits off Norway on 16 February. The same two Russian command and control aircraft flew exactly the same routes that they did the day before: one over Moscow and one out to Tyumen and back.[142] The Russian leadership was signaling that its nuclear command and control systems was survivable in the face of cruise missile attack. Not coincidentally, multiple Russian information operations conduits suddenly extolled the virtues of the Russian ballistic missile warning system.[143] The simultaneous NC3I posture over the United States included three E-6Bs rotating singly throughout the day and an E-4B aloft all day, and a similar posture the next day. However, the USAF MQ-4 UAV that had been monitoring Ukraine's borders was replaced with an RAF RC-135 Rivet Joint.[144]

At 1630 EST 16 February, the USAF High Frequency Global Communications System (HFGCS) issued what appeared to be Emergency Action Messages (EAMs) every half hour, a significant increase in frequency from similar traffic over the past two days. These continued all night.[145] EAMs can, of course, be sent for communications checks as well as for training purposes and do not necessarily mean war is imminent. Immediately after this, amateur radio operators posted on social media what appears to be a properly formatted Red Dash Alpha message on the USAF HFGCS.[146] This suggests there was some change in the posture of U.S. strategic forces. Two hours later, early on the morning of 17 February, a Cobra Ball aircraft lifted off from Eielson AFB in Alaska and an E-4B departed its base at Lincoln.[147] This was an hour or so after Russia issued new PRIP/NAVWARNs over the Barents Sea and the Kura Test Range in Kamchatka for the dates 20–22 February.[148]

Meanwhile, in Europe, the Norwegian minister of defense canceled an appearance in Munich and flew home "to be available for decision-making, due to the current situation."[149] The B-52Hs at RAF Fairford appear to have taken off around this time, while an RAF RC-135 Rivet Joint and an MQ-4 Global Hawk flew along Ukraine's borders with Belarus and Russia as well as occupied Ukraine.[150] Later in the day, an E-4B Nightwatch lifted off and conducted an infinity loop over the SAC Museum at Ashland, Nebraska.[151] There was a burst of EAM traffic at this time. NORAD then reposted a 24 February 2020 post which depicted two CF-18 interceptors and a KC-135 tanker operating in the Arctic with a vague statement about protecting approaches to North America, suggesting that a similar operation was in progress.[152]

Russian activity during the course of the day included an announcement that two Tu-22M3 Backfires escorted by Su-35s were flown over Belarus as part of the Allied Resolve exercise.[153] In terms of visible nuclear command and control activity, two Tu-214PU command post aircraft were aloft flying from Moscow to Ulyanovsk and Tyumen and back to Moscow. Notably, an Il-96–300PU command post aircraft, the one that approached the Ukrainian border in late 2021, conducted a Kozelsk Loop flight.[154]

The ongoing concerns about Russian ballistic missile test launches account for some of this activity. The Grom 2022 exercise was believed to be imminent and then the date for it shifted at the last minute to 20–22 February, thus the launch and recovery of the Cobra Ball. Indeed, Blinken announced that the Russians were again considering a false flag attack that could be "a fabricated so-called terrorist bombing inside Russia, the invented discovery of the mass grave, a staged drone strike against civilians, or a fake—even a real—attack using chemical weapons."[155] Curiously, a Russian information operations conduit released a story claiming that Ukraine had a plan to let the Zaporizhzhia nuclear power plant deliberately spew radioactive material or otherwise deploy radioactive material to contaminate Russian forces in occupied Ukraine.[156] So now the possibility of a false flag incident involving nuclear or biological material emerged. It is likely the B-52s, the E-4B, and the NORAD activity plus other activity yet to be revealed was to underline the counter-information operation regarding the latest Russian false flag plan.

On Twitter, an account established in early January dedicated to Cold War Soviet nuclear weapons imagery suddenly produced more contemporary material.

The author off-handedly identified the location as Belgorod-22, object 1150, seventeen km from Ukraine. He claimed "NT-236 supercontainers" (nuclear weapons transport containers) were stacked near the groomed and protected bunkers, and supplied comparative Google Earth imagery to imply that nuclear weapons had recently been deployed to the site. The account ceased to function in May 2022 but continued to contain a mix of historical and contemporary Russian nuclear handling and storage information.[157]

On 18 February, multiple Russian information conduits simultaneously announced that the Grom 2022 strategic nuclear forces exercise would occur on 19 February and that Putin would lead it.[158] Putin spokesman Dmitry Peskov leaned heavily on the existence of "the black suitcase" and "the red button" and the fact that only the head of state could order the launch process. He also noted, "The fact is that a number of exercises are underway now, and these are actions that are absolutely transparent and understandable for specialists from other countries."[159]

There were, however, other events in play. Pictures taken of BMD-2 airborne mechanized combat vehicles used by Russian airborne forces and rigged for airdrop suddenly appeared on Western social media. These pictures were associated with airfields at Pskov (Ostrov) and Tula (Klokovo), the home stations of the 106th Guards Airborne Division and the 76th Guards Air Assault Division, respectively.[160] The activation and preparation of two airborne divisions at this point in the crisis would have given the NATO intelligence apparatus pause.

It is therefore not a coincidence that two B-52Hs departed RAF Fairford and headed for Sweden in the early afternoon. These two aircraft ostensibly conducted training with Swedish forces, but the proximity of their orbit areas to Russian Iskander missile sites in Kaliningrad is most interesting.[161] The flight time for forty cruise missiles launched from two B-52Hs to their targets would have been about fifteen minutes. In the early evening Germany increased the readiness levels of its commitment to the NATO Rapid Reaction Force.[162] At this point there were three E-6B Mercuries and forty-three tankers visible over the United States, an increased American posture.[163]

In an unexpected move, the Special Flight Detachment at Chkalovskiy Air Base scrambled one of two Tu-214PU-SBUS Ministry of Defence command post aircraft, which proceeded to orbit northeast of Moscow. This was followed by

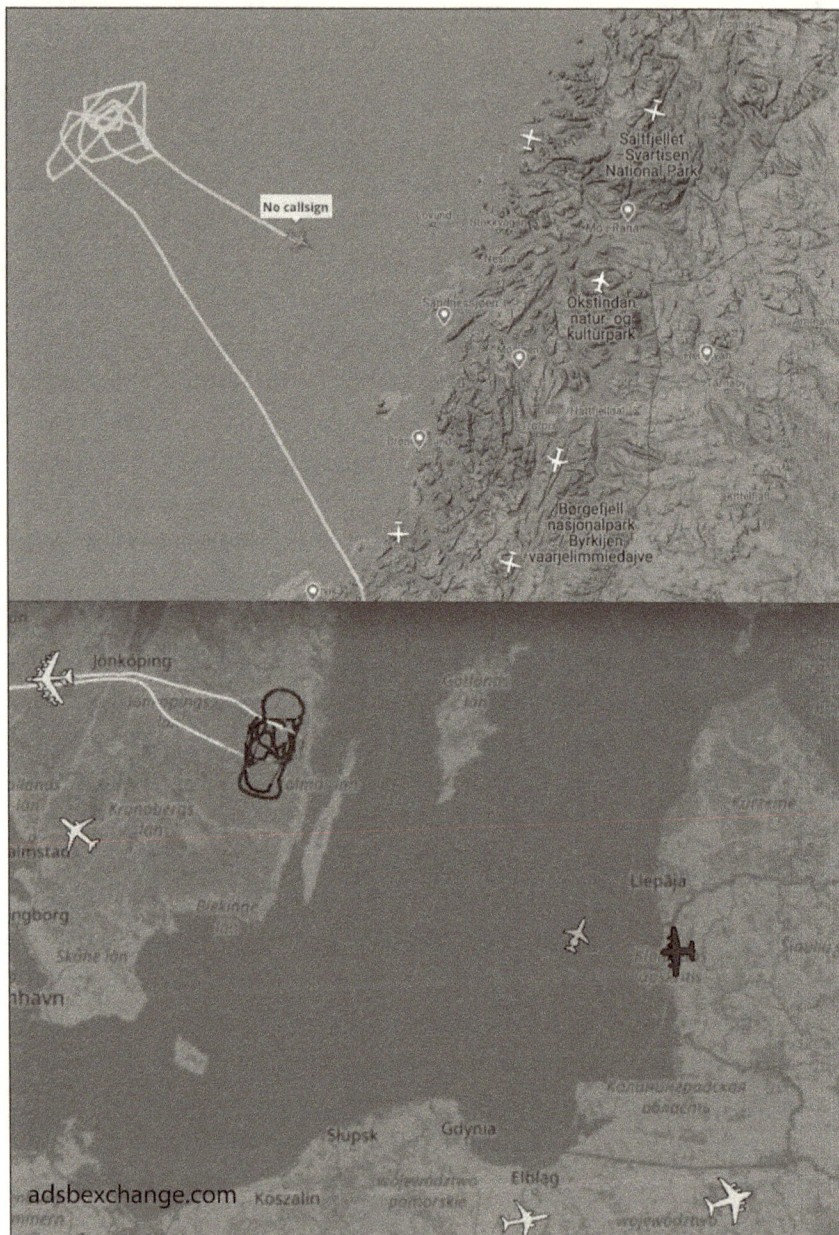

PHOTO 16 ♦ As the crisis deepened, B-52s deployed to a point off Norway on 15 February and again on 16 February 2022 (*top*). On 18 February they deployed to orbits over Sweden, thus putting Russian Iskander missiles in Kaliningrad and south of St. Petersburg within ten to fifteen minutes of flight time. *flightradar24, ADS-B Exchange*

an Il-96–300PU presidential command post aircraft from Vnukovo, which then undertook a Kozelsk Loop.[164]

The blending of preparations for the next phase of the invasion of Ukraine, implying that the Baltic countries were at risk from Russia, and preparations for a strategic nuclear exercise that might serve as cover for nuclear weapons use at this point in the crisis underscores how dangerously irresponsible the Putin regime was behaving. Any one of these events decontextualized from a major international crisis would be understandable but not in combination.

Grom 2022 was conducted on 19 February with Putin at the helm and Belarusian strongman Lukashenko at his side. The main attractions included a Sineva SLBM launch from the Delta IV–class ballistic missile submarine, *Karelia*; a RS-24 YARS mobile ICBM launch from Plesetsk; and cruise missile launches from Tu-95MSs. Kh-47s were launched from MiG-31Ks, while Iskandar ballistic missiles were fired in the Astrakhan region. Ships from both the Northern and Black Sea Fleets fired Kalibr missiles. Numerous Russian information conduits carried stories on the Grom exercise, and Russian Ministry of Defence provided extensive video coverage to them. This was all transmitted via social media to Western outlets.[165] The inclusion of dual-capable systems in Astrakhan 2022 was a clear message that Russia had the ability to operate in the sub-strategic nuclear engagement zone.

The Russian over-the-horizon radar system 29B6 Konteyner situated in Mordovia was also exercised during Grom 2022. Konteyner is an early warning system geared toward detecting ballistic missile attack but also for the detection of stealth aircraft and cruise missiles.[166] This aspect is strange and suggests that Russian authorities may have been concerned of the outside possibility of a western strike while the exercise was in progress.

The American reaction to Grom 2022 started with the launch of the Constant Phoenix aircraft from the United Kingdom with accompanying tankers to an operating area over the Barents Sea, and then a Cobra Ball aircraft was deployed over Alaska. An E-4B Nightwatch also lifted off in the central United States. Prior to the Sineva SLBM launch, two E-6B Mercuries deployed: one over the Gulf of Mexico and another off the Oregon and California coast. These were augmented or rotated with another E-6B out of Albuquerque and another from a base in California. EAMs relating to the overwater aircraft were picked up by the amateur radio community. In Europe, the Fairford-based B-52s launched and proceeded to orbit Romania along with several tankers, while RC-135 Rivet Joint and E-8 JSTARS flew routes parallel to the Russian-Ukraine border.[167]

The Russian plans for Ukraine were completed on 18 January and were printed and delivered to units on 22 February. However, materials captured later by Ukrainian forces depict 20 February as the original start date of the operation. In theory, the Putin regime would have wanted the planned false flag, the Grom 2022 exercise, and the assault to occur in a logical sequence, with the false flag acting as justification, Grom 2022 acting as a deterrent signal to the United States and its NATO allies, and then the assault itself. The Americans clearly thought Grom 2022 was going to start on 17 or 18 February, but then Russia suddenly changed its navigational warnings in the Barents Sea and on the Kamchatka range to 20 February that day. Russia then conducted Grom 2022 a day earlier, 19 February. The original sequence was probably supposed to be Grom 2022, then the false flag, and then the invasion. It appears as though the revelations of the "false flag" and possibly other factors threw off the timeline, which delayed Grom 2022 and the Russian invasion operation.

After the excitement over Grom 2022 died down, Russia declared a NOTAMS for the Sea of Azov and moved part of the Black Sea Fleet through the Kerch Strait on 20 February. There was also a formal Ministry of Defence announcement that Russian troops would not depart Belarus after the joint exercise "because of tensions in the Donbas." This was in parallel to massive cyberattacks directed at Ukrainian institutions. Russian nuclear command and control aircraft activity consisted of an Il-96–300PU command post that conducted a routine Ufa run to check RVSN communications.[168]

There were two manufactured public spectacles mounted by the Putin regime on 21 February. The first was a Russian Federation Security Council meeting "to discuss the current developments in Donbass" with an aim toward Russia recognizing "the independence and sovereignty of the Donetsk People's Republic and the Lugansk People's Republic," which were artificial constructs controlled by the Putin regime after the invasion of 2014. The televised meeting was designed to provide legal, diplomatic, historical, and strategic legitimacy for further Russian intervention in Ukraine. The excuse mechanism was that there was an alleged humanitarian catastrophe involving 800,000 Russians in the two pseudo-statelets brought on by Ukrainian shelling and the use of neo-Nazi sabotage groups.[169] Defense Minister Sergy Shoigu asserted that Ukraine was pursuing a nuclear capability, both domestically and with NATO help.[170] The situation in Ukraine was firmly blamed on the United States: "They are hiding their specific goal, which is none other than the collapse of the Russian Federation."[171]

Having achieved his "consensus," Putin then gave a televised speech to the Russian Federation four hours later. The alleged "situation in Donbass has reached a critical, acute stage." Putin framed Ukraine as "an inalienable part of our own history, culture and spiritual space . . . [a] people bound by blood" in a space where people "have called themselves Russians and Orthodox Christians," in other words clear and direct references to Eurasianism. The "threat" to these people consists of Kyiv's "parasitic attitude" enabled by the "rise of far-right nationalism, which rapidly developed into aggressive Russophobia and neo-Nazism." Putin went on to reference "color revolutions," claiming that "a role in this was played by external forces, which used a ramified network of NGOs and special services to nurture their clients in Ukraine and bring their representatives to the seats of authority."[172]

Putin then focused on Ukraine's alleged nuclear ambitions, reiterating what Shoigu briefed on in the earlier televised spectacle and in particular mentioning the Tochka-U missile system: "they can do more, it is only a matter of time." However, Putin elaborated on what he viewed as the wider threat. He specifically expressed concerns about the potential use of precision weapons from "NATO warships" in the Black Sea against Russia. He asserted that the BMD launchers in Poland and Romania could "be used for Tomahawk cruise missiles-offensive strike systems." According to Putin, "the flying time of Tomahawk cruise missiles to Moscow will be less than 35 minutes" if stationed in Ukraine. Putin conjured up fear of "NATO's tactical aviation . . . including precision weapons carriers [that] will be capable of striking at our territory to the depth of the Volgograd-Kazan-Samara-Astrakhan line" and "the deployment of reconnaissance radars on Ukrainian territory will allow NATO to tightly control Russia's airspace to the Urals." Pausing for effect, Putin said, "It is like a knife to the throat. . . . Russia has every right to respond in order to ensure its security."[173]

The Russian National Security Council spectacle was designed by Putin to implicate his national security leadership in what was about to happen, and to present a consensus to the Russian audience vis-à-vis absorbing Donetsk and

Luhansk into Russia so they could be "defended" in a preventive war. The speech
to the Russian Federation was designed to appeal to two groups: those who had
been immersed in Russian ideology over the past fifteen years and understood
its language and, second, the armed forces and security services who had a vested
interest in maintaining the security apparatus of the state in its broadest sense.

Presenting the existence of the conventional OTR-21 Tochka-U battlefield
missile systems in the Ukrainian inventory as a strategic threat to Russia pushed
the absolute limits of combat potential reasoning to the edge of reality. The SS-21
Scarab dated back to the Soviet period of the late 1980s and came in two models:
"A" with a range of 70 km and "B" with a range of 120 km.[174] There was no evidence
that Ukraine possessed any nuclear warheads. The assertions that Putin made
regarding Tomahawk ranges and timings amount to sophistry in that he knew
full well that such systems did not need to be deployed on Ukrainian soil to pose
a threat to the Russian Federation. It is with some irony to note that USAF B-52s
were orbiting Poland near Kaliningrad during the Security Council melodrama.[175]
Putin's discussion of nuclear matters was designed as an integral component of
the excuse mechanism to justify the next phase of the invasion of Ukraine.

There were some indicators throughout the day that events had been put in
motion by the Putin regime. Armored columns started to move into Donetsk and
Luhansk after Putin's speech.[176] At the same time, the Ukrainian cell network
opposite the occupied territories was suddenly jammed.[177] Also during Putin's
speech, the Russian Ministry of Defence released high-grade footage of a Tu-95MS
exercise, with shots lingering on cruise missiles loaded on pylons, in contrast to
similar footage deployed earlier in February from the same unit.[178] Concurrent
with this was the evacuation of U.S. State Department personnel by CV-22B
Osprey aircraft from Lviv to Poland.[179] The American command and control
aircraft deployment during the day included three visible E-6B Mercuries.[180]

The next day, 22 February, Russian information operations ramped up its
emphasis on Ukraine's alleged nuclear capability but altered course slightly. RT
trotted out disgraced American UNSCOM inspector Scott Ritter to provide
skewed commentary accusing Ukraine of pursuing weapons acquisition for
"nuclear blackmail" purposes and attacking the realities of the Budapest memo-
randum.[181] Gazeta.ru reanimated the Cold War–era Colonel General Viktor Yesin
to walk back the idea Ukraine might have nuclear warheads for the Tochka-U

missiles. Instead, the idea that Ukraine may have constructed radiological material dispersion warheads was raised.[182]

That afternoon, a Tu-214SUS special command and control aircraft lifted off from Vnukovo and headed to Tyumen to check in with the 33rd Guards Missile Army and other RVSN formations to the east. At this point a pair of B-52s were orbiting the North Sea and the WC-135 Constant Phoenix aircraft had just conducted a Baltic run off Kaliningrad, presumably monitoring Russian nuclear forces there. An RAF Rivet Joint and an American Rivet Joint had been operating over Ukraine all day, while two Russian A-50U AWACS aircraft were up over Belarus keeping an eye on them and Ukrainian airspace. An Il-96–300PU presidential command post aircraft conducted a Kozelsk Loop at this time, again to check in with RVSN forces there and signal that the Russian nuclear forces were in some form of higher readiness.[183]

The heightened tension also expressed itself over North American airspace. One E-6B Mercury that was up in the morning was replaced with two E-6Bs, and then three E-4B Nightwatches also lifted off. Additionally, a Cobra Ball aircraft positioned itself southwest of the abandoned ABM site in North Dakota. As Grom 2022 was long over, this move was unusual. This posture was maintained into the late evening of 22 February.[184]

The afternoon of 23 February brought with it a significant step-up in surveillance activities in the region: an RAF Rivet Joint collected on targets in Belarus, while an American Rivet Joint and an MQ-4 Global Hawk flew routes over Ukraine all day collecting on Russian forces over the border. They were joined by an E-3B Sentry AWACS aircraft orbiting western Poland.[185]

Over North America, there were three E-6B Mercuries up in the morning (late afternoon in eastern Europe). By noon, or 1900 hours in eastern Europe, those three aircraft were replaced with four other E-6Bs and an E-4B Nightwatch. Two E-6Bs proceeded toward the ICBM fields in the northern United States (these were likely ALCS-equipped), one headed for the Atlantic coast, and the other to the Pacific coast for ballistic missile submarine communications. The E-4B orbited southwest of Washington, D.C.[186] At 1646 hours Eastern European Time (EET) observers reported on social media that Tu-95MS Bear aircraft at Engels Air Base could be heard running up their engines, while amateur radio enthusiasts reported that the LRA communications network was active.[187] The shift in the American nuclear command and control posture vis-à-vis TU-96MS Bear operations is not coincidental and strongly suggests there were other movements of Russian strategic nuclear forces that remain outside of the public domain.

This may also have been related to the deployment of Russian ballistic missile submarines. There are usually five of these based out of Gadzhiyevo near Murmansk. Two of these had deployed for Grom 2022 before 19 February but returned by 23 February. Two ballistic missile submarines and possibly two *Oscar* II–class guided missile submarines, departed sometime on 23 February. This was a significant proportion of the Northern Fleet's strategic nuclear capability.[188] Russian ballistic missile submarines can, incidentally, hit their targets from their berths. Submarines deploy for survivability and stealth. This deployment, naturally, generated apprehension and uncertainty within the NATO intelligence community. It is logical to conclude that the U.S. nuclear command and control forces, and probably other forces including bombers, were deployed to increase readiness and to signal the Russian leadership.

By 2221 EET, increased military activity in the occupied areas was reported, right before the Ukrainian internet opposite those areas suffered an outage. A Russian Il-22PP electronic warfare aircraft was also spotted over the occupied territories.[189] This was followed by a tremendous cyberattack on Ukrainian institutions which rapidly expanded to Latvian and Lithuanian systems by 0040 hours EET.[190] Viasat's KA-SAT communications service network was then hit with a cyberattack designed to disrupt Ukrainian military communications.[191] The Konteyner OTH radar was also "swamping out" high-frequency communications in Europe.[192] Early in the morning Russian issued a NOTAMS for the Ukrainian-Russian border region and Eurocontrol ATC started diverting civilian air traffic away from Ukraine.[193]

Around 0600 EET, Tu-95MS and Tu-160 bombers launched waves of Kh-101 and Kh-555 cruise missiles against the Ukrainian national air defense system. These were followed by coordinated launches of Kalibr cruise missiles from naval units in the Black Sea, and Iskander ballistic and cruise missiles from land positions. Those were immediately followed by strikes from Su-34 Fullback strike aircraft to a depth of three hundred km from Russian-controlled territory against air defense systems associated with Ukrainian ground forces in order to clear the way for the insertion of Russian airmobile and airborne forces in the Kyiv region.[194]

With the first missile strikes, Putin publicly announced that the assault was prompted by "tragic events taking place in the Donbass" and the "expansion of the NATO bloc to the east, bringing its military infrastructure closer to Russian borders." He decried the collapse of the Cold War order that the Soviet Union presided over, justified the assault by comparing it to NATO Operation Allied Force, and decried the use of force in Iraq, Libya, and Syria. Putin then likened all of this activity to the functional equivalent of Nazi German designs on the Soviet Union in the 1940s. In his speech, Putin issued a nuclear threat, reiterated his concern about NATO infrastructure, and decried NATO as an instrument of American power. He blamed the West for supporting "extreme nationalists and neo-Nazis in Ukraine." Given the alleged danger to the people in the so-called Donetsk People's Republic and Luhansk People's Republic, "I decided to conduct a special military operation" to "protect people who have been subjected to bullying and genocide." He then used Article 1 and Article 51 of the UN Charter to justify this.[195]

The visible response of American nuclear forces to developments on 24 February initially included the deployment of B-52s from RAF Fairford to orbit areas over Poland opposite Kaliningrad and over Sweden. These covered Russian nuclear forces in that enclave and south of St. Petersburg. This move was accompanied by the deployment of four E-6B Mercuries over the United States to include ALCS-equipped aircraft in proximity to the ICBM fields, and a Cobra Ball over North Dakota. Later in the day, E-4B code named Order66 took off from Lincoln, Nebraska.[196] The call sign of this E-4B Nightwatch referenced *Star Wars Episode III: The Revenge of the Sith*, a film in which a secret order is given to senior military personnel by the emperor to execute his opposition.

Russian moves did not correspond to patterns established up to this point in the crisis. A VIP Tu-214 flew to Nur Sultan but was likely a diplomatic flight. An Il-96–300PU command post aircraft flew from Moscow to Dushanbe, while a Tu-214PU-SUS command post aircraft flew from Moscow into uncontrolled airspace in Kazakhstan.[197] These last two moves suggest that the Putin regime dispersed nuclear command and control resources into CSTO countries that they knew would not be targeted by the United States. The Kazakhstan deployment of this unique aircraft is significant in that it was geographically positioned to communicate with the three mobile ICBM divisions of the 33rd Guards Missile Army in eastern Russia while operating inside Kazakh airspace.

It was the French, however, who responded publicly first. Foreign Minister Jean-Yves Le Drian was asked by media if he thought Putin was making a nuclear threat; Le Drian said, "Yes, I think that Vladimir Putin must also understand that the Atlantic alliance is a nuclear alliance."[198]

Conclusions

The year began with British signaling using a ballistic missile submarine to underscore resolve during a period of intense international diplomacy in the face of the massive Russian military buildup opposite Ukraine. Unexpectedly, there were several major and concurrent destabilizing events in January in addition to Ukraine: the unrest in Kazakhstan, North Korean flirtations with intercontinental delivery systems, and Russia's use of hybrid methods against NATO members and neutral Baltic countries as a possible distraction from the Ukraine operation. The American response involved exercising its nuclear command and control apparatus, revealing the location of submarines, and exposing Russian plans for a false flag event.

The Russian response was increasing the visible training tempo of its strategic missile forces, and establishing a Russian nuclear umbrella over Belarus. These moves were supported with enhanced information operations. Russian LRA was also employed to increase the pressure on NORAD, especially during an internal crisis in Canada that was aggravated by Russian hybrid methods. Other Russian military activities made it clear by the end of February that there was a coordinated effort to pressurize and distract NATO by operating in its rear area. These activities were backed by significant strategic nuclear force movements, including the Grom 2022 exercise. Collectively, these activities amount to implementation of the Russian definition of deterrence, that is, compellence. The message was, stay out of the Ukraine situation.

Deploying British and American ISR aircraft over Ukraine throughout this period was by no means the equivalent of deploying B-52s over Mariupol and P-8s over the Sea of Azov for deterrence signaling. There were no attempts at this point to establish that Ukraine was under an American, British, or French nuclear umbrella in the way Russia blatantly did so over Belarus. When it came to hybrid methods directed at NATO members, however, the American response was proportionate to the activity and employed assurance signaling. This included, for example, the deployment of B-52s off Norway, over Sweden, and then Poland to cover Russian nuclear systems in Kaliningrad pointed at NATO countries.

Probably nothing short of blatantly introducing nuclear-capable systems or military forces from NATO in or over Ukraine would have deterred Russia from mounting the next phase of its invasion. And that would have given the Putin regime confirmation of its worldview as well as the casus belli it wanted. Indeed, years of successful Russian information warfare directed against NATO members ensured there could be no prompt response along these lines. By exercising its strategic nuclear forces away from the conflict area and by conducting limited signaling with theater forces like the F-16 DCA units away from the conflict area, the United States used assurance signaling that it was prepared to respond if the Russian leadership chose to expand operations outside of Ukraine or involve the smaller NATO members closer to Russia. In the end however, any indications of a Russian military move against NATO members during this period was a distraction from the main event, though they likely had contingency plans to conduct "humanitarian interventions" in the Russian-speaking areas of the Baltic countries if the opportunity arose to exploit the situation. The Putin regime retained the initiative in part due to confidence, even though they had been thrown off by the false flag revelations. That confidence existed because of its strategic nuclear forces and their unbridled use for signaling and coercion. Strategic nuclear arsenals canceled each other out, Russian theater nuclear forces offset NATO theater nuclear forces, and there was no deterrent umbrella over Ukraine, thus making eastern Europe safe for conventional warfare. None of the Russian activity stopped vital military aid from being delivered to Ukraine in the weeks before the 23–24 February assault. Russia's coordinated activities did, however, set up the Western information ecosystem and make it susceptible to "nuclear fear" information operations.

CHAPTER 6

◆ ◆ ◆

Abyss Creep

February–April 2022

Russia set the stage for nuclear confrontation forty-eight hours after the start of the operation to seize Kyiv. Despite the public focus on the Russian conventional force buildup on the periphery of Ukraine, this was a supporting effort to the Kyiv coup de main. Starting in December 2021, Russian airborne forces (VDV) worked out a plan involving the 9th Directorate of the 5th Service of the FSB that was nothing short of the "repression of Ukrainian civil society," that is, the liquidation of the Zelensky government.[1] On 14 February, one week before the start of operations, Rosgvardia units, including their SOBR special operations units deployed to Belarus for Allied Resolve, discovered that they were going to be part of a Kyiv occupation force.[2] Seven days later, the VDV force earmarked to go in on 24 February received the finalized plan. The day before the operation was mounted, supporting Russian motor rifle and Rosgvardia units received their final orders. The personnel involved in the coup de main believed "this was a 'Crimean scenario,'" that is, as in 2014 with "minimal resistance."[3]

At 0549 hours EET on 24 February, amateur radio enthusiasts reported significant interference in high-frequency communications which they attributed to the Russian Konteyner over-the-horizon radar system.[4] What they heard was the initial electronic warfare attacks designed to disrupt Ukrainian air defenses around Kyiv, particularly at the Hostomel airfield fifteen kilometers from Kyiv. The cruise missiles arrived shortly after this.[5]

Thirty-four helicopters, including twenty-four Mi-8 transports carrying several hundred VDV personnel escorted by ten Mi-24 and Ka-25 attack helicopters, departed their base in Belarus and headed for Hostomel. Despite attack runs against Ukrainian defenses, eight helicopters were shot out of the skies. About eighty members of the assault force were killed. The rest of the landing force was engaged by Ukrainian defense forces on arrival and a desperate battle was on.[6]

At 1000 EET, eighteen four-engine Il-76 transports loaded with around 2,000 VDV personnel from the 76th Guards Air Assault Division, or two battalion tactical groups, departed the air base near Pskov, and headed south with the intention of landing at Hostomel to expand the airhead.[7] While they were in the air, the Ukrainian 4th Rapid Reaction Brigade, working with local defenders and supported by Mi-24P gunships, Su-24 strike aircraft, and MiG-29 fighters contained and pushed back the VDV.[8] The Il-76 force from Pskov turned back.[9] This force later diverted to Gomel airfield in Belarus where they disembarked two battalion tactical groups and moved south to join the fighting near Chernahiv.[10] Ukrainian special forces, possibly from the 3rd Special Purpose Regiment, landed by Mi-8 helicopter, infiltrated into the area, and called down artillery fire to damage the Hostomel runway.[11]

Russian Spetsnaz, some wearing Ukrainian uniforms, initiated operations in Kyiv itself. These were multiple assassination and saboteur groups supported by an agent network in the city: these teams had been inserted on 18 February to kill the Ukrainian political and military leadership. Acting on precise information, Ukrainian forces killed or captured these teams before they could achieve their objectives. One of the teams was intercepted as it prepared to attack the State Emergency Services headquarters building. Dead Russian operators were found equipped with AS-Val suppressed assassination weapons.[12] Other Russian teams, using ambulances to move around the city, were apprehended over the next two days.[13]

As the back-and-forth fighting raged at Hostomel on the morning of 25 February, the Rosgvardia SOBR unit that was supposed to support or reinforce the teams in Kyiv bypassed the fighting and made its way down the E373 highway. The unit was ambushed and destroyed in detail, including its commanding officer, on the bridge over the Irpin River. There were only three survivors out of eighty personnel.[14] A second Russian air assault, this one supported by ground forces that fought their way to points north of Hostomel from Chernobyl, arrived on the scene, and by 1400 EET the Russian Ministry of Defence claimed it now controlled the airfield. Elsewhere, a platoon of infantry from the Russian 74th Motorized Brigade was cut off and captured. They told their Ukrainian interrogators that they were unaware that they were being sent into Ukraine for combat operations.[15] Other prisoners told their captors that "we were told Zelensky had already signed a decree of capitulation" and had been told merely to drive to Kyiv.[16] A cell phone from a dead Russian soldier texting his mother: "We were told that they would greet us, but they threw themselves under our vehicles and did not let us pass."[17]

The Russian supporting efforts made varying degrees of headway in the face of stiff Ukrainian resistance. A traitor compromised the Ukrainian defense plan for Kherson, which facilitated a rapid Russian advance north from Crimea. A Russian amphibious landing between Melitopol and Mariupol threatened the defense lines there. Over in Donbas, Russian armored columns led by Russian vehicles painted white with OSCE (Organization for Security and Co-operation in Europe) markings on them were used to infiltrate Ukrainian positions. Kharkiv was bombarded. Armored columns spidered their way along the road network into northeastern Ukraine. But none of this mattered if Kyiv was not taken. And the Russian army was not psychologically or logistically prepared for a protracted campaign.

At 1450 EET 25 February, Putin's spokesman, Dmitry Peskov, announced through TASS that Russia was willing to send a team to Minsk to negotiate with Ukraine.[18] This was after Putin called an emergency meeting with senior Russian leaders.[19] It was at this point the Putin regime internally acknowledged the failure of the plan to seize Kyiv and liquidate the Ukrainian leadership. The public version of the meeting appears to have been a rewritten script for one of the false flag scenarios justifying the assault. Putin asserted that the Ukrainian armed forces was "infiltrated with nationalist elements" and that these "Banderites and neo-Nazis" were deliberately siting rocket systems in Kharkiv to attack the Donbas and "force return fire by Russian strike systems against residential quarters." This was "being done on the recommendations of foreign consultants, primarily American advisors."[20] Journalist Andrew Roth noted with alarm that "the speech seemed to be ripped from an alternate reality—or from the Second World War."[21]

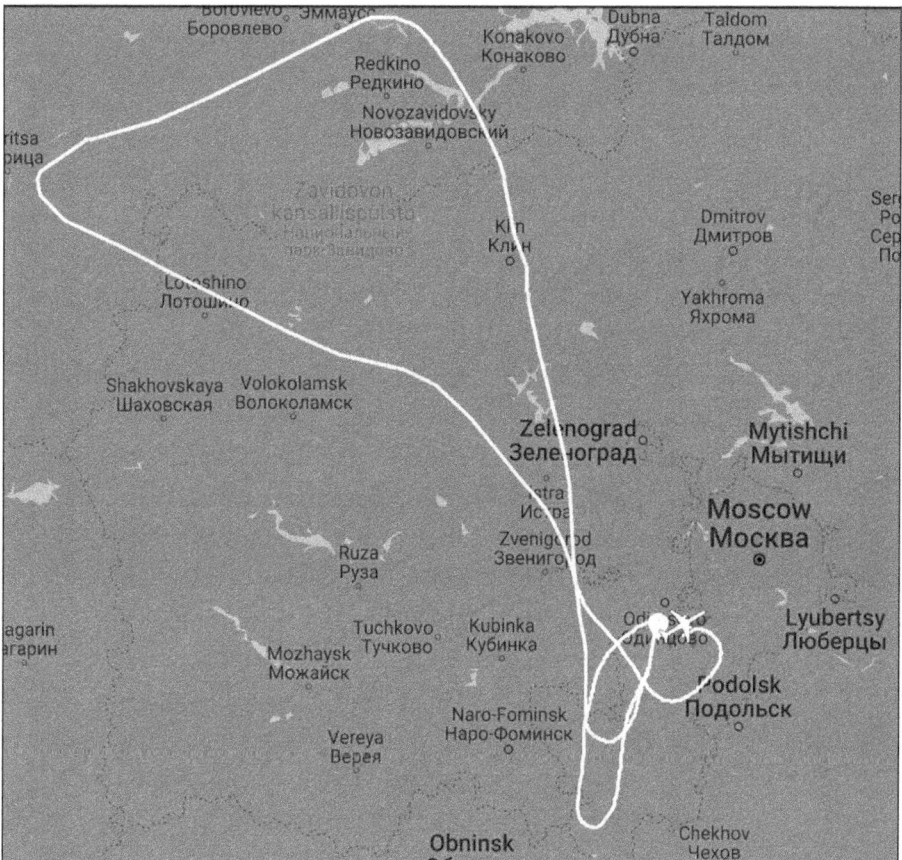

PHOTO 17 ◆ The Tver Loop, this one conducted by a Tu-214 PU command post aircraft. The Tver Loop replaced the Kozelsk Loop immediately after the latest phase of the Russian invasion of Ukraine in late February 2022 and exploits the strategic communications facilities northwest of Moscow. *flightradar24*

Turning to the Russian command and control aircraft plot, a Tu-214 PU command post took off from Vnukovo, headed to Tver, looped around and returned in an hour. This inaugurated the "Tver Loop," which replaced the Kozelsk Loop communications check and signaling mechanism. The tip of the Kozelsk Loop was likely deemed to put the aircraft in harm's way given the distance to Ukraine. A Tu-214 SUS special communications aircraft repeated its previous flight paths taking it into Kazakh airspace where it was less vulnerable and could communicate with the eastern RVSN missile divisions. These moves should be categorized as ongoing general deterrent signaling.[22]

Two hours later, Russian information outlets released video of four YARS or Topol-M mobile ICBM launchers being blessed by an Orthodox priest before departing Vladimir on the highway toward Moscow.[23] The Russian Ministry of Defence tweeted that they were being deployed and prepositioned for the 9 May Moscow Victory Day Parade.[24] A Russian information conduit specializing in aerospace affairs noted that "this happened against the backdrop of a statement by the Russian Foreign Ministry that Ukraine refused to conduct negotiations, which raised a number of questions about whether these strategic missile systems would be used in a special military operation in the Donbass."[25] The possibility exists this movement was in fact prepositioning for the 9 May parade and was also used to backstop Russian diplomatic efforts in the information environment.

American NC3I movements were restrained. On 25 and 26 February, six E-6Bs were sequentially aloft in singles or pairs throughout each day, with a single trailing wire antenna deployment in the Pacific on each day. The following day, this dropped to four rotations of single aircraft during the course of the day.[26]

An emergency meeting of the UN General Assembly that evening voted to condemn Russia for the invasion while the merits and impact of sanctions against Russia filled the global information space. A press release at 0048 EET 26 February by the General Staff of the Armed Forces of Ukraine stated that "the enemy has not achieved the previously set strategic goals." Ukrainian SBU operations around the country moved to dismantle a number of Russian-controlled cells at the oblast level that were to act as "People's Republic" governments modeled on the Donetsk and Luhansk puppet governments once the Russians had liquidated the government in Kyiv.[27] The Putin regime's gamble to seize Ukraine by coup de main had failed less than forty-eight hours after it was launched.

After the First 48

As the fighting raged on 27 February, Putin held another public spectacle, this time with only two other actors: resigned-looking Defense Minister Sergy Shoigu and General Valery Gerasimov, whose theories on hybrid warfare were not faring well. These were the three men who had the authority to collectively release nuclear weapons. The president and defense minister provide legal authorization with two separate codes to the chief of the General Staff (CGS), who digitally combines the codes with his system. This permits the CGS to then interact with the nuclear force commanders to issue preliminary or direct commands. Preliminary commands consist of various states of readiness, targets, and destruction criteria, while the direct command is essentially the launch order.[28] Presenting this troika in such a public way was a clear signal to nuclear cognoscenti in the West.

Putin referenced "unfriendly economic actions" and "officials of the leading NATO countries [who] are indulging in aggressive statements directed at our country," presumably a reference to the French foreign minister but also interpreted by some to mean statements by UK minister of defence Liz Truss.[29] What he said next is open, quite literally, to interpretation. The English translation released by Putin's office says: "Therefore, I order the Defense Minister and Chief of the General Staff to put Russian Army's deterrence forces on high combat alert."[30] However, in the original Russian, Putin used повышенную боевую готовность. This can translate to "increased combat readiness" or "special regime of combat duty." That is very different from "high combat alert." However, nearly every Western media outlet had "nuclear forces" and "high alert" in their headlines that day.[31] TASS followed up ten minutes later but used the term "special service regime" to describe the actions of the Russian deterrent forces. TASS added a helpful blurb about the purpose of the strategic deterrent forces, to deter aggression against Russia and to "defeat any aggressor in a war by using various types of weapons including nuclear ones."[32]

The details of the specific readiness system that Russia currently uses for its strategic nuclear forces, the equivalent of the U.S. DEFCON system, are not in the public domain. Historically, there were four Combat Readiness levels:[33]

- Combat Readiness 4: Constant twenty-four hours
- Combat Readiness 3: Increased seven to two hours
- Combat Readiness 2: Increased, First Degree one hour
- Combat Readiness 1: Complete (or Total) twenty-two to thirty minutes

So did Putin increase combat readiness from nothing to the equivalent of CR 4? Or had he selected one between the equivalents of CR 4 and CR 1? Again this was opaque to the media but not to specialists in the field. These levels were not necessarily progressive. Unlike the Cold War DEFCON system, Russian levels are not intended as signaling mechanisms. They merely establish the fact that nuclear formations and units meet certain readiness standards set by the supreme authority at a given time. What we do not know is whether there were preliminary commands associated with the Combat Readiness change. If there were, and they were detected by the U.S. intelligence apparatus, then the signal would have been a strong one. If there were none, then Putin's statement was designed as part of a larger information operation to use nuclear fear to support his objectives in Ukraine and elsewhere.

Forty minutes after the Putin statement and Shoigu and Gerasimov essentially said, "Yes, sir!" in front of the cameras to convey docile subservience. The U.S. response was low key, despite the media freak-out. White House spokesperson Jen Psaki noted this was a consistent pattern for Putin and that Russia was under no external threat from any source.[34]

The Russian negotiation team took off from Moscow for Belarus, while Turkey closed the Bosporus and blocked Russian naval forces from passage into the Black Sea. The supporting Russian information effort employed Dmitry Kiselyev from Rossiya 1 TV, known as "Putin's mouthpiece," to announce to the country and the world: "Our submarines alone can launch more than 500 nuclear warheads, which guarantees the destruction of the U.S. and NATO for good measure.... Why do we need the world if Russia won't be in it?"[35] This was immediately amplified all over Twitter. There were, however, conflicting signals. Russian ambassador to the UN Vasily Nebenzya replied to a press conference question, "On the use of nuclear weapons, God forbid it." Moscow, he said, was exercising "a kind of deterrence."[36]

The immediate U.S. military response remains blurred. There is a belief that the United States warned Russia against nuclear weapons use "through government channels."[37] An E-6B was over the Pacific Ocean with its trailing wire deployed and there were EAMs.[38] Half an hour after this, Senator Marco Rubio from the Senate Intelligence Committee tweeted that "Russian military leaders should think very carefully before following orders they recently received."[39,40]

There was a pause for several hours, then the LRA communications system came alive late in the evening of 28 February.[41] This coincided with Shoigu going on television to report to Putin that he had carried out his orders and "the duty shifts of the control centers of the Strategic Missile Forces, the Northern and Pacific Fleets, as well as the long-range aviation command, began to carry out duty with reinforced personnel."[42] Notably, Gerasimov was absent from this announcement which led to speculation that he had been removed. In all likelihood, Gerasimov had moved to a bunker to underscore the seriousness of the situation.

After this, four E-6B Mercuries and an E-4B Nightwatch were visible on ATC systems over the United States.[43] Pentagon spokesperson John Kirby noted that Russian nuclear weapons activity was "as unnecessary as it was escalatory. . . . The Secretary is comfortable with the strategic deterrent posture of the United States"[44] without elaborating on what U.S. actions had been taken.

Russian and Israeli media conveyed that ballistic missile submarines of the Northern Fleet "carried out combat training activities."[45] The Tu-214VPU command plane associated with the Interior Ministry took off from Vnukovo and orbited for an hour near Moscow while a Tu-214PU command post conducted the new Tver Loop strategic forces communications check. A Tu-214SUS also conducted a Tver Loop later on.[46] It also appears as though the 29th Guards Missile Division around Irkutsk carried out a readiness exercise. The United States, however, maintained one E-4B and four E-6Bs up throughout the day. Of note, the Cobra Ball aircraft deployed to Japan overflew Hiroshima while squawking on ATC systems.[47]

It was at this point the Putin regime decided to overtly bully the Baltic neutrals either to distract NATO or maintain some form of initiative somewhere other than Ukraine. An aircraft identifying itself as an Ilyushin Il-114LL was detected flying collection loops paralleling Finland.[48] The Il-114LL was a civilian airframe performing "complex radar, photography, and thermal vision mapping of ground and sea surfaces."[49] Its nearly indistinguishable military version conducts "reconnaissance and cartographic missions."[50] Then the GPS interference started. Not only did this affect civil life on the ground, it also affected Finnish air travel, which was shut down in some border areas for two weeks.[51]

The next day, 2 March, the U.S. ambassador to the United Nations decried Putin "threatening to invade Finland and Sweden."[52] Within hours, Swedish Gripens intercepted two Su-24 strike aircraft escorted by two Su-27 fighters off Gotland in Swedish airspace. This initiated a Swedish debate as to whether the Su-24s were carrying nuclear weapons or not.[53] The Swedish air force asserted that "the incursion had been done deliberately to signal Sweden" and "we see it as a deliberate act."[54] Pictures of the Su-24s showed they were armed with stores. They could have been carrying either training rounds or conventional weapons.[55] It is extremely unlikely that actual nuclear weapons would have been released by the Putin regime for this task, but the signaling was clear.

That day, a probable Russian social media information operations conduit released footage from VKontake on Twitter that depicted a 12th GUMO nuclear weapons convoy of sixteen vehicles observed near the Mozarhaisk-10 nuclear weapons depot west of Moscow heading west. This footage showed dedicated nuclear weapons transport and monitoring vehicles escorted by security vehicles.[56] Phone footage released on a Russian TikTok channel depicted a similar 12th GUMO convoy also heading west, except this one was equipped with a crane used for missile warhead loading.[57] The location was noted as Velikly Novgorod, which happed to be midway between the Novgorod-17/-18 nuclear weapons storage site and the town of Luga, south of St. Petersburg. Luga was an area suspected of hosting an Iskander missile formation. It is highly likely that the footage was meant to be discovered by Western open source enthusiasts and amplified in the infosphere.[58]

American moves included eight sequential flights of E-6B aircraft throughout the day on 2 March over the United States, with trailing wire antenna deployments by two aircraft over the Atlantic and Pacific.[59] Importantly, Secretary of Defense Lloyd Austin postponed a previously scheduled Glory Trip Minuteman III launch. Austin noted that "Russia's nuclear submarines and mobile missile launchers reportedly staged drills and units of [the RVSN] dispersed intercontinental ballistic missiles in forests in eastern Siberia to practise secret deployment," thus confirming that Russian strategic nuclear movements over the past forty-eight hours were much more extensive than understood in the public domain. The postponement was "an effort to demonstrate that we have no intention in engaging in any actions that can be misunderstood or misconstrued."[60]

Russian NC3I activity increased on 3 March, possibly to underscore an "interview" with Sergy Lavrov on RT which was rebroadcast globally. Lavrov stated that "the thought of nuclear war is constantly running through the minds of Western politicians, but not the minds of Russians. . . . If a real war is unleashed against us, this must be a concern for those who are hatching such plans. And I believe those plans are being hatched."[61] The Interior Ministry Tu-214VPU was launched and orbited southwest of Moscow, while a Tu-214PU command post flew a route from Moscow to Chelyabinsk, Magnitogorsk, Yekaterinburg, and then Moscow. An Il-96–300PU presidential command post conducted a Tver Loop, while another flew to Ufa and back to Moscow. The aircraft routes, numbers, and types indicate this was a significant strategic nuclear forces communications check.[62] This activity was concurrent with the operations of two E-4B Nightwatches and seven E-6B Mercuries over the course of the day, with the two E-4Bs up at the same time with two E-6Bs. One of the E-6Bs conducted a trailing wire antenna deployment over the Gulf of Mexico.[63]

A tweet from USSTRATCOM noted, "Shown here moored in Saipan late last month, the USS *Emory S. Land* is one of two U.S. Navy submarine tenders providing . . . support to SSBNs."[64] The other was the USS *Frank Cable*.[65] This was an extremely subtle and technical signal. It implied that U.S. SSBN's could be kept on station significantly longer than the usual 110-day deterrent patrol, thus maintaining or even expanding target coverage. During the Cold War, U.S. Navy submarine tenders had the ability to carry reloads of submarine-launched ballistic missiles for the SSBN force. Mobile and operating from discrete forward locations, the tenders could reload SSBNs for the trans- and post-SIOP periods. In the post–Cold War environment, the ships were reconfigured to support SSNs and SSGNs carrying Tomahawks, with each tender carrying seventy-two Tomahawk reloads. Though the two ships no longer have the ability to reload complete SLBMs, they likely have the ability to fill up empty slots in the Trident II D5s with reentry bodies, or alter load-outs as required at out-of-the-way sheltered locations using their cranes.[66]

This was the same day the Ukrainian negotiating team was believed by some to have been "tactically" (nonlethally) poisoned on their way to talks with Russian counterparts in Turkey.[67] If this was the case, there was mixed messaging going on.

In addition to underscoring Russian diplomacy, another possibility was that the so-called Anonymous hacker collective announced that they had penetrated ROSCOSMOS and shut down its control center.[68] If this is accurate—and given the incendiary statements by ROSCOSMOS boss Dmitry Rogozin threatening war that were permitted on practically every Russian information outlet, it probably was—the degradation of Russian satellite communications used for strategic nuclear forces may have forced the Russians to deploy their command and control aircraft as a precautionary measure.[69]

In February 2022 Russia was down to two optical reconnaissance satellites and a radar imaging satellite, all nearing end of life. The status of the forty-seven communications satellites was unclear, likewise with the six of the required ten ECS missile warning satellites. As for GLONASS, the Russian GPS equivalent, twenty out of twenty-four were operational, but many of these were also nearing end of life. One assessment was that "half of the GLONASS satellites can fall out [of] orbit at any moment."[70] Indeed, Russian strike aircraft being used against Ukraine after this date were seen to have commercial Western GPS systems taped to panels in their cockpits. Captured Russian troops also had commercial GPS.[71] The situation was so bad that Russia fell back on using terrestrial LORAN navigation system, of which three stations were positioned to ensure coverage of Ukraine.[72] A crash program led to repopulating their constellations throughout 2022, but for these early weeks and months, Russia's space capability was problematic.

On the Kola Peninsula, Norwegian sources noted a surge of surface vessels from the Northern Fleet, as well as heightened air defense readiness and heightened base security around nuclear weapons.[73] At the same time, Putin met with his Security Council and asserted that the objective of the Special Military Operation was to protect Russians in Donbas and provide security for Russia. Employing Eurasian ideological language, Putin asserted that "Russians and Ukrainians are one nation" and that people had been "duped by nationalist Nazi propaganda." Ukraine had to be "denazified and demilitarized" so that Russia could not be threatened with nuclear weapons "on our border."[74]

Despite the reduction in Russian nuclear rhetoric, an Il-96–300PU presidential command post aircraft proceeded to Washington, D.C. and New York to evacuate Russian diplomatic personnel.[75] The following events then occurred simultaneously over a one-hour period on 4 March. An Il-96–300PU presidential command post aircraft conducted a Tver Loop while a Tu-214PU conducted a flight from Moscow-Chelyabinsk-Magnitogorsk, having taken off at 1230 EET. Once both planes were in the air, the 39th Missile Division at Novosibirsk and the 29th Missile Division at Irkutsk deployed mobile ICBMs from their bases into the forests. An E-4B Nightwatch then came on strip alert at Lincoln, Nebraska, and five E-6Bs sequentially rotated throughout the day, with one conducting a trailing wire antenna deployment over the Pacific. A Cobra Ball aircraft also deployed to North Dakota.[76]

A pair of B-52Hs flew over Czechia at 1140 EET and then moved to Romania and orbited close to Moldova, bringing the Moscow area and everything in between in theoretical ALCM range. At some point in the afternoon the Carrier Strike Group based on the USS *Truman* moved into the northern Aegean Sea, conveying a similar message. During the night an E-4B Nightwatch call sign Truce18 staged at Abilene, Kansas.[77]

The same account used to channel the 12th GUMO nuclear weapons convoy footage noted on 5 March that "several TOPOL-M, YARS missiles may have non-strategic warheads for preventive tactical strikes." The account noted that "VNIIEF [Sarov or Arzamas-16] has developed three modifications" that were tested eight times that may have been nonstrategic warhead tests. A detailed list of Topol-M flights was attached.[78] These data suggest that three different reentry vehicles were tested, with the last one tested six times in 2007–2008. This suggests the fine-tuning necessary to handle nonstrategic targeting.[79]

The next day, an article appeared on a specialist website dealing with Russian issues. This article highlighted "escalate to de-escalate" doctrine, whereby low-yield weapons are used as a "shot across the bow" to demonstrate Russian resolve and compel opponents to back off. Sections of Russian declaratory strategy and policy were quoted.[80]

These pieces, when combined with the earlier Topol-M mobile ICBM movements out of Vladimir, might have been part of a Russian information operation to convince analytical elements in the West that the Putin regime was considering the use of sub-strategic weapons in Ukraine. It might have been designed to enhance a nuclear fear information line and pressure Western governments to force Ukraine to the bargaining table, or to intimidate the West and Ukraine in a broader sense. The fact that there were ongoing negotiations in Turkey supports that argument.

Russian information outlets, meanwhile, ramped up justification for the Ukraine operations for their domestic audience, this time focusing on nuclear weapons development in line with Putin's and Shoigu's statements. A compilation of these included references to uranium mining; the alleged transfer of polonium to Ukraine from the United States; the allegedly sinister role that Zaporizhzhia nuclear power plant was playing; and the alleged destruction of nuclear weapons design data by the Kyiv government.[81]

The deployment of the guided missile destroyers USS *Donald Cook* and USS *Forrest Sherman* into the Baltic on 7 March does not appear to have been a scheduled move: Exercise Cold Response 2022 would not start until 14 March and BALTOPS 22 was not scheduled until June. With ninety and ninety-six vertical launch cells, respectively, such a move could have been calculated to appeal to Russian paranoia over cruise missile systems in close proximity to its territory, especially Kaliningrad, but also missile sites south of St. Petersburg, thus extending some kind of deterrent umbrella over the Baltic countries. And it did, with the Russian National Defense Control Center using Russian information outlets to publicly acknowledge the ships' presence.[82] B-52Hs continued to operate over Romania, while on 8 March two F-16s out of Aviano Air Base scrambled and made a beeline for Romania before turning around.[83] Two E-6B Mercuries lifted off. One headed for the North Dakota–South Dakota border, while the other flew up to the Montana-Wyoming area. This was clearly a visible demonstration of E-6B ALCS capabilities and the Minuteman III ICBM fields, while at the same time it was a step down from the four E-6B/one E-4B posture.[84]

The U.S. intelligence community's assessment was that Putin's statements on raising Russian nuclear force readiness were "extremely unusual," something not seen since the 1960s. The intelligence community leaders, however, minimized the situation by stating that they had not seen "force-wide nuclear posture changes that go beyond what we have seen in prior moments of heightened tension during the last few decades." It was limited to "[deterring] the West from providing additional support to Ukraine." When pressed they conceded that "we have not yet seen posture changes that are beyond" what were seen in 2014 or 2016.[85]

At this point the arena had been demarcated. NATO members had made it clear that NATO territory would be defended but were clearly deterred from projecting a "no-fly zone" over Ukraine or from providing combat aircraft to replace Ukraine's losses.[86] NATO members were not deterred from supplying Ukraine with other conventional weapons systems, however, and it was these that were killing Russian armored vehicles and their crews at unprecedented rates.

Seven Days in March

At this point, Russian forces were bogged down on all axes of attack in Ukraine. Putin's regime launched allegations that chlorine had been delivered to the

Kharkhiv front by a Ukrainian "nationalist group," who were going to employ it on a civilian population and "accuse Russia of allegedly using chemical weapons."[87] The U.S. administration responded from the White House level, countering that this was being done to "justify a false-flag operation or [Russia] using chemical or biological weapons in Ukraine themselves."[88] Russian rhetoric continued to escalate at the UN, with the White House keeping pace. The Chinese leadership then started to endorse the Russian position using its information apparatus.[89] The informational space groundwork had actually been laid on 1 March, with an article in *VPK*. This article implied that nerve agents should be used to clear built-up areas in order to avoid "Stalingrads" and there were precedents to do so. The article implied the West would not be able to deter a chemical attack because it lacked the capability to do so.[90]

At the same time, Ukrainian intelligence revealed that the Putin regime had planned or was planning a false flag attack on Chernobyl nuclear power plant using the frozen corpses of dead Ukrainian troops held in refrigerated trucks in Belarus. In their assessment, "having failed to get the desired outcome from a ground military operation and direct negotiations, Putin is ready to resort to nuclear blackmail of the international community for concessions to support Ukraine."[91]

At this point, U.S. nuclear command and control posture was at an elevated level, with twelve NC3I aircraft moving on 9 March, six on 10 March, and eleven on 11 March. Qualitatively, this included trailing wire deployments over Lake Superior, maintaining three E-6Bs aloft simultaneously on two days, and maintaining two E-4Bs aloft at once on one day.[92] Corresponding Russian moves had one command plane fly to Kazan or Ulyanovsk and back and another conducting a Tver Loop per day.[93] The seriousness of the chemical false flag situation was underscored by President Biden's point-blank statement on 11 March that "Russia will pay a severe price if they use chemicals."[94] The number of E-6Bs up on 12 March included five E-6Bs and an E-4B sequentially over the course of the day and one trailing wire deployment over the Pacific. There were no corresponding Russian flights.[95]

The next two events occurred simultaneously. The first was either a purge or an attempt to preempt a putsch in Moscow. The FSB's 150-man organization for collecting on and subverting Ukraine led by Sergey Beseda was more or less incarcerated en masse along with its boss on corruption charges. Then the deputy commander of Rosgvardia, Roman Gavrilov, was ousted. Gavrilov had previously been part of the presidential protection service, the FSO. His boss, Viktor Zolotov, disappeared for some days after being overheard bad-mouthing Putin at church. The deputy chief of the General Staff, Igor Kostyikov, was also fired. There was widespread speculation that Shoigu and Gerasimov had also been purged, but this turned out to be inaccurate.[96] Ukrainian information operations poured gas on the fire by releasing information that led to headlines like "Russian elites planning to overthrow Putin."[97] In any event, the political situation in Moscow was destabilized.

The second was the Russian missile strike against the Yavoriv Training Centre, a NATO Partnership for Peace facility near the Polish-Ukrainian border on 13 March. The attack was conducted by Tu-95MS strategic bombers based at Engels firing an estimated thirty cruise missiles.[98] This is significant in that it would have been easier and quicker to strike the Yavoriv facilities with Iskander ballistic missiles stationed in Belarus. Indeed the Iskander brigade near Brest could have conducted the strike. The use of a dual-capable asset like a Tu-95MS and the proximity of the target to Poland was an explicit Russian warning that they were prepared to escalate actions against those who support Ukraine. This strike kicked off a series of measures throughout the rest of March designed to intimidate Poland, the Baltic countries, and NATO; a Russian television program even featured several retired generals presenting in detail with electronic maps how they would retake the Baltic countries and invade Bornholm island.[99]

With a destabilized political situation in Moscow and increased pressure against Poland, two scheduled exercises started on 14 March: NORAD's Noble Defender 2022 and NATO's Cold Response 2022. Noble Defender exercised all three NORAD regions and involved the defense of the Alaskan North Slope and the eastern approaches to North America, including Thule Air Base. B-52s were used to simulate Russian bombers, which were intercepted by RCAF and USAF aircraft.[100] The possibility that the B-52s conducted other operations in the Arctic before penetrating the NORAD air defense system should be considered. Notably, interceptor aircraft were armed in the pictures used on social media posts and a connection was made between Noble Defender and the ongoing ICEX.[101]

Cold Response was a previously scheduled large-scale NATO exercise in Norway. Though it officially began on 14 March, elements of II Marine Expeditionary Force (II MEF) started operating in Norway during December 2021, with forward elements flown in on 13 January 2022. The exercise involved 45,000 personnel from 23 countries, including Sweden and Finland, implementing a multiphase plan that involved securing command of the sea, projecting air operations to protect a land deployment, and the land defense of northern Norway. Amphibious operations by a composite French-Dutch-U.S. force figured prominently, as did the USS *Kearsarge* LHA with a Marine Expeditionary Unit and F-35B strike aircraft embarked. This was on par with ACE Mobile Force exercises during the Cold War. Cold Response was a threat to Northern Fleet bases around Murmansk, whose naval infantry brigade was currently getting chewed up in Ukraine. ICEX was also underway concurrently with Cold Response and Noble Defender, as were port visits by the guided missile destroyer USS *Forrest Sherman* to Stockholm and other scenic Baltic locations.[102] This state of affairs likely caused unease in Russian military quarters regarding the correlation of forces in the Arctic while their operations were increasingly problematic in Ukraine.

British defense intelligence warned on 14 March that Russia could employ "a faked attack . . . a staged discovery of agents or munitions, or fabricated evidence of alleged Ukrainian panning to use such weapons" accompanied by "extensive disinformation."[103] The same day Noble Defender and Cold Response kicked off, Russia appears to have staged a Tochka-U missile strike on Donetsk which killed civilians, apparently after having gathered local people for a rally near where the missile allegedly hit. There were, however, no chemical aspects of this event.[104] Russian information outlets then focused on alleged American biological warfare labs in Georgia and asserted that if Russian inspectors were not allowed in, then cruise missiles should be fired at that country.[105] The U.S. posture shifted from one E-6B Mercury at 1600 EET, to four E-6B Mercuries by 2146 EET and a Cobra Ball over North Dakota. Russian movements involved a Tver Loop, but it was conducted by an Il-96–300PU presidential command post.[106]

The chemical weapon false flag issue remained active on 16 March, with National Security Advisor Jake Sullivan having a telephone conversation with Nikolai Patrushev. Sullivan "warned [Patrushev] of the consequences of any possible Russian decision to use chemical or biological weapons in Ukraine."[107] The move from public comments by a White House spokesperson to a phone call from the national security advisor indicated how seriously the Biden administration took the situation.[108]

At 1400 EST Volodomyr Zelensky addressed the U.S. Congress via remote means for what amounted to an epic Churchillian speech. For several hours prior to this, eight E-6B Mercuries (half of the fleet), one E-4B Nightwatch, ten E-3C/E-3TF AWACS aircraft, a Cobra Ball over North Dakota, and more than a hundred tanker aircraft were aloft in the largest demonstration of visible American nuclear command and control signaling seen this far.[109] The Russian side responded with a Tu-214SUS specialized command plane on a Tver Loop, an Il-96–300 on a Moscow-Ufa run, and an Il-96–300PU presidential command post run from Moscow to Novosibirsk and back. A Tu-214PU out of Tyumen orbited near Yekaterinburg, with the 42nd Missile Division nearby, and then the aircraft flew to Orenberg (home of the 31st Missile Army) and then back to Moscow.[110] These aircraft were in the air simultaneously, which suggests a precautionary posture or at the very least a communications check vis-à-vis the RVSN forces. Some B-52Hs were seen returning to RAF Fairford late that night, which suggests movements in northern Norway in support of Cold Response, in line with previous exercises (or both), or possibly a flight into the Arctic Basin.[111]

The events of 17 March remain murky. On the Baltic front, the Russian foreign ministry used social media to reach out to ethnic Russians in Latvia to acquire abuse stories that could be used in a Sudetenland-like scenario. A Polish defector was murdered in Belarus, while the Polish rail system was hit with a massive cyberattack. These appear to have been fragments of a stepped-up information campaign against the Baltic countries that revealed itself later but were connected to Russian threats regarding Polish military support to Ukraine.[112]

The movements of Russian nuclear command and control aircraft this day were extensive and were almost like a delayed but escalatory response to the 16 March armada put up by the United States. In addition to a Tver Loop and a Novosibirsk run by a Tu-214PU and an Il-96–300PU, respectively, a Tu-214SR communications relay aircraft conducted orbits near Tyumen and Omsk, locations associated with the 33rd Guards Missile Army and its subordinate divisions. And in a move not yet seen, an Il-96–300PU departed Moscow for the north Arctic coast, flying east and then disappearing into airspace where ATC was minimal or nonexistent known as the Siberian Black Hole. It is possible it was testing communications with the submarine force, or it could have been deployed to the area to make it more difficult to target.[113] A Russian special detachment aircraft heading to China took a U-turn and returned to Moscow. It was reported that Sergy Lavrov was aboard, which suggests there were urgent matters to discuss in Moscow.[114]

The American command and control activities that day consisted of five E-6Bs and one E-4Bs up sequentially over the day. But there was one exception. Departing Patuxent River NAS, the E-6B call sign Mass97 flew to an elliptical orbit point west of the Ravenrock NMCC Complex and Camp David, where it orbited for nearly four hours while the Russian aircraft were airborne.[115] The numeral 97 is the atomic number associated with Berkelium, which "has no commercial . . . or practical use" but was a radioactive material produced in a cyclotron in 1949 by Glenn Seaborg, the discoverer of plutonium.[116]

A 2021 U.S. State Department report concluded that despite the destruction of declared stockpiles, "the United States does not believe Russia has declared all of its CW stockpile . . . and development facilities."[117] Indeed as of 2017 the Kizner CW depot near Kazan was still operational, and it previously was assessed as containing "small artillery shells" loaded with VX, sarin, and soman nerve agents.[118] Nuclear Biological and Chemical Defense vehicles had been photographed on trains in Belarus in January and again on 9 March, while Ukrainian intelligence reported that vaccination of Russian troops with a sarin antidote was underway for GRU special detachments.[119] Dead Russian personnel in some sectors were found equipped with protective equipment. All Russian tubed and rocket artillery, of course, was theoretically capable of firing chemical munitions.

In addition to the ongoing fighting all over Ukraine, 18 to 20 March was relatively quiet when it came to overt visible signaling, with only three E-6s aloft sequentially on 19 March and one E-6B and one E-4B on 20 March. The USAF E-4B's call sign was Kruegr81, referencing the fictional serial killer Freddy Krueger from *A Nightmare on Elm Street*.[120] A Russian information conduit sowed ambiguity by noting the fact that there were nine Tu-160 Blackjacks and twenty Tu-95MS Bears at Engels Air Base, but "the reasons for the presence of such a large number of strategic aviation assets . . . remains unknown." Further, "a number of assumptions have been made that strategic aviation can be used to strike at military facilities of the Armed Forces of Ukraine, however this information is refuted by the lack of relevant statements from the Russian Ministry of Defence."[121] That is, outside observers should assume that these forces had purposes other than targeting Ukraine.

The Putin regime ramped up on the chemical weapons issue in the information environment, once again claiming that Ukrainian neo-Nazis were preparing a chemical attack provocation.[122] This was designed to legitimize Lavrov's statement that "any foreign supplies to Ukraine containing military equipment will be considered legitimate targets for Russian strikes."[123] Pro-Russian influencer George Galloway posted on social media, "It is now clear that the US is about to stage a false-flag #WMD incident in Ukraine. Fasten your seat belts."[124] President Biden opined that Russia's focus on imagined Ukrainian WMDs indicated Putin was considering the use of biological and chemical weapons, a move that would provoke a very serious NATO reaction.[125]

USSTRATCOM tweeted that B-52 bombers were supporting "air-to-ground integration training over Germany, Romania, and the Black Sea."[126] Air-ground training over the Black Sea was an interesting turn of phrase if one were signaling about the possibility of, say, striking chemical weapons storage areas in Crimea with conventional weapons.

Force de Dissuasion

It was at this point all eyes shifted to Poland. Russia initiated a new information campaign on 16 March asserting that Poland should be partitioned and "de-nazified."[127] In a rambling Telegram post five days later, Medvedev decried "Russophobia" in the Polish elites stimulated by "senile" Western leaders. After this, Polish media reported that a Russian Tu-22M3 Backfire violated Polish airspace. The Polish Ministry of Defence refused to confirm.[128] The next day on 22 March, clouds of smoke were seen emanating from the Russian embassy in Warsaw. Social media compared this to what the Russian embassy did in Kyiv before the assault in February, implying war was imminent.[129] As it turned out, Poland expelled forty-five out of sixty Russian embassy staff in response to their security services uncovering a spy ring.[130] Rossiya1, however, chimed in to ramp up the nuclear fear line, discussing a "possible nuclear strike on Europe, Russian invasion of Poland and Lithuania."[131] Over the course of these days, there was what can only be described as a rhetorical offensive with multiple statements from Lavrov, Peskov, and Deputy Foreign Minister Sergey Ryabkov, all emphasizing Russian declaratory nuclear weapons policy, made for a variety of audiences.[132]

This time, France came to the fore. The French nuclear deterrent forces include the Strategic Oceanic Force, consisting of four *Triomphant*-class ballistic missile submarines, each of which has sixteen SLBM launchers. Three submarines are on an operational cycle with one at sea all the time on a seventy-day patrol and the fourth submarine is in overhaul.[133] On 11 March, regional French media reported that a second French SSBN had put to sea from the base at Île Longue. Having two SSBNs deployed, apparently, had not been done since NATO INF affair in 1981.[134] Then the number increased to three.[135] One commentator suggested this was a cautionary measure to prevent them from being destroyed in one strike or damaged by sabotage.[136]

Right on the heels of this came an announcement from the French Ministry of Defence that a successful test of the Air-Sol Moyenne Portée Amélioré (ASMPA) nuclear air-to-surface missile was conducted on 24 March.[137] The significance of the ASMPA test lies its relationship to French nuclear doctrine. This missile carries a 300 kt yield warhead and is launched from Mirage 2000NK3 aircraft or Rafale Marine aircraft from the French aircraft carrier force. French nuclear doctrine has a "nuclear warning" component in which nuclear weapons like the ASMPA are employed before the decision to fire the ballistic missile force is made in order to get the aggressor to back off.[138]

A French nuclear flourish was certainly an interesting development. Operating outside of NATO, France possesses the ability to extend deterrence over whoever it chooses using the *tous azimuth* doctrine established during the Cold War and more finely tuned into the twenty-first century. The *force de dissuasion* had the ability to saturate Russian ABM defenses and fulfill a range of target options that contributed greatly to overall deterrence. As a NATO member, France can also activate Article 5 if attacked by Russia. Knowing that France has this capability, a disproportionate percentage of the Russian strategic forces must be kept available to cover targets in France, thus placing some stress on the Russian capacities if they are not to reduce target coverage of the United Kingdom and the United States. It is clear that France was deliberately signaling during this period, but not using the social and mainstream media informational space to enhance the distribution of the message. This message was directed at Russia's dedicated intelligence apparatus to keep the band narrow and focused on the top of the Putin regime. The French leadership clearly believed they were subject to, or about to be subjected to, Russian threats against vital French national interests.

The numbers of nuclear command and control aircraft remained constant on 23 March with one Russian aircraft in the Moscow area and another on an eastern run, and in the United States a pair of E-6B Mercuries up on either coast.[139] American media reported that Secretary of Defense Lloyd Austin and JCS Chairman General Mark Milley had for the past week attempted to contact their opposite numbers, Shoigu and Gerasimov, but had received no response.[140] To dispel rumors of Shoigu's incapacitation, the Russian Ministry of Defence released a proof-of-life video.[141]

Ex Cold Response continued in Norway while other non-exercise assets conducted operations against Russian targets. A Combat Sent aircraft deployed from RAF Mildenhall orbited off Murmansk on an eleven-hour mission.[142] In terms of the Russian concept of combat potential, this is an activity that would take place immediately before an operation was executed. Combined with the presence of the Cold Response forces, this would have generated uncertainty in the Russian command structure regardless of U.S. or NATO intent.

Russian NC3I movements on 24 March reflected this uncertainty. One of two Tu-214SUS special communications aircraft conducted a Tver Loop, while a Tu-214PU command post flew to Ulyanovsk and later back to Moscow. During this time a Tu-214SR communications relay aircraft moved back and forth to St. Petersburg. In another unusual move later in the day, one of two Tu-214PU-SBUS Ministry of Defence command posts departed Chkalovskiy Air Base and headed northeast away from Moscow where it shut its transponder off.[143]

On the evening of 24 March, President Biden stated that NATO would "respond in kind" if Russia used WMD in Ukraine.[144] Given that NATO nations did not possess chemical or biological stockpiles, this statement implied that nuclear weapons would be employed in such a scenario. In a tweet after the speech, Rubio remarked, "To force Ukraine into a deal he can claim as victory Putin needs battlefield momentum. The danger is that [Putin] has no economic or diplomatic cards to play, his conventional forces are stalled, & cyber, chemical, biological, & non-strategic nukes are his only escalation options."[145] The situation in Ukraine on 25 March supported that assessment.

As for American deterrent moves, social media revealed that the USAF had been quietly building up an F-35 force at Eielson AFB in Alaska opposite Siberia, and imagery of forty-two F-35As operating and supporting tankers there emerged on 25 March.[146] (The F-35A was certified for B61 nuclear weapons delivery later in 2022, and there is a large groomed and maintained Weapons Storage Area at Eielson AFB which could easily be used to store weapons flown in from, say, Fairchild AFB's nuclear storage area in an emergency.[147]) B-52s continued to conduct operations throughout the day, this time over Denmark and specifically the Skagerrak Straits. A VC-25A, Air Force One, departed Brussels and landed at a base in Poland so President Biden could meet with U.S. troops before proceeding to Warsaw for the night. While he was in transit, the Russian Ministry of Defence held an information operation spectacle. None of the senior Russian military officials were present. The gargantuan map on display was deceptive in that blobs of color indicated control over territory that was contested, especially by stay-behind forces. The spokesmen lied and asserted that there was never any intent to seize Kyiv or the other major cities, that this was a distraction from the real aim, which was "liberation" of the Donbas. This fooled nobody except the ideologically committed in the West. These were excuses so that Russian forces could withdraw from the failed northern and western directions and regroup in the south and southeast directions.[148] None of this was backed up with any significant visible strategic forces movement. To some observers this looked like an attempt to freeze the conflict in place, as the Russians had in 2015, so they could keep grinding away and try again later. Others believed that the Putin regime would try to create a divided Korean or Germany scenario.

In complete contrast, the Biden speech in Warsaw on 26 March was a clear statement of intent. Biden leaned on the "four freedoms," the basis of the Second World War effort and the basis of the postwar order as expressed in the UN Charter, as justification for American and NATO policy. The United States and its allies had a "sacred obligation" under NATO's Article 5 to defend NATO territory "with the full force of our collective power," implying nuclear weapons use if required. He asserted that American forces in Europe were there to defend NATO members and would not be committed to engage in conflict with Russian forces.

Turning to Russia, Biden made it clear he believed that "this war has already been a strategic failure for Russia." Russia complained about NATO forces on their borders, but they had 100,000 U.S. troops there now that were not there before.[149]

Dmitry Medvedev was earmarked to deliver the definitive Russian response. Echoing the rhetorical offensive three days before, Medvedev called attention to a "special document on nuclear deterrence" but elaborated that only if Russia or an ally were attacked, even with conventional weapons, Russia was "entitled to use nuclear weapons."[150] Right after Medvedev's statement and Russian Ministry of Defence spokesman Igor Konashenkov's redeployment announcement, Tu-214SUS and Il-96–300PU command and control aircraft maneuvered in the Moscow area, with another Tu-214PU on a trip to Ulyanovsk and back. The ground units involved included the 42nd Missile Division equipped with YARS mobile ICBMs and the 13th Missile Division at Dombarovskiy, controlled by the 31st Missile Army at Orenberg. The significance of the 13th Missile Division was that it was equipped with SS-18 silo-based ICBMs, with one regiment equipped with the new Avangard hypersonic glide vehicles.[151] These forces, which are the geographically farthest away in Russia from North America, were targeted against North America.

One possible engagement scenario has the hypersonic vehicles (that defy interception) go first to generate EMP, then follow that attack with ICBMs, bombers, and SLBMs. This flourish was probably a response to Biden's Warsaw speech, but it may also have been a face-saving maneuver given that this was the day Russian troops retreated from their salient north and east of Kyiv, from Chernihiv, Sumy, and Kharkiv.

None of this, however, deterred Western countries from pouring military support into Ukraine or the United States from providing a forward presence: Secretary of Defense Austin publicly announced that the USS *Harry S. Truman* Carrier Strike Group would extend its deployment in the Mediterranean.[152] Indeed, the discovery of the mass murder of civilians by Russian forces in the Bucha area increased the resolve of all pro-Ukrainian parties. Debates still lingered on providing jet aircraft and longer-ranged missiles to the Ukrainian air force, and a public debate over "offensive vs. defensive" weapons systems was in play, stimulated by the White House press secretary's comments.[153] The simplistic argument that emerged, which was not new and dated back to when the Trump administration supplied Javelin anti-tank missiles to Ukraine in 2018, asserted that deployment of "offensive weapons" to Ukraine would "escalate" the conflict and thus bring the world closer to nuclear war because Russia would then use nuclear weapons against Ukraine and those supporting her.[154] What an "offensive weapon" was remained ill-defined. However, Ukraine could not recover territory with "defensive" systems, which suggests the argument could be used to support a "frozen conflict" akin to that of 2014–22 and solidify Russian territorial gains.

Russian command and control movements dropped off after the RVSN exercises and returned to one aircraft conducting a Tver Loop and another conducting a long-range flight to the east and back to Moscow. What appeared to some to be a sinister development, one of the Tu-214PU-SBUS Ministry of Defence command posts was seen orbiting Kazan. This was likely the test of a communications or avionics upgrade.[155] U.S. command and control movements also dropped off, with the exception of the flights of six E-6B Mercuries and two E-4B Nightwatches up to and during President Zelensky's speech to the UN on 5 April, and then maintaining six E-6Bs sequentially aloft each day over the next two days.[156]

During this period the Putin regime increased pressure on the Baltic front again, with a special focus on the neutrals. Initially this took the form of cyberattacks against the Finnish Ministry of Defence immediately after a meeting with NATO.[157] Eventually, however, the sortie rate of Russian aircraft over the Baltic, and then the Black Sea, significantly increased to the point where NATO air policing forces started to reinforce with additional aircraft by 8 April.[158] This increased sortie rate was accompanied by a pair of Tver Loops in one day, including a Tu-214SUS special communications aircraft.[159] There was no corresponding U.S. activity until 11 April, when six E-6Bs were up sequentially, one conducting trailing wire deployment over the Pacific, an E-4B, and a Cobra Ball over North Dakota.[160] Exercise Northern Viking had been underway since 2 April, but this was conducted geographically between Iceland and Tromsø, Norway, with additional B-52 operations over the Skagerrak Straits. A Combat Sent aircraft was also deployed off Murmansk.[161]

It was at this point that specialist observers noticed several line items in the U.S. Department of Defense budget related to upgrading nuclear weapons storage infrastructure in Europe. In addition to the DCA bases in Belgium, Germany, Italy, the Netherlands, and Turkey, B61 storage and support areas in the United Kingdom were scheduled for refurbishment. These storage areas had been in caretaker status after the last weapons were removed in 2008. The first F-35 fighters arrived in December 2021, and these were aircraft capable of using B61-3 and -4 weapons. This was done with full knowledge that U.S. budgetary processes were heavily monitored by adversaries. As such, this was more along the lines of programmed general deterrence rather than crisis deterrence despite media coverage of it. It did, however, signal increased long-term interest in improving deterrence in the NATO area.[162]

The next day Dmitry Peskov asserted in comments that Sweden and Finland joining NATO was destabilizing, but he relied on veiled threats that lacked spine. At this point Putin was meeting Lukashenko at the future Vostochny Cosmodrome in the Russian far east and had dragged five of his available nuclear command and control aircraft with him.[163]

There were multiple moving parts and Putin had to remain in communications with them. There was the response to Sweden-Finland-NATO situation. But there were also events in Ukraine that could escalate. Ukrainian forces continued to resist in the Azovstahl complex in Mariupol. Russian ground command-ers publicly stated that chemical weapons should be used to exterminate the Azovstahl defenders, and this was rebroadcast by Ukrainian media outlets.[164] There was another purge underway in Moscow, including not just FSB officers associated with Ukraine operations but also Putin's key ideologist, Vladislav Surkov, who was accused of "embezzlement."[165] Then there was another foreign ministry assertion that Ukraine was trying to acquire nuclear weapons, followed by accusations that Ukrainian "nationalists" were about to engage in chemical terrorism.[166] Unconfirmed reports that Russia had employed chemical weapons appeared in Ukrainian social media and then in mainstream Western media.[167]

This produced a quandary for the Biden administration. The Obama admin-istration had lost international credibility by establishing "red lines" and then not standing up to chemical weapons use in Syria, which in turn gave the green light for continued use by the Assad regime. Biden had given a "red line" earlier on this matter regarding Ukraine.[168] It would take time to verify what was hap-pening in Mariupol.

That day, the guided missile submarine USS *Georgia* with its 154 Tomahawk cruise missiles surfaced off Souda Bay, Crete, and the event was tweeted by the U.S. Department of Defense to make sure observers were paying attention to the proximity of the submarine to southeastern Ukraine.[169] There was also another Combat Sent run off Murmansk.[170] As for nuclear command and control aircraft, two E-4B Nightwatches and an E-6B Mercury were up during the course of the day.

While the drama over Azovstal played out, Ukrainian forces tracked and sank the cruiser *Moskva* in the Black Sea west of occupied Crimea on 13 April. The destruction of the Black Sea flagship, which was a dual-capable vessel, was a major material and informational space loss for the Putin regime. The Russian foreign ministry decried the attack and insisted that the United States restrain weapons provision to Ukraine or there would be "unpredictable consequences for regional and international security."[171] There were also warnings by Medvedev that Russian would deploy nuclear and hypersonic weapons if Finland and Sweden joined NATO, a strange threat given what was stationed in Kaliningrad.[172] The only visible movement of Russian command and control aircraft were two flights of one of the Tu-214PU-SBUS Ministry of Defence planes to orbits northeast of Moscow, which was unusual, though the U.S. posture consisted of the baseline two command and control aircraft aloft during the day and was nothing out of the ordinary.[173]

Statements that day amounted to Director of Central Intelligence William Burns noting that "none of us can take lightly the threat posed by a potential resort to tactical nuclear weapons or low-yield weapons.... We're obviously very concerned."[174] He was contradicted by Ukrainian military intelligence chief Kyrylo Budanov, who asserted that Putin's threats were "blackmail" and that "the use of nuclear weapons is technically difficult."[175]

The next day the Russian command and control posture returned to the familiar Tver Loop by a Tu-214SUS and a Tu-214PU run to Tyumen, where it orbited between the 42nd Missile Division and the 33rd Guards Missile Army. There were unconfirmed reports that a Tu-214PS-SBUS, the same aircraft that conducted orbits northeast of Moscow, flew with transponder off to Murmansk, suggesting a visit to the Northern Fleet by either Shoigu or Gerasimov.[176] In an interview that day, Zelensky, contradicting Budanov, "warned . . . that the whole world should be prepared" for Russian chemical or nuclear weapons use.[177] The Russian Ministry of Foreign Affairs obliged on 16 April with a strange accusation: Germany, apparently, was working closely with the United States on a network of thirty biological laboratories in Ukraine, implying this was a weapons program.[178]

This could be interpreted as the Putin regime providing information space lay-down and possibly justification for nuclear weapons use against Germany, the United States, as well as Ukraine if they needed to threaten to do so. The U.S. response was to cycle three B-52Hs from their base at RAF Fairford over Romania and Poland.[179] Russia continued to apply pressure to the Baltic front: RAF and Luftwaffe Typhoons operating from Estonia conducted multiple intercepts of Russian aircraft that day.[180]

Pressurizing Poland

This was the prelude to a period of intense activity that started on 17–18 April. Why, exactly, is difficult to determine. The focus of activity shifted to Poland and to some extent Lithuania. A former U.S. ambassador to Poland noted that the Polish leadership was undertaking continuity of government measures. Polling noted that 65 percent of Poles were concerned that Russia would use nuclear weapons against Poland, and 77 percent thought they would be used against Ukraine.[181] Border issues between Belarus and Poland were aggravated again, with Poland coming under attack from Amnesty International for its alleged treatment of "refugees" brought in by the Lukashenko regime and pushed onto Polish terri-tory.[182] Social media influencers claimed that Lukashenko was going to push back using military force.[183] Indeed the Belarusian 38th Brest Independent Guards Air Assault Brigade conducted an airborne exercise near the border using Il-76 aircraft.[184] The Polish air force leadership noted that Russian A-50 AWACS aircraft were up "at all times" and that there were alert scrambles against Russian aircraft taking runs at the Polish border from Belarus three times a day.[185] Secretary to the Belarusian Security Council Alexander Volfovich asserted on TV that there was a military buildup in Poland, Lithuania, and Latvia, saying, "It could be a sign that Western countries want to start a war with Belarus," and if they did, "there will be destruction, death and explosions on their territory as well."[186] RT then conducted what amounted to "supporting fires" in the information space, asserting that Belarus was preparing to "respond to provocations from Poland and Lithuania."[187]

This burst of activity most likely came out of the Putin-Lukashenko meeting at Blagoveshchensk. One purpose was to distract NATO but more important to put pressure on Poland to back off facilitating Ukrainian resupply. There was definitely some abnormal movement in the Russian strategic command and control realm on 18 April. A Tu-214PU command post aircraft flew to Saratov from Moscow, looped around, and flew to a point between Kaluga and Ula, where it orbited. This flight permitted communications with the 60th and 28th Missile Divisions, including the Kozelsk ICBM field. There were also two Tver Loops flown, one of the two Tu-214SUS special communications aircraft and a presidential Il-96–300PU command plane. These flights were sequential over a six-hour period, which suggests an exercise of the RVSN and LRA communications systems which was meant to be visible.[188]

American and allied activity that day does not appear to have been specifically geared to respond to these events. In Gibraltar, close-up pictures of Tomahawk cruise missiles being loaded aboard the *Astute*-class submarine HMS *Audacious* circulated throughout British media, which used headlines like "Britain's most powerful submarine arrives in Gibraltar hours after Putin's threats."[189] The *Astute*-class vessels can carry up to thirty-eight weapons and were a proven capability, having fired Tomahawks against targets in Libya in 2011.[190] Finally, on the evening of 18 April, two E-4B Nightwatches scrambled from their dispersal field, with one headed for the ICBM fields in the northern United States.[191]

Questions about possible Russian nuclear weapons use in Ukraine were expressed to Lavrov on 19 May during a press conference in Moscow, but he deflected them with nonanswers.[192] Russian nuclear command and control flights included a Tu-214PU to Saratov, where it landed and then flew to Moscow, where it looped, as one did previously. A Il-96300PU presidential command post did a Tver Loop but unusually was in the air three times as long. These two flights overlapped, so once again it was some form of strategic forces communications check.[193]

Back in Washington, D.C., the U.S. Department of Defense expressed the view that "we've seen no indications that the use of nuclear weapons is in play or is imminent in any way."[194] This was reflected in the flight of one E-4B Nightwatch and two E-6B Mercuries, which represented a steady state at this point in the crisis.[195] The U.S. intelligence apparatus had to have known that preparations were underway by Russia to test the RS-28 SARMAT heavy ICBM on 20 April.

This event can be assessed in several ways. One, it can be overestimated and fall into the category of Russian *wunderwaffen* spectacles to awe both international and domestic audiences as to the power of Russian nuclear might but ultimately are irrelevant outside that context. Indeed RS-28 was a development of the Cold War–era R-36 ICBM. Some noted that this test had been scheduled for three years and had not met the timings for one reason or another.[196] Two, any launch of a ballistic missile system with intercontinental range from Plesetsk poses a danger if not a threat to North America, as it has since the 1960s. And with the advent of hypersonic glide vehicles, the reaction time of North American defense systems was greatly reduced, therefore vigilance was advised. Three, the United States had publicly forgone the scheduled Glory Trip Minuteman III launch to reduce international tension. The Putin regime was not reciprocating in kind, which itself was a signal that the Putin regime was not interested in the niceties of dialogue.

The U.S. Department of Defense made soothing noises the day of the test: the secretary was briefed "two or three times a week," the United States had "not seen any indication Russia has made any moves to prepare nuclear weapons for use during the war." Departmental sources had told media that "U.S. officials are more concerned about the threat of Russia using them than at any time since the Cold War."[197]

The Russian command and control system was in the same steady state it had been the previous two days: a Tu-214SUS conducted a Tver Loop, while a Tu-214PU flew to Saratov and back, with loops near Moscow. This posture was maintained during the SARMAT launch process.[198] The visible American command and control system moves consisted of an E-4B Nightwatch, which departed its base at Lincoln and flew to Lake Superior where it proceeded to conduct loops consistent with trailing wire antenna deployment for communicating with ballistic missile submarines. Three E-6Bs lifted off, and then these three rotated later on with three more. This occurred during the course of the SARMAT launch process. Unusually, not one but two Cobra Ball aircraft were at this point over the Bering Sea, keeping tabs on the missile and whatever vehicle(s) emerged from it.[199]

Operations on 20–21 April started off with a routine Rivet Joint flight off Murmansk to keep an ear out for the Northern Fleet, while Sergy Shoigu, clearly concerned about NATO activities in Norway, Sweden, and Finland, announced that the region would receive an influx of military and civil resources.[200] By 0900 EST, an E-4B Nightwatch was over Lake Superior conducting maneuvers consistent with trailing wire antenna operations, with a tanker on the way to replenish it. Four E-6B Mercuries were visibly aloft over California, Oklahoma, and the central United States, making headway for the coasts. At 1109 EST, a second E-4B Nightwatch lifted off out of Warner-Robins AFB, while the president's VC-25A was airborne from Andrews AFB to Wright-Patterson AFB. Fifty tankers of all types were visible. By late afternoon, two more E-6Bs and a third E-4B conducted a rotation with the other aircraft and the posture returned to its day-to-day routine. Of note, none of the aircraft call signs involved mass murder, serial killers, or radioactive isotopes.[201]

For comparative purposes, three E-4Bs and one E-6B plus forty-three tankers had been up on 11 January when the Russians launched the large field training exercise; two E-4Bs, with one over Lake Superior, and two E-6Bs during the false flag affair on 12 January; one E-4B and five E-6Bs during GROM 2022; three E-4Bs and two E-6Bs on the eve of the invasion on 22 February; one E-4B and four E-6Bs the day of the invasion and the same package the day Shoigu announced an increase in readiness and again with Lavrov's threats on 1 March. The exception was the deployment of eight visible E-6Bs and a hundred tankers when Zelensky addressed Congress on 16 March. Otherwise, the baseline was three aircraft on any one day.

There was one threat publicly expressed by a significant Russian personality, in this case ROSCOSMOS boss Dmitry Rogozin. Rogozin asserted that any tampering with Russian satellite constellations could trigger war. In a 21 April tweet, Rogozin deployed a bogus statement he attributed to British prime minister Boris Johnson that basically said that the United Kingdom could use nuclear weapons without consulting NATO.[202]

There were significant cyberattacks underway against Estonia and the NATO Cyber Defence Centre.[203] The Five Eyes intelligence sharing community also issued, depending on the sources, an "advisory"[204] or an "alert."[205] This alert made its way into the public domain in the United Kingdom through IT websites, where principles were warned that "Russian-aligned hacking groups" had "pledged allegiance to the Russian government."[206] The British version of the warning noted that "the Russian government is exploring options for potential cyberattacks against critical organizations such as the [National Health System], nuclear power stations and parts of the civil service." The ostensible reasons for this activity were payback for the global sanctions regime targeting Russia, as well as to interfere with those countries providing direct military support to Ukraine by targeting critical infrastructure.[207] If a demonstration like the one on 21 April was conducted in response to a Russian or allied cyber threat, that threat operation would have to have advanced its timeline toward execution or there was a reconnaissance pathway leading to a viable destructive outcome that was uncovered.

In what would increasingly become a staple of the conflict in the information space, Russian state TV personalities on evening talk shows that night were "giggling uncontrollably while discussing nuclear strikes against the continental territory of the United States." Others crowed that there would be "war against Europe and the world once the operations against Ukraine has [sic] been concluded."[208] Commentary like this appeared on state-controlled media, therefore it had official imprimatur. Peter Pomerantsev's commentary on the matter is straightforward: "The Kremlin . . . uses information with a military mindset to confuse, dismay, divide, and delay."[209] As outrageous as it appears to those of us in the West, whenever such commentary was deployed, it was in support of a nuclear fear information line that had both general and specific objectives.

In effect, when the Russian momentum in Ukraine stalled out, the Putin regime attempted to regain the initiative elsewhere. In this case, there was a concerted effort to pressurize Poland and NATO because their military material support to Ukraine was seen by the Putin regime as a *schwehrpunkt*. This effort included the strike at the Yavoriv training center using dual-capable strategic systems; the new wave of Belarus' migration warfare; the deployment of Belarusian airborne forces to the border; accusations of German and Polish support for the nonexistent Ukrainian biological warfare program; and the possibility of cyberattacks against nuclear power plants in NATO countries. These were all backed with constantly menacing commentary from senior regime figures and supported with information operations via Russian "media."

Two items fall out of this. First, this approach could not work without being backstopped by a credible strategic nuclear force. Second, it is clear that the Putin regime continuously sought to manufacture or generate situations that constantly permitted them the "legal" option to use nuclear weapons, thus making their use credible and therefore generating nuclear fear information lines that could be exploited in a variety of venues. In Soviet and then Russian society, the demand for accountability at all levels so authorities knew where to lay blame was so acute that absolutely everything had to be justified and documented—even mass murder. In this case, Putin had to retain the "legal" credibility to threaten credible nuclear weapons use to both his domestic and international audiences.

Conclusions

The failure of the Russian coup de main on Kyiv immediately led to posturing using nuclear forces. The global backlash against Russia, both in terms of shocked rhetoric and the various sanctions regimes, was the trigger for an increase in Russian nuclear forces readiness. This was probably done to backstop diplomatic negotiations designed to extricate the Putin regime from the situation or freeze the conflict in place as had happened in 2014–15. On another level, there was a symbiotic relationship between Russian nuclear movements and verbalizations in the information space in order to compel the international community to acquiesce to the Putin regime's objectives. These moves were met with discrete movements of the American nuclear command and control apparatus, which backstopped a dramatically increased conventional arms flow by NATO members into Ukraine.

When the Putin regime needed to regain momentum, it pressurized the Baltic neutrals with a combination of conventional forces and nuclear force intimidation in a failed attempt to bully them off the NATO membership path. This was counter-productive and eventually led to Finland joining NATO with the co-commitment effect of expanding Russia's geographic frontage with NATO, the reduction of which allegedly was the justification for invading Ukraine in the first place.

There was a constant drumbeat of nuclear threats emanating from the Putin regime. When combined with the increased Russian readiness levels, this could not go unanswered, and as a result there were a plethora of moves made by American strategic nuclear forces and propelled into the information space via social media and other methods. Russian threats, however, did have an effect on deterring what amounted to "extended conventional deterrence" over Ukraine when it came to a no-fly zone and the provision of combat aircraft.

With Russian forces bogged down, the Putin regime once again planned to use false flag attacks involving chemical weapons or interference with nuclear power stations with the objectives of pressurizing NATO members to stop providing military support, or possibly even to lay the groundwork for limited battlefield nuclear or chemical weapons use to restore momentum on the battlefield. There were direct responses by the president of the United States to these moves and these were supported by the manipulations of elements of strategic nuclear forces.

The Putin regime then shifted focus onto Poland using hybrid methodology, but by this time France had had enough and signaled using its strategic nuclear forces. At the same time, NATO forces exercising in Norway generated combat potential that could be used against the Kola Peninsula, something Russia could only counter with nuclear weapons use as the conventional ground forces had been destroyed in Ukraine. This offset the pressure on the Baltic neutrals. The Putin regime, taking all of this into account, then announced a change in its objectives in Ukraine and withdrew from the northern part of the country while maintaining operations to the east and south. This was accompanied by Russian strategic nuclear force movements.

Frustrated, the Putin regime applied pressure to the Baltic front, threatening Sweden and Finland with nuclear weapon deployments, and in the Black Sea using airpower against Romania to heighten tension. Then it was Poland again, with Lukashenko's Belarus in the lead. In all cases, the United States provided assurance signaling that also backstopped public statements and conventional efforts in both the Baltic and the Black Sea.

Russian nuclear posturing did not stop aerial intelligence collection in support of Ukraine by NATO and allied members. It did not stop support from Poland that was absolutely vital to the Ukrainian effort. The Russian version of deterrence had its limits. The United States, on the other hand, was able to extend deterrence to the Baltic neutrals while ensuring deterrent coverage for the frontline NATO countries and sustaining the Ukrainian military aid effort. More important, the Putin regime did not employ chemical weapons as part of a false flag operation, even though it planned to do so. It did not interfere with nuclear power plants, and there is some evidence that it was preparing to do so. And it did not use chemical or battlefield nuclear weapons against Ukrainian military units, even though it hinted that it might. If the United States, France, the United Kingdom, and NATO could not provide classical "extended deterrence," they could provide some form of "virtual deterrence."

CHAPTER 7

◆ ◆ ◆

Ramp-Up

May–September 2022

As winter gave way to spring, and spring turned into summer, the situation in Ukraine stabilized. The remnants of Russian forces from the Kyiv, Chernahiv, and Konotop fronts pulled back to refit and regroup before redeployment to the southern fronts. Ukrainian forces continued to rearm as weapons and logistics continued to pour in through Poland. The situation on the Kherson front was far from static, with constant pressure applied to all Russian forces that were north of the Dnipro River. Ukrainian forces held in the Zaporizhzhia direction, all the way along the Donetsk direction to Kharkiv. The Azovstal defenders continued their Stalingrad-like resistance in Mariupol, generating more Russian casualties and drawing in increasingly scarce resources.

The Putin regime maintained pressure on the strategic nuclear front, including threats over nuclear weapons use, but even though this appeared to slack off during the summer, it never really fully dissipated. Russia attempted to maintain strategic momentum by reapplying pressure on NATO's northern flank and on the Baltic front with a variety of activities designed to disrupt and distract. The summer was also a period of strategic nuclear force demonstrations, many designed to support general deterrence, with others directly applicable to the ongoing crisis. The problem of food scarcity driven by Russia's war on Ukraine also came to the fore, with an emphasis on activities in the Black Sea. In August,

the Putin regime had determined new angles of attack, but by the time of the
Ukrainian Kharkhiv offensive in September and October, chemical weapons
provocations and nuclear threats were back on the table.

Into Spring

On 21 April, an Il-96–300PU presidential command post conducted a Tver Loop;
Tu-214PU command post conducted a run to St. Petersburg and back, while a
Tu-214SR communications relay aircraft flew out to Kazan, orbited, and returned
to Moscow. In essence, there was nothing dramatic about this series of flights
beyond the baseline RVSN communications checks.[1] When queried in a Pentagon
press briefing, John Kirby stated, "We have no reason, at this point, to change
our own nuclear posture."[2] There was a single E-6B aloft that day.[3]

The Putin regime ramped up an information operation on 23 April, probably
to enhance its position during ongoing negotiations. This messaging asserted that
Poland had a biological weapons program (because Poland had one in the 1920s)
and was working with their "accomplice," Germany.[4] Tracked back, almost all
information lines went to a "briefing" by General Igor Kirilov (later assassinated
in December 2024), who commanded the Russian forces associated with bio-
chemical defense. He asserted that "the United States is preparing provocations
to accuse Russia" of using chemical and biological weapons.[5] The scenario was
that Azovstal defenders in Mariupol were going to use WMD, apparently against
themselves, which "would give NATO the excuse to attack Russia" with WMD.[6]
Furthermore, Azerbaijan, Georgia, and Moldova were said to have partnered
with Ukraine in this operation, and other possible target areas included Poland,
Belarus, and Moldova.[7][8]

Starting on 26 April and lasting for four days, there was a significant increase
in Russian flights testing defensive responses in NATO-controlled airspace over
the Baltic and the Black Sea. There were no flight plans filed, transponders were
turned off, and the crews refused to converse with air traffic control.[9] The visible
Russian command and control plot remained similar, but the aircraft types shifted
to Il-96–300PU presidential command posts instead of Tu-214PUs.[10]

On 27 April, Putin bent reality by implying Russia intervened in Ukraine to
prevent a larger conflict from unfolding. Ukraine was going to mount an assault
on Donbas and the Russian attack preempted it. Specifically, the "Kiev regime's"

nuclear weapons program and "the deployment of a network of Western biolabs on Ukrainian territory . . . has confirmed that our reaction to those cynical plans was correct and timely."[11]

There was yet another slight shift in the visible Russian command and control posture: in this case, a Tu-214PU-SBUS Ministry of Defence command post conducted a run to Omsk and back, while an Il-96–300PU took Putin to Sochi, and a Tu-214PU command post went back to Moscow.[12] NATO intercepts over the Baltic and the Black Sea remained at increased levels. Putin's remarks were dismissed as "irresponsible rhetoric" by Secretary of Defense Lloyd Austin.[13] There was no change to the U.S. visible nuclear command and control posture.

The next day, Baltic and Black Sea intercepts remained at elevated levels, but now there was suddenly discussion in social media about a possible planned Russian intervention in Moldova using amphibious forces. Footage of Topol-M TELs in the Moscow area also appeared, while Margarita Simonyan at RT asserted that Russia had two options: defeat in Ukraine or World War III. She predicted that Putin would use nuclear weapons.[14] There was, however, nothing unusual regarding the visible Russian command and control structure.

On 29 April, things changed. Events that day included Norway's decision to close its border to Russian ships and vehicles; the Swedish interception of Russian aircraft on the edge of Swedish airspace; and ridiculous Russian reports that a retired three-star Canadian general, Trevor Cadieu, was trapped in the Azovstal siege.[15] None of this explains the sudden deployment of four E-6B Mercuries, an E-4B Nightwatch, and a hundred visible tanker aircraft over the United States that day. None appeared to be conducting specialized activities, like ALCS or trailing wire antenna operations.[16] This was likely related to the launch of a Russian Angara-1.2 rocket from Plesetsk carrying a military payload. The launch inclination was directed over North America instead of to the southeast of Plesetsk.[17] Lavrov asserted that "a nuclear war must never be launched as there could be no winners."[18] In a departure from the past three days, Russian command and control aircraft altered their patterns. An Il-96–300PU and a Tu-214PU deployed to Sochi, while a Tu-214SR conducted something similar to a Tver Loop but with an additional flight loop northwest of the usual location. And finally, a Tu-214PU command post aircraft flew into and landed in Crimea, something that had not been seen before.[19]

This all coincided with the launch of a Russian satellite.[20] The flight path of this Plesetsk-launched system took it over Alaska and northwestern Canada with obvious implications for NORAD. This doesn't explain the significant increase in U.S. posture on 30 April. This consisted of five E-6B Mercuries aloft, one E-4B Nightwatch, and more than ninety visible tankers. This time one E-6B was over Montana testing its ALCS, and another over the Gulf of Mexico using its trailing wire antenna.[21] When queried several times during a press conference, Pentagon spokesman John Kirby reiterated previous points that nuclear rhetoric from the Putin regime was irresponsible.[22]

Admiral Charles Richard's testimony on U.S. nuclear readiness on 4 May was eye-opening. Richard was blunt: "We are facing crisis deterrence dynamics right now that we have only seen a few times in our nation's history.... The nation and our allies have not faced a crisis like Russia's invasion of Ukraine in over 30 years. President Putin simultaneously invaded a sovereign nation while using thinly veiled nuclear threats to deter U.S. and NATO intervention.... My ability to maintain strategic deterrence is limited." Specifically, Richard said, "The war in Ukraine and China's nuclear trajectory, their strategic breakout demonstrates that we have a deterrence and assurance gap against the threat of limited nuclear employment."[23]

The bulk of visible military moves in late April–early May outside of Ukraine were in and around Norway, with a visit of Tomahawk-capable submarine HMS *Ambush*; the USS *Kearsarge* amphibious group with its stealthy strike aircraft; a division-sized Polish exercise opposite Belarus and Kaliningrad; and ongoing air intercepts over the Baltic.[24] On 4 May, Russian nuclear forces in Kaliningrad, the Iskander units, conducted a snap electronic launch exercise and deployed the TELs to "avoid a possible retaliatory strike." Russian activity was not accompanied by a visibly significant shift in its command and control posture.[25] The United Kingdom responded by publicly stating that it would assist Sweden and Finland regardless of their NATO status.[26] In the United States, the Order66 E-4B was aloft, along with five E-6Bs rotating during the course of the day that suggests the Kaliningrad exercise was signaling and this was the response.[27]

Despite the lack of significant visible U.S. nuclear posture changes since 30 April, the Putin regime suddenly ramped up its nuclear rhetoric on 5 May. Russian ambassador to the United States Anatoly Antonov issued provocative statements to *Newsweek*, quoting extensively from Russian declaratory nuclear policy. He characterized the situation "as one of the most dangerous moments since the Cuban Missile Crisis." NATO politicians "clearly [do] not take the nuclear threat seriously." He emphasized that the use of any WMD against Russia and its allies would be grounds for retaliation.[28] So were the DPR and LPR "allies"? Part of Russia? What about the newly occupied Ukrainian oblasts? Was a conventional assault against them a threat to "the very existence of the state"? *Newsweek* released the interview on 5 May. The same day a Ministry of Defence command post Tu-214PU-SBUS flew to Novosibirsk and back, while an Il-96–300PU presidential command post conducted a Tver Loop, and another repositioned itself in the Moscow area. A third Il-96–300PU also moved to a dispersed location.[29]

The next day, the Russian operational commander in Mariupol made claims that Ukraine was preparing to use chemical weapons in his area of operations. These statements made their way into American media, which immediately interpreted that this was a Russian attempt at false flag activity.[30] However, testimony later that week in the Senate Select Committee on Intelligence amped up the fear factor in the public domain. DNI Avril Haines noted that Putin could not afford to lose in Ukraine but might change his definition of victory if his nuclear saber rattling succeeded in deterring the West from further support of Ukraine.[31]

Haines compared the situation to posture changes seen in 2014 over Crimea and 2016 regarding Syria.[32] Secretary of Defense Austin "had consistently requested for weeks to no avail" communications with his opposite number, Sergy Shoigu. This eventually occurred on 13 May, possibly due to concurrent events.[33] The first was the buildup and movements of U.S. forces on the northern and Baltic flanks. Part of a USAF A-10 squadron deployed to Andoya Air Base in Norway on 6 May. U.S. Army forces deployed to Finland on Exercise Arrow 22 with a mechanized formation, while the USS *Kearsarge* amphibious group moved into the Baltic Sea. On 7 May, a battalion from the 4th BCT flew into Bardufoss, Norway on eight C-17s directly from Alaska over the Arctic. U.S. Special Operations Forces then conducted a landing on an expedient Latvian airstrip (a highway) during Trojan Footprint on 12 May, followed by nine A-10s to Latvia on an ACE operation from the United States.[34]

Another event appears to have been an unscheduled USSTRATCOM exercise on 11 May that involved two E-4B Nightwatches and five E-6B Mercuries, with three of the E-6Bs and an E-4B operating simultaneously over water with their trailing wire antennas. Unusually, a number of B-52Hs were visible over the United States.[35] Finally, on 13 May, a C-17 was tracked over the Atlantic to RAF Lakenheath, which hosts an interim nuclear weapons storage facility and where security was observably tightened, and then on to Volkel Air Base in the Netherlands. Volkel hosts NATO dual-capable aircraft and American B61 nuclear weapons.[36] This last move, however, was more likely a scheduled event. It was not accompanied by any information operation support. There were four E-6Bs and an E-4B up.[37]

Visible Russian command and control activity during this period consisted of combinations of Tu-214PUs and Il-96–300PUs moving back and forth from Moscow to Sochi, with a Tu-214SUS on strip alert at Sochi. Another Tu-214PU flew an unusual loop near St. Petersburg. A second Tu-214SUS conducted a Tver Loop. On the day of the phone call, an Il-96–300PU conducted a Tver Loop, while a Tu-214SR communications relay aircraft flew out to Krasnoyarsk and back to Moscow. The Interior Ministry Tu-214VPU conducted communications checks in the Moscow area and then the St. Petersburg area. Similar movements took place on 15 May as well. In other words, these were important, and in some cases uncharacteristic, movements indicating significant strategic nuclear forces communications checks.[38] Russian Ministry of Defence also announced that the SSBN *Knyaz Oleg* used a torpedo to break through the ice at a location in the Arctic.[39]

All of this activity was followed by a phone call between General Milley and Gerasimov. Visible U.S. activity at that point consisted of a one E-4B Nightwatch and two E-6Bs over the United States, while the Russians deployed two Il-96–300PUs on Tver Loops, a Tu-214PU-SBUS on a run to Novosibirsk and back, and a Tu-214PU at Sochi. According to a Pentagon spokesman, the call "didn't specifically solve any acute issues or lead to a direct change in what the Russians are doing or what they are saying."[40]

Summer: June–July

Russia focused on the northern and Baltic flanks with the same objectives: bully Baltic countries, continue to pressure Poland and stop Ukrainian resupply, and pressurize NATO. Defender Europe 22 was a U.S. Army exercise held in late May involving the deployment of artillery systems from prepositioned stocks in

Germany to exercises held in Latvia, Lithuania, and Slovakia.[41] During the course of Defender Europe 22, HIMARS battlefield missile systems were deployed on Bornholm island, within range of Kaliningrad and its Iskanders. At the same time, discussions were underway about the provision of HIMARS systems to Ukraine, which would provide its forces with deep-fire capabilities and significantly alter battlefield dynamics.

The Putin regime mounted an information operation on 1 June to deter the provision of HIMARS to Ukraine, using Bornholm as a mechanism. Russian diplomats in Vienna protested the HIMARS deployment. This was proliferated by TASS, which leaned on the word "escalation" in social media, while the 54th Missile Division with its mobile ICBMs were publicly exercised.[42] Medvedev followed up bombastically on 3 June just to make sure the message was clear: those countries providing such weapons "could trigger retaliation" and "the Horsemen of the Apocalypse are already on their way and all hope is with Lord God the Almighty."[43] This line was amplified by Vladimir Solovyov on Rossiya1; he also shifted into religious apocalyptic language.[44]

The Belarusian 38th Air Assault Brigade conducted an exercise on 4 June, the day NATO exercise BALTOPS 22 kicked off. This was part of a larger Belarusian mobilization exercise that started on 6 June, the motive of which was to draw Ukrainian resources away from the fighting in the southwest while at the same time generating uncertainty in Poland and Lithuania.[45] That day, the Russian Ministry of Defence announced that a second regiment of Avangard hypersonic missiles had achieved "combat duty" with the RVSN. This was amplified by TASS for the international audience.[46] Tu-214PU and Il-96–300PU aircraft both conducted Tver Loops that day, which was unusual.[47]

In addition to linking two nuclear force events to events on the Baltic front, there was intense Russian activity during BALTOPS 22, including buzzing the USS *Kearsarge*. A pair of Russian corvettes also violated Danish territorial waters near Bornholm.[48] In turn, Lithuania decided to close the railway corridor to Kaliningrad as part of the EU sanctions regime. Nikolai Patrushev flew into the enclave personally to deliver a threat: these were "hostile actions" and Russia "would respond with unspecified measures that will have a significant negative impact of the population in Lithuania."[49]

This threat was accompanied by three days of unusual Russian command and control activity in what can only be described as a major unscheduled exercise of the whole RVSN communications system. A Tu-214PU-SBUS disappeared into the Siberian Black Hole, while the other one looped back and forth between Moscow and Kazan. Il-96–300PU presidential command posts and Tu-214SUS communications planes conducted repeated Tver Loops, including two on one day in conjunction with the Interior Ministry's Tu-214VPU repeatedly orbiting Podelsk between the Chekhov and Sharapovo bunker complexes.[50]

The day after this exercise, Lukashenko publicly asked Putin to reequip Belarusian aircraft to carry nuclear weapons and transfer nuclear-capable Iskander-M missiles to Belarusian control. He asserted that the closure of the corridor to Kaliningrad "was akin to declaring some kind of war." Putin agreed that "this would be done."[51]

American moves during this period were initially low key. There was a series of Trident II D5LE launches from the ballistic missile submarine USS *Kentucky*, with three shots on 15 June and another on 17 June. These were billed as Commander Evaluation Tests of life-extended SLBMs to make sure they worked, and USSTRATCOM insisted they were not related to ongoing international events which suggests general deterrence signaling. The missiles were launched off the U.S. West Coast toward Kwajalein Atoll impact area, but it was unclear in the public domain if they involved dummy MIRVs or not.[52] The U.S. Air Force also released information on five successful tests of a B61–12 gravity bomb from B-2A bombers, noting that they incorporated the Radar Aided Targeting System, which assisted in giving the weapon a precision capability.[53] The only visible move of nuclear forces was the day after Patrushev's threat, when B-52Hs from Barksdale AFB deployed on an ACE exercise to the former SAC Chennault AFB at Lake Charles, Louisiana. This included the 2nd Munitions Squadron, the nuclear weapons handling unit.[54]

In Europe, the U.S. announced a significant force buildup on 29 June. This included establishing the Cold War–era V Corps headquarters in Poland; a mechanized brigade in Romania; multiple rotations of armored, aviation, and Special Operations Forces (SOF) in the Baltic countries; and two F-35 squadrons to be based in the United Kingdom, presumably at RAF Lakenheath with its nuclear storage capability.[55]

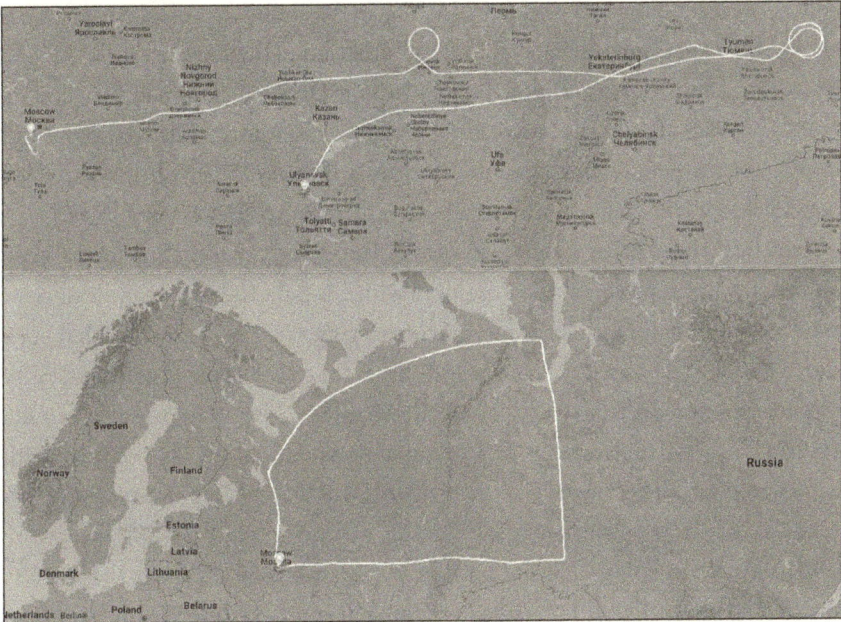

PHOTO 18 ♦ Flight tracks of Russian nuclear command and control aircraft during an unscheduled nuclear forces exercise, June 2022. The Tu-214SR is communicating with the 14th Missile Division and the 33rd Guards Missile Army. The Tu-214PU-SBUS is flying a route that permits communications with ballistic missile submarines in the Arctic. *flightradar24*

In return, the Putin regime ramped up intimidation against NATO's northern flank. As part of the ongoing sanctions regime against Russia, Norway refused transit of Russian materiel to Svalbard. The issue was raised at the diplomatic level by Russia, complete with a threat of unspecified "retaliatory measures."[56] The threat was accompanied by the unusual movements of one of two Ministry of Defence Tu-214PS-SBUS command planes in the Arctic Basin.[57] Additionally, there was a massive wave of cyberattacks directed at Norwegian internet facilities.[58] It is unclear if there were other military movements, but the next day U.S. Naval Forces Europe tweeted that the SSBN USS *Rhode Island* was visiting Faslane, Scotland.[59] This visit was followed by the arrival of an unnamed *Ohio*-class SSGN and a *Virginia*-class attack submarine, also at Faslane on 6

July.[60] The messaging here was that NATO had the ability to project power across the conventional-nuclear spectrum if the situation with Norway escalated. The Russian response was to deploy the 39th Missile Division equipped with mobile YARS missiles around the Novosibirsk area on an exercise, while command and control aircraft conducted a communications check with that unit. Medvedev was also deployed to make statements about Russia's nuclear capabilities.[61]

Much had been made about the *Belgorod,* a modified *Oscar*-class guided missile submarine equipped with the Poseidon, which is believed to be a large nuclear powered autonomous underwater vehicle carrying a nuclear warhead. It has been described as a response to U.S. withdrawal from the ABM Treaty, with the ability to generate nuclear tidal waves.[62] A U.S. Air Force ARRW hypersonic missile test on 8 July failed, the latest in a string of booster failures. That day, Russian information outlets crowed that the *Belgorod* was declared operational and this made its way into Western media.[63]

That Russian euphoria did not last long. A pair of ARRW tests were successfully conducted on 12 July, with the boosters going beyond Mach 5.[64] The Russian foreign ministry lashed back that day, asserting that the United States was "provoking escalation" through Ukraine and "unleashing a violent hybrid confrontation with Russia." Production of the RS-28 SARMAT was authorized and the information was released by a jubilant Dmitry Rogozin.[65] Russian command and control aircraft were maneuvered, notably a Tu-214SR relay aircraft that orbited near the 60th Missile Division with its Topol-M silos at Tatishchevo and the 42nd Missile Division near Nizhniy Tagil with its mobile YARS missiles.[66] The 35th Missile Division at Barnaul was exercised soon after on 15 July.[67]

Russia continued its operations on the northern flank. NATO reported that a Royal Norwegian Air Force P-3C Orion and a DA-20 Falcon intercepted Russian Su-33 Flanker and MiG-29K Fulcrum fighters in the "High North."[68] This was followed by a pair of Tu-160 Blackjacks escorted by MiG-31 Foxhounds that appeared over the Barents.[69] These had not been seen in some months and their appearance underscored increased Russian interest in the region.

Russian activities appeared to have no effect on Norway's sanctions regime and the Putin regime was unwilling to escalate further. Indeed, Russian capabilities to execute certain types of operations in the region decreased with the destruction in detail of the Northern Fleet's sabotage-reconnaissance unit in Ukraine in

July.[70] Similarly, the Putin regime's behavior did not deter Sweden and Finland from signing accession protocols to join NATO in July 2022. On the Black Sea, Russia was unable or unwilling to establish a blockade of Ukrainian trade in the face of international pressure related to global grain supply and world hunger. The Black Sea Grain Initiative was signed on 22 July, with Russia lobbing a Kalibr cruise missile at the Port of Odesa the next day.[71] Again, the Putin regime was unable to generate compellence in these three areas despite its signaling efforts, both conventional and nuclear.

Summer: August

A meeting of the Non-Proliferation Treaty review conference in early August provided a platform for nuclear-capable states to air their positions. The United States, the United Kingdom, and France released a joint statement condemning Russia's use of its nuclear forces for "military coercion, intimidation, and blackmail." Putin himself denied this, and his representatives fell back on Russian declaratory policy and strategy, emphasizing the lines about using nuclear weapons to protect sovereignty while leaving it undefined.[72] The French elaborated on their declaratory policy, specifically that the *force de dissuasion* was "aimed at permanently guaranteeing the country's decision-making autonomy and freedom of action . . . including against any attempt at blackmail that may be made in the event of a crisis."[73]

During the conference, there was no letup in aerial activity over the Baltic. There were three NATO scrambles against Russian aircraft in one day, with subsequent interceptions daily to 4 August.[74] The U.S. secretary of defense delayed a Minuteman III Glory Trip operation the day the conference ended, ostensibly due to Nancy Pelosi's visit to Taiwan and associated tensions with the Beijing regime.[75]

The RVSN held an exercise on 8 August. The visible aspects of it consisted of a Tu-214SUS conducting a Novosibirsk run and an Il-96–300PU conducting a Tver Loop. However, a Ministry of Defence Tu-214PU-SBUS command aircraft took an unusual route into the Arctic, which suggests communications tests with ballistic missile submarines. Russian sources claimed an RC-135 Rivet Joint was also present near Novaya Zemlya shadowing the Tu-214PU-SBUS.[76] U.S. NC3I activity consisted of four E-6Bs and an E-4B throughout the day, with a trailing wire antenna deployment over the Pacific. B-52s lifted off and headed for the Canadian border near Michigan, and KC-135 tankers were positioned off each coast.[77]

What appears to have been another RVSN exercise or communications check was run on 14 August. An Il-96–300PU departed Moscow and then appeared northeast of Irkutsk near Belaya air base, the 29th Missile Division and the Irkutsk-45 nuclear storage depot. It proceeded to Krasnoyarsk, near the Gladkaya missile field and the Krasnoyarsk-26 nuclear weapons depot. It then returned to Tyumen, Kirov, and back to Moscow.[78] An RAF Rivet Joint operating off the Kola Peninsula was challenged by Russian MiG-31 interceptors and allegedly forced to leave the area.[79] U.S. movements also included a protracted E-6B flight that did a communications check with ALCS over Montana, recovering to Fairchild AFB.

Shoigu made a public statement on 16 August and dismissed any claims about possible chemical weapons use as "absurd" and part of Western hybrid warfare against Russia. Any claims that Russia was preparing for nuclear weapons use were "provocations" and emphasized Russian nuclear weapons existed to deter attack.[80] Simultaneously, there was significant command and control activity on the same scale as 14 August involving Tu-214SR communications aircraft, Tu-214SUS and Tu-214PU-SBUS command aircraft, as well as an Il-96–300PU on a Tver Loop, essentially the footprint of a communications exercise. This was repeated with Tu-214SUS and Tu-214PU-SBUS aircraft the next day.[81]

American activity during this period consisted of an ACE exercise with B-52Hs to the former Loring AFB in Maine, the first time B-52s had been there in twenty-nine years. The delayed Minuteman III launch from Vandenberg AFB took place on 17 August. The next day, with commander USSTRATCOM present, a BTF of four B-52Hs departed Minot AFB, flew "above the Arctic Circle," exercised with Norwegian and Swedish fighters, and landed at RAF Fairford.[82]

The arrival of the BTF coincided with a flight of three MiG-31Ks carrying Kh-47 Kinzhal missiles that violated Finnish airspace. The BTF B-52Hs then flew to the Swedish Arctic region where they delivered conventional precision guided munitions "400 miles from the Russian border."[83] In a move not seen for some time, two Tu-95MS Bears escorted by Su-30SM fighters were announced by the Russian Ministry of Defence as flying over neutral, international waters of the Sea of Japan. The presence of a U.S. Air Force tanker flying around St. Lawrence

Island suggests that there was additional Russian activity near Alaska as well as NORAD surveillance.[84] In what appears to be a supporting information effort, three E-6Bs were aloft at once, were replaced with four others, one conducting a trailing wire antenna deployment over the Pacific, and those four were replaced with three more E-6Bs. The possibility exists that these flights were a readiness exercise while the Beijing regime conducted a launch of "technology demonstration" space vehicles.[85] A similar deployment took place on 25 August when the Beijing regime launched a "reusable Suborbital Carrier." Four E-6Bs were aloft sequentially, followed by three E-6Bs with two of them conducting trailing wire antenna operations over the Atlantic and Pacific at the same time.[86]

The USAF continued to emphasize readiness with its information activities. Barksdale-based B-52Hs conducted an ACE deployment to Fairchild AFB, which was unusual because it was an active base instead of an abandoned SAC one.[87] Another ACE, Patriot Fury, was conducted over a six-day period starting 26 August. This involved the dispersal of several C-130J Super Hercules from Dyess AFB, presumably in support of dispersed B-1B operations in the continental United States. The 171st Aerial Refueling Wing conducted a Nuclear Operational Readiness Inspection with eight KC-135 tankers on 27 August.[88] As part of its information activities, the USAF also released imagery of a nuclear weapons movements exercise involving the 5th Bomb Wing's 705th Munitions Squadron at Minot AFB on 30 August.[89]

The *Marshal Ustinov* is a Russian *Slava*-class cruiser, the sister ship of the ill-fated *Moskva*. It is equipped with sixteen P-500 Bazalt (SS-N-12 Sandbox) missiles capable of carrying conventional or 350 kt-yield nuclear warheads.[90] The *Ustinov* deployed to the Mediterranean in early February, where it conducted operations with its sister ship *Varyag* shadowing the USS *Harry S. Truman* carrier strike force. On 24 August, it passed through the Straits of Gibraltar into the Atlantic with the *Udaloy*-class antisubmarine destroyer *Vice Admiral Kulakov*, having been replaced with a *Yasen*-class guided missile submarine.[91]

Six days later the *Ustinov*, the *Kulakov*, and a Russian tanker (the *Vyazma*), shadowed by Portuguese forces, proceeded to the Irish Sea and entered the Irish Exclusive Economic Zone but not Irish territorial waters. Speculation emerged that a Russian submarine was accompanying the surface task force. The Irish Defense Forces, possessing no ability to engage the ships, could only monitor the situation.[92] This was not merely the case of the Putin regime bullying another neutral, however. The P-500 range is five hundred kilometers, and there is no visual method available to determine whether the sixteen missile cannisters carry conventional or nuclear rounds. This could only be determined from observation of the loading procedure or by sensor systems that can detect the presence of special nuclear material. From this position in the Irish Sea, the *Ustinov* had the ability to strike NATO headquarters in Brussels, the French SSBN facilities, or anywhere in the United Kingdom. The potential target of most concern, however, was the United Kingdom's only ballistic missile submarine base at Faslane.

As of January 2022, one British *Vanguard*-class ballistic missile submarine was undergoing a multiyear refit, leaving three at sea. At least two were kept on patrol until May. In June another *Vanguard* came out of refit and began reacceptance trials. Its replacement did not enter refit until 2023. In late August 2022, at least one SSBN was at sea on patrol. Once the *Ustinov* approached the British Isles, one *Vanguard* departed Faslane on 26 August. It was followed by another on 31 August.[93] One of the *Vanguards* returned to Faslane on 5 September after the Russian ships departed, and it was replaced with another on 6 September. As with the French, the precrisis baseline deployment of a British SSBN was one on patrol, one in refit, and two preparing for patrol. Two or three SSBNs on patrol clearly suggests increased concern over the international situation.

On 30 August the Royal Navy frigate HMS *Lancaster* arrived on station and shadowed the Russian task group, supported by a U.S. Navy P-8A Poseidon out of Iceland. The frigates HMS *Westminster* and HMS *Richmond* moved in later. The probability that a Royal Navy submarine was also in the area is high. The next day, the *Ustinov* task group turned to the southwest and headed through the English Channel, where it was escorted by the French Navy.[94] Russian domestic coverage of the *Ustinov* affair emphasized Russian freedom of movement to contrast the stalled-out situation in Ukraine.[95] There was no visible significant shift in Russian command and control aircraft movement patterns.[96]

Into the Fall: September

The key event driving nuclear force maneuvering in September was the Kharkiv offensive. After shaping operations near Kherson started on 29 August, Ukraine unleashed a counteroffensive east of Kharkiv on 5–6 September. Two days later,

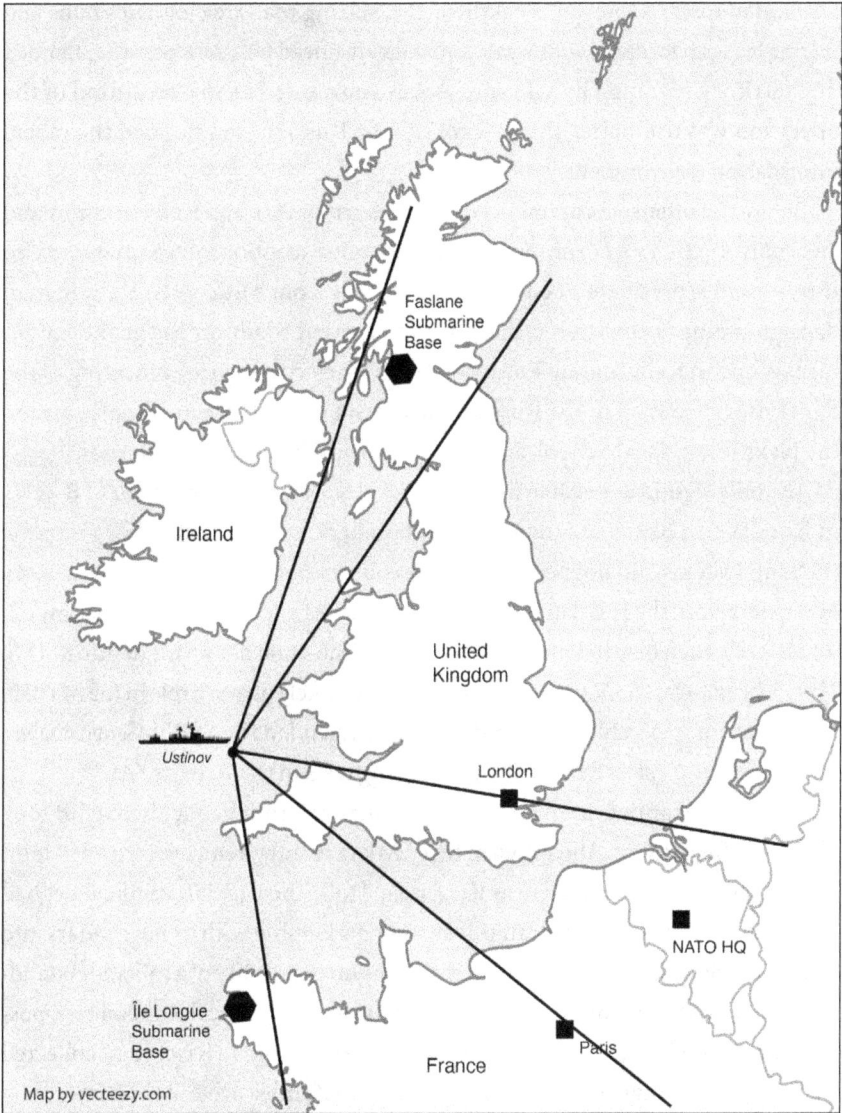

Map labels (within image): Faslane Submarine Base, Ireland, United Kingdom, Ustinov, London, NATO HQ, Île Longue Submarine Base, France, Paris, Map by vecteezy.com

MAP 6 ◆ Range gates for P-500 missiles on the Russian cruiser *Marshal Ustinov* while positioned in the Irish Sea in late August 2022. This demonstrated Russian combat potential against British and French ballistic missile submarine bases. *Created by the author*

Ukrainian forces achieved breakthroughs, seizing rear-area logistics hubs and forcing Russian forces to withdraw from their advanced positions on 10 September. The territory regained by Ukraine was as substantial as the execution of the operation was dramatic. The perception of a Russian rout flooded the global information environment.[97]

Before the offensive commenced, the patterns of U.S. and Russian command and control activity were unremarkable with one exception: on 1 September two Il-96–300PUs presidential command posts flew from Moscow to Kaliningrad, departing nine hours later.[98] These aircraft carried Vladimir Putin and entourage. While in Kaliningrad, Putin chaired (via teleconference) a meeting of the Supervisory Council of the Russian Movement of Children and Youth, visited the Nakhimov Naval School, and conferred with the enclave's governor.[99]

The only significant visible movement of U.S. bombers was a flight of B-52Hs to Estonia on 2 September in which the bombers remained off public air traffic tracking systems and popped up over the country before orbiting about seventy kilometers from the Russian border (seven hundred kilometers from Moscow).[100] The B-52Hs then flew to Poland and landed.[101] The same day an unidentified U.S. Navy *Virginia*-class SSN arrived in Faslane, which likely played a role in the *Marshal Ustinov* event.[102] Visible Russian movements included the SSBN *Generalissimo Suvorov* Borei-A–class submarine undergoing tests in the Barents Sea.[103]

A previously announced test of a Minuteman III from Vandenberg AFB took place on 7 September. Aboard were three Mk 12 reentry vehicles with W78 Joint Test Assemblies that landed near Kwajalein Atoll. The Russian Pacific Fleet had deployed the *Marshal Krylov* range instrumentation ship with its huge radars into the Bering Sea three days before, so there was no possibility of a misunderstanding.[104] The possibility of some kind of unrealized symmetrical Russian test, possibly using the *Generalissimo Suvorov*, must have existed, as a Rivet Joint collected not only off Murmansk but also toward Novaya Zemlya around this time.[105]

This does not explain the significant command and control movements on 8 September, which included a Tu-214PU to Novosibirsk, two Il-96–300PUs to Vladivostok, one Tu-214SR in the Siberian Black Hole, and another Tu-214SR to the Far East and back.[106] This package coincided with a Putin visit to the Far East, suggesting that Putin was unaware of the ramifications of the Ukrainian counteroffensive forty-eight to seventy-two hours after it was launched. It is clear,

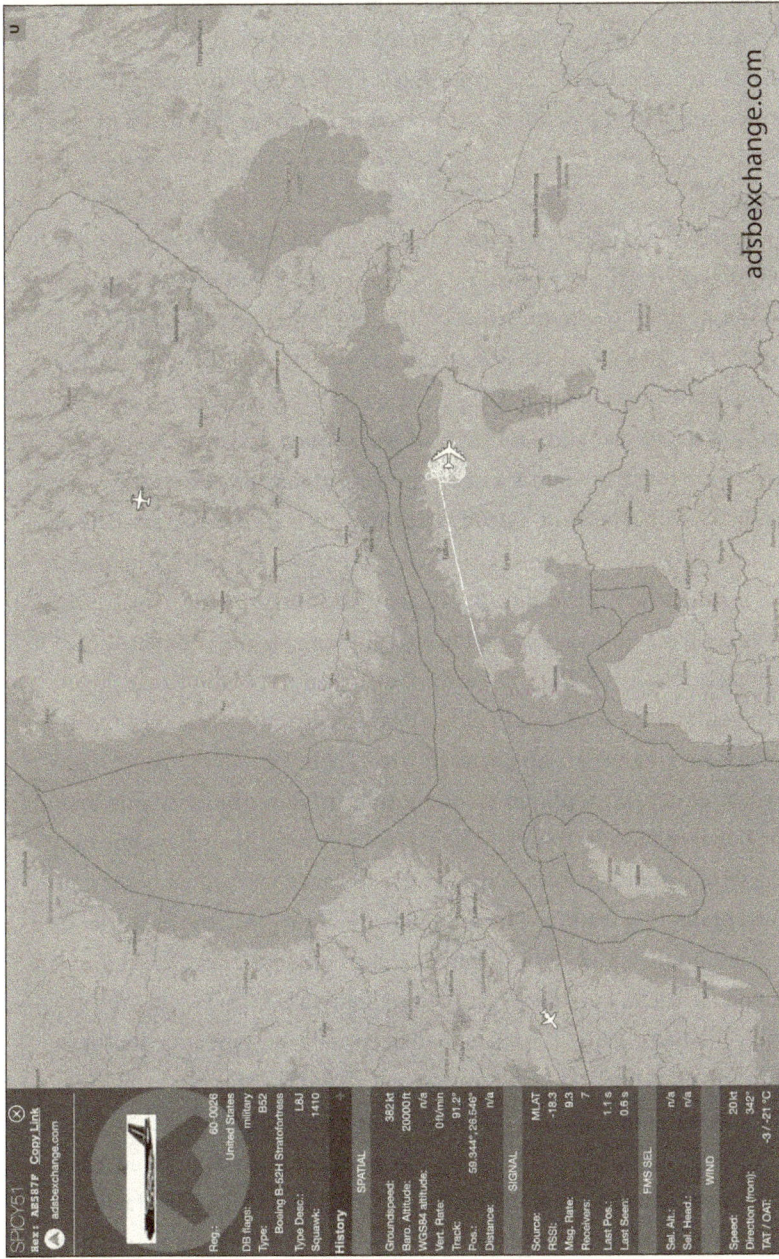

PHOTO 19 ◆ B-52 operations over Estonia, 2 September 2022. Note Moscow targets are in range if the B-52s were loaded with ALCM, thus demonstrating combat potential to Russian leaders during the crisis. This is a significant departure from B-52 flights to Norway and Sweden earlier in the crisis. *ADS-B Exchange*

however, that the Russians became aware on 10 September when a presidential command post Il-96–300PU repositioned in the Moscow area, while another flew to Sochi accompanied by a Tu-214PU and a Tu-214SUS. The repositioned aircraft then flew to Sochi. In this case, it was likely Putin relocated out of Moscow to confer with a select group of advisors on what to do about the situation. An RAF Rivet Joint conducted collection activity off Sochi while these meetings were in progress.[107]

Visible U.S. activity that day included three E-6Bs and one E-4B sequentially deployed, a tanker surge of forty-one aircraft, then two E-6Bs with one over the Atlantic conducting trailing wire operations.[108] The only other overt Russian moves were a flight of two Tu-142 Bears that "entered and operated in the Alaskan and Canadian air identification zones" and were tracked by NORAD. A corresponding series of flights of Il-38 Mays and Tu-142 Bears were conducted over the Barents Sea. Together this has the appearance of a large antisubmarine exercise.[109]

The first of what would become a steady stream of nuclear threats started on 13 September with Medvedev threatening NATO members with "earth burning and concrete melting" for having the audacity to arm "Ukrainian Nazis."[110] Russian command and control activity could best be described as slightly enhanced, with two Tver Loops by Il-96–300PUs and a Tu-214PU run to Novosibirsk.[111] U.S. activity, however, was unusual. A trio of E-6Bs were aloft simultaneously, with one of these conducting an ALCS check-in with the ICBM fields. These were replaced with a sequential deployment of another three E-6Bs, with one over the Atlantic, but another flew to Lake Michigan and conducted what appeared to be trailing wire antenna patterns. These were replaced with yet another pair of E-6Bs, one conducting trailing wire antenna operations over the Atlantic.[112]

The same day, NORAD's Noble Defender kicked off. U.S. Navy SEALs also publicly dropped in on St. Lawrence Island to establish a UAV base and presumably other surveillance means. The island had been an important contributor to early warning and hosted a large radar station and a SIGINT station throughout the Cold War. On 22 June 1955, Soviet MiGs notoriously shot down a U.S. Navy P2V Neptune patrol aircraft over the island.[113] It would not take a paranoid mind to see St. Lawrence Island as a base to raid the Russian bomber base and facilities 522 km away at Anadyr. Overall, Noble Defender "set the stage for a demonstration of layered air defense" as part of "NORAD's mission of deterring aggression."

KC-135 tankers deployed to forward operating locations at Joint Base Elmendorf-Richardson and Moose Lake, Alaska, and Cold Lake, Alberta; they worked with RCAF and USAF interceptor aircraft.[114]

On 13 September, Global Strike Command mounted Exercise Prairie Vigilance to "conduct strategic bomber readiness operations." Public affairs highlighted the role of the 705th Munitions Squadron at Minot AFB, in addition to the B-52H operations of the 5th Bomb Wing.[115] The same day, the B-52H BTF at RAF Fairford conducted operations over Norway with Royal Norwegian Air Force aircraft while an RAF Rivet Joint collected off Murmansk. The exercise was billed as "integrated air defense takedown" training.[116] Noble Defender, Prairie Vigilance, and the BTF were scheduled events that demonstrated ongoing collective defense, deterrence, and collective offense if required.

There was a discernable spike in nuclear rhetoric from Russian information outlets on 15 September. Russian Telegram propaganda channels, using Alex Wallerstein's Nukemap application, modeled nuclear strikes on Kyiv, Vinnytsia, and Lviv and distributed the results widely.[117] On Rossiya1, the dialogue between Vladimir Solovyov and Margarita Simonyan was equally disturbing. Simonyan decried the length that war would take now "3 months, 3 years, or 30 years." Solovyov replied, "Our other choice? Reduce the whole world to dust. Just not yet," quoting a line from the cult Russian military satire film *Demobilized*. Simonyan's response? "And we will go to heaven."[118]

In this light, Putin's response to "questions" posed by Russian "reporters" on 16 September is interesting. Russian forces had by this point struck Ukrainian infrastructure targets with cruise missiles. Putin referred to these as "warning shots," and if the situation on the ground continued to deteriorate for Russian forces, he stated that "our response will be more impactful."[119]

Using selective high-resolution video deployed all over the information environment, the Russian Ministry of Defence showed the operations of the *Oscar* II–class SSGN *Omsk* and *Yasen*-class SSGN *Novosibirsk* launching cruise missiles at a target in either the Chukchi Sea or the Bering Strait on 16 September in exercise Umka 2022.[120] The Oscars carry twenty-four P-700 Granit cruise missiles with either a conventional or nuclear (350 to 500 kt) warhead. The *Yasens* carry 32 Oniks or Kalibr cruise missiles, which also are nuclear capable. These missiles range out to 625 km for Granit, 300 km for Oniks, and 1,500–2,500 km for Kalibr, which essentially put every possible target in Alaska in range of one system or another.[121] For example, Oniks could be used to destroy outer early warning, Granit to destroy air defenses, and Kalibr for high-value nuclear-capable and BMD systems.

There was, of course, more to Umka 2022 than exciting footage of nuclear-capable systems moving in the direction of Alaska. A Tu-214PU flew into remote airfield near Norilsk in northern Siberia before the cruise missile launches which suggests higher-level coordination of the exercise.[122] Umka 2022 also involved the range instrumentation ship *Marshal Krylov* either as a command platform, launch verification system, missile defense warning system, or all three. The ballistic missile submarine *Knyaz Oleg*, with its sixteen Bulava missiles with a range of eight thousand kilometers, each potentially carrying ten MIRVd warheads yielding up to 550 kt per vehicle. The *Knyaz Oleg* reportedly conducted a "trans-Arctic under ice" from the Barents Sea to the Bering Sea and then sailed to the submarine base near Petropavlovsk.[123]

Umka 2022 was a sophisticated intimidation operation. In the public space, missiles launched in the direction of the United States (Alaska) were perfect fodder for "copium" (informational opium to feel better) for a Russian population given the extant circumstances of the Kharkhiv offensive. The addition of a third *Borei*-class SSBN to the Russian Pacific Fleet, not to mention its new under-ice capabilities, was clearly directed at the U.S. administration. The Kharkiv counteroffensive generated significant angst in the Putin regime, which responded by firing off missiles against Ukrainian infrastructure targets. This was framed by Putin as an implied "shot across the bow" before employing battlefield nuclear weapons to contain the offensive. That threat was backed up by Umka 2022, which represented significant combat potential off Alaska with SSGNs and North America with SSBNs, all supported with information operations.

That night, President Biden warned Putin against any WMD usage in response to Ukrainian advances, as that would provoke a strong NATO response.[124] These remarks appear to have been unscripted. The number of U.S. NC3I aircraft aloft that day was elevated: an E-4B and an E-6B, followed by the same E-4B and then three new E-6Bs, then the E-4B and an E-6B, which were both replaced with two E-6Bs, providing continuous coverage for nearly a twenty-four-hour period.[125] The next day—19 September—the RVSN's 42nd Missile Division exercised

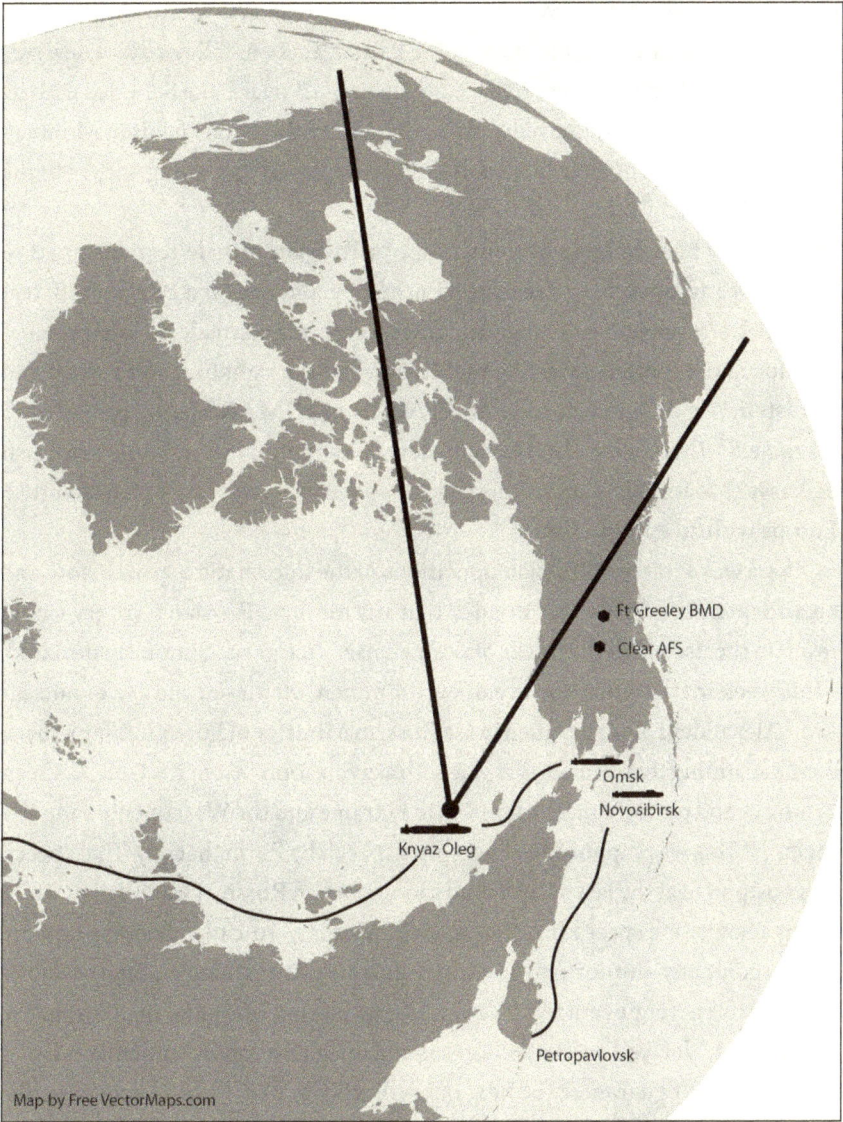

MAP 7 ◆ The Russian ballistic missile submarine *Knyaz Oleg* off northern Russia during Umka 2022, with the range gate of its sixteen Bulava submarine-launched ballistic missiles (8,000 km) each potentially carrying ten MIRVd warheads yielding up to 550 kt per vehicle. *Created by the author*

their mobile missiles near Sverdlovsk.[126] The U.S. response was to progressively deploy an E-6B with antenna out over the Pacific. This was followed by a new pair of E-6Bs, which in turn were replaced with an E-4B over Lake Superior with its trailing wire antenna deployed; two E-6Bs skirted the ICBM fields in Montana and Wyoming. Right after this another E-6B deployed its trailing wire antenna over the Pacific.[127]

This may have been in anticipation of a Russian missile test designed to underscore an upcoming speech by Putin. Speculation on a Burevestnik test circulated starting 17 September, and there were remarks made on 19 September by information outlets asserting that Russia had the right to strike Poland and Ramstein Air Base in response to ATACMs or HIMARS being provided to Ukraine.[128] There were also Russian assertions routed through their Swedish embassy that RAND Corporation had supposedly developed a plan "to crush Europe with an energy crisis."[129]

There was a flurry of Russian operations in the information space before and on 20 September. Medvedev implied that any move on Russian territory would result in the use of "all the forces of self-defense." Margarita Simonyan declared, "This week marks either the eve of our imminent victory or the eve of nuclear war." Alexander Dugin weighed in via Russian Ministry of Foreign Affairs social media, claiming that "Russia is at war with a civilization that fights God. . . . Open Satanism and outright racism flourish in Ukraine and the West openly supports them."[130] This was supposed to backstop a speech by Putin that day. That speech was postponed at the last minute, and the airtime on Rossiya1 was filled by a film about a Soviet-era spree killer nicknamed "Mosgaz," so-called because he wore the gas company's uniform to gain entry into his victims' houses, a not unsubtle analogy to current events.[131] There was speculation on social media that the speech was "delayed as the demonstrative nuclear weapon launch failed and the RVSN can't guarantee the next one will work better."[132] In anticipation, the United States deployed a Cobra Ball aircraft to North Dakota and an E-6B with ALCS capability over the ICBM fields in Wyoming and Montana.[133]

Putin delivered his speech on 21 September. It announced a partial mobilization of Russian reserve military forces. Putin then asserted that the Zaporizhzhia nuclear power plant was being shelled by Ukraine with Western encouragement, and he decried statements by NATO countries "on the possibility and admissibility of using weapons of mass destruction—nuclear weapons—against Russia." In the latest bluff, Putin leaned forward and looked into the camera, "In the event of a threat to the territorial integrity of our country . . . we will certainly make use of all weapon systems available to us. This is not a bluff. . . . Those who are using nuclear blackmail against us should know that the wind rose can turn around."[134] If Putin believed that he and Russia were on the receiving end of "nuclear blackmail," this suggests that whatever Western powers were doing in this regard was working.

Shortly before Putin's speech, an E-4B Nightwatch took off from its dispersal base at Lincoln, Nebraska, in the middle of the night. Around 0600 that morning, one E-6B flew out over the Atlantic Ocean to communicate with the ballistic missile submarines assigned to the East Coast. Another E-6B orbited the Oklahoma City area, while a Cobra Ball took off from Eielson AFB and orbited above Alaska.[135] A KC-135 assigned to NORAD to support the RCAF, call sign Molson01, took off from Fairchild AFB and connected with RCAF CF-18s over Vancouver Island on what was labeled a "NORAD-tasked mission."[136]

A Rivet Joint operated off Sochi, and an MQ-4 kept an eye on occupied Crimea.[137] At 0642 Eastern Daylight Time (EDT), a Tu-214SR communications aircraft took off from Vnukovo, headed toward St. Petersburg but then entered an orbit in the Luga area. This aircraft held this orbit for the next seven hours, an unprecedented deployment of this asset. At that time, two B-52Hs from the BTF departed RAF Fairford and were already on station in northern Sweden conducting "integration" with Swedish defense forces. A Tu-214PU then launched from the Moscow area, headed for the St. Petersburg area. The E-4B Nightwatch and the two E-6B Mercuries landed before 0909 EDT, while the B-52Hs flew out over the Norwegian Sea to refuel from KC-135s. A third E-6B Mercury took off from Oklahoma City and flew around the Midwest. Eventually the Tu-214SR broke off and headed for Moscow.[138]

At that time, a tweet appeared that Lithuania had activated its Rapid Response Force "to meet the threat from Russian-occupied Koenigsberg."[139] And then reports came in that unidentified UAVs were spotted "at the Johan Sverdup, Gullfaks C, Snorre A, Gina Krog, Heidrun, and Kristin oil gas installations," that is, nearly every important Norwegian energy facility.[140] At this point, two Russian Navy tugs escorting Russian ship "Special Vessel SS-750" with its remotely

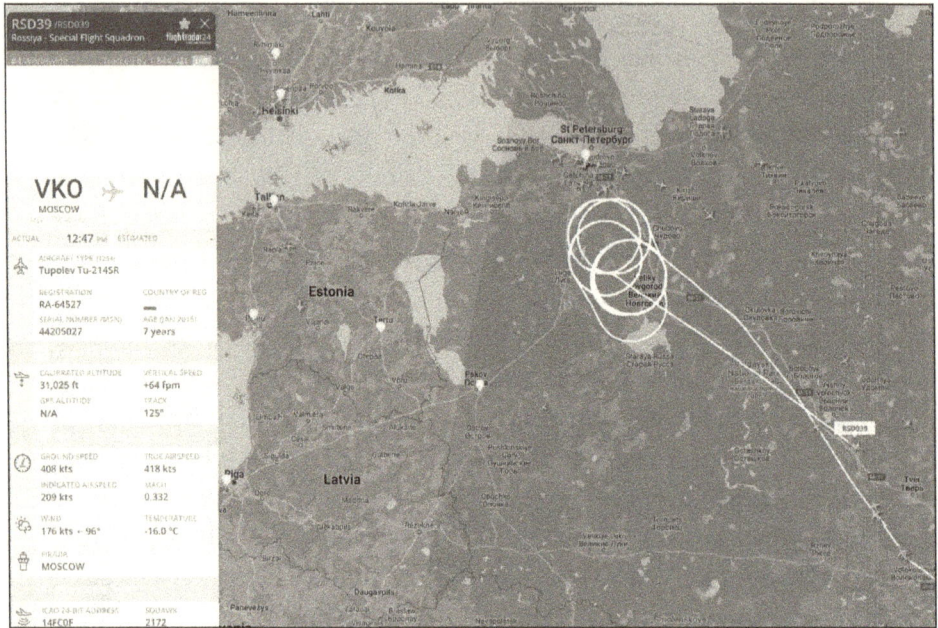

PHOTO 20 ◆ Tu-214SR strategic communications relay aircraft orbits Iskander missile positions south of St. Petersburg while Putin announces mobilization and makes threats to use nuclear weapons. *flightradar24*

operated vehicles and mini-subs, approached Bornholm island, where they were intercepted and photographed by the Danish Navy surveillance ship *Nymfem*.[141]

The activities of the Tu-214SR aircraft are crucial in this narrative. The flight, its route, location, and duration, was exceptional in all three categories during the crisis. There were indications in the information space that nuclear weapons had been deployed to this area, which hosted an Iskander missile unit. The Tu-214SR started its orbits roughly at the same time the B-52Hs came on station, almost as though it was timed to do so given the transit time from RAF Fairford to Sweden. By orbiting this exact area for seven hours, the Putin regime threatened Finland, Sweden, and Norway, the three Baltic countries, and underscored Putin's messaging: "You threaten me in Ukraine, I can threaten you in Scandinavia."

The West's response was multifaceted. Stoltenberg shot back that there should be "no misunderstanding in Moscow about the seriousness of using nuclear weapons. . . . Nuclear war cannot be won by Russia." Biden rebuked Putin for making nuclear threats.[142] The next day Blinken made it clear that the U.S. interpretation of the Russian threat was that occupied and illegally annexed Ukrainian territory constituted "Russian land" and that it would be "protected" with nuclear weapons. Blinken tried to clarify the U.S. understanding that Putin would use nuclear weapons against Ukraine if it tried to regain occupied territories militarily, while at the same time taking Russia to task for making nuclear threats.[143] Medvedev issued another missive stating that nuclear weapons could and would be used to protect all territories under Russian control, and hypersonic attacks could be made against the United States and Europe.[144]

While Blinken was speaking, an E-4B Nightwatch had been operating west of Lincoln for an hour or so. Then five E-6B Mercuries lifted off with one over the Atlantic communicating with submarines; another orbiting the Midwest; a third flying northwest of San Francisco headed for the Pacific Ocean; a fourth over northern Arkansas; and the fifth orbiting Schriever Space Force Base, just southeast of Colorado Springs, where U.S. military GPS satellites are controlled. More than forty tankers were visible. B-1B bombers were also visible on ATC tracking sites.[145]

NATO Air Policing forces intercepted Su-24s over the Baltic. These were fully armed with Kh-25ML penetration missiles. An unannounced Russian exercise run out of Kaliningrad had fifteen aircraft, including Su-24s and Su-30Ms in the strike role with Su-27s escorting, while four missile boats conducted "anti-convoy" attack profiles. The presence of the USS *Kearsarge* amphibious task group in the area made the Russian exercise less than coincidental.[146]

Both sides reduced their NC3I presence with one U.S. E-6B up while a pair of Russian aircraft headed to Sochi from Moscow.[147] Activity focused on Kaliningrad that day, where a Rivet Joint flew all the way around the enclave twice and stood off orbiting over Polish airspace. It was at this point four Russian wide-body commercial aircraft and an Il-76 transport, all flying in trail, landed in Kaliningrad.[148]

The nuclear call and response continued on 24 and 25 September. Lavrov reiterated everything Putin said and once again leaned on Russian declaratory policy and strategy. Sullivan's response went further. He revealed that "the US government had communicated directly, privately, and at a very high levels to the Kremlin that any use of nuclear weapons will be met with catastrophic consequences for Russia, that the US and our allies will respond decisively, and we have been clear and specific about what that will entail." This strongly suggests that at some point direct communications had been made with Putin, and the specific U.S. response to Russian nuclear weapons use was conveyed to him. The fact that the Putin regime continued with its course of action, which seems to have been preparation to use battlefield nuclear weapons against Ukraine, indicates that the American signaling was not getting through. Visible command and control moves over these days show seven E-6Bs sequentially deployed and conducted trailing wire antenna deployments over the Gulf of Mexico and the Atlantic up on 24 September, which was reduced to five flights the next day with three of those simultaneous movements of the E-6Bs.[149]

At 1203 GMT 26 September, two explosions occurred under the Baltic Sea. These damaged the Nord Stream 1 undersea gas pipeline at the intersection of the Swedish and Danish Exclusive Economic Zones (EEZ), and Nord Stream 2 inside the Danish EEZ, permitting gas to leak to the surface where it was photographed by Danish surveillance aircraft. Seventeen hours later, a second pair of explosions was detected, one at each site. These detonations completely destroyed the pipelines. The explosive charges were at the time estimated to have been 150–500 kg of explosive power each.[150]

The sabotage generated massive amounts of "white jamming"—blasting audiences "with so much cynicism about everyone's motives" in order to "persuade them that behind every seemingly benign motivation is a nefarious, if impossible to prove, plot so that they lose faith in the possibility of an alternative," a concept invented by a Russian media specialist.[151] These acts had the effect of completely drawing the attention of the global information space away from what was happening in Ukraine, that is, the rout of the Russian army east of Kharkiv. The event also took attention away from the phony "referendums" that Russia was holding in occupied Ukrainian territory, which in theory gave the Putin regime the "right" to "defend" them with nuclear weapons. It also had the effect of protecting Gazprom from massive financial losses it was taking in the sanctions environment as the Russian corporation could claim force majeure.[152]

On 27 September, Medvedev asserted that Russia could use nuclear weapons and NATO wouldn't do anything about it.[153] Russian military discussion groups on Telegram then increased their chatter level over the iodine pills, chemical suits, and other radiation effects mitigation equipment that was apparently being issued—this was conveyed into Western social media.[154] That day, intelligence collection aircraft once again circumnavigated Kaliningrad and continued to do so for the next several days. An unprecedented four E-4B Nightwatches, the whole fleet, were aloft during the course of the day: two in morning to mid-afternoon spelling off with two in the evening. The first pair's call signs were Order66 and Omega07. Four B-52Hs using Fear call signs were visible operating from northern tier bases in the United States, again highly unusual.[155] Russian visible moves amounted to a pair of Tver Loops with Il-96–300PUs.[156]

The war of words continued from 28 to 30 September but was lowered to the State Department–foreign ministry level.[157] Intelligence missions were flown by Rivet Joints around Kaliningrad and off the Kamchatka Peninsula, and an SSGN, the USS *Florida*, was seen visiting Gibraltar. The visible command and control plot was reduced to a single aircraft on each side.[158]

On 29 September an RAF Rivet Joint conducting surveillance of occupied Crimea over international waters was buzzed by a pair of Russian Su-27 Flankers. One of these aircraft fired an air-to-air missile at the aircraft to no effect. This incident was downplayed by all players and the "facts" were not released for several weeks.[159] The same day, a hacker group claimed to have accessed and taken down the Russian GONETS satellite communications network, which remained down until 4 October. GONETS handled communications for, among many clients, the FSB, ROSCOSMOS, and possibly organizations involved in cruise missile activities.[160]

A Royal Navy *Vanguard*-class SSBN was seen deploying from Faslane, but this likely was part of the normal deterrent patrol schedule as another arrived at the base around the same time.[161] A Russian Delta IV SSBN also departed its base near Olenya Guba this day. This activity made its way into social media sometime after the event, and the sailing took place without information operations support from higher levels of the Russian government. This could have been a signal using the intelligence channel. Five Russian SSBNs were determined to be in port, though they had the ability to conduct alert operations from their berths.[162] RVSN also exercised the 42nd Missile Division with its mobile ICBMs but again not specifically for public consumption.[163] The U.S. response was comparable to the previous mobile ICBM exercise, that is, a Cobra Ball aircraft over North Dakota, two pairs of E-6Bs and an E-4B over the course of the day. One E-6B did a check with its ALCS system near the Montana ICBM fields.[164]

Speculation continued in social media on the possibility of the Putin regime employing battlefield nuclear weapons in Ukraine. A "defense source" in Moscow, through an observer of the crisis, said that "Putin had three options after Kharkiv: retreat, mobilize, or use nukes. He opted to mobilize, which suggests that tactical nuclear weapons use might not be imminent." The Polish foreign minister opined that Putin's threats were confined to Ukraine, not NATO, and if there were a response, it should be a conventional one.[165]

Putin finally announced the annexation of Donetsk, Luhansk, Kherson, and Zaporizhzhia on 30 September, littering his speech with references to Hiroshima, Nagasaki, the firebombing of Dresden, and destruction of Cologne by the RAF. Putin asserted that there was a precedent for nuclear weapons use as established by the U.S. in World War II. Since this so-called "precedent" existed, Russia also had the right to use nuclear weapons.[166] Sullivan's response suggests there may have been indicators of consideration or preparations detected by the intelligence apparatus.[167] The visible U.S. command and control plot consisted of an E-4B Nightwatch (call sign Hulk21), two E-6B Mercuries, and a Cobra Ball near the abandoned ABM site in North Dakota, an unusual posture more in line with preparations for a Russian missile test.[168]

The threat to the Norwegian oil and gas fields increased on 30 September with the movement of several Russian Navy auxiliary vessels near facilities in the North Sea, followed by reports of unidentified UAVs. A Russian "trawler" was also spotted "fishing" near the Svalbard submarine cables. Norway deployed F-35s over the platforms to observe and deter, and increased the security presence on all gas and oil platforms. Sweden, which had lowered the reaction time for its critical infrastructure protection units to two hours, deployed them for "exercises" at this time.[169] Also on NATO's northern flank, an Israeli private intelligence organization announced that there was an "irregular presence" of Tu-160 Blackjack and Tu-95MS Bear aircraft at Olenya air base on the Kola Peninsula.[170] Without a baseline of how many bombers were normally deployed there, it was difficult to determine whether this was a significant increase or not and how it related to the crisis.

Olenya air base had been an LRA base since it was constructed during the Cold War, and the Olenegorsk nuclear storage site that supported it was seventeen kilometers away. It is safe to assume these facilities retained the ability to handle nuclear weapons for both types of aircraft.

Conclusions

The Putin regime once again employed a combination of information operations and nuclear force movements to gain momentum. In this case, the chemical weapon theme remained in heavy play, but the "accomplices" broadened out to include NATO and non-NATO states in the region, not just Ukraine. This was backed up with significant pressure brought to bear against countries surrounding the Baltic Sea and then the Black Sea. There was now a shift in theme: Russia might employ low-yield nuclear weapons against Ukraine to "defend" areas in Ukraine it occupied. The dual Russian themes of Ukraine possessing all three types of WMD ricocheted back and forth in the information environment, as well as ominous but unspecified threats as to what Russia might do. At the same time, NATO nations applied pressure on the northern flank to bolster Norway in the face of Russian provocations and gray-zone activities. These were backstopped with significant U.S. nuclear force movements, but outside of the northern flank area. Following this, NATO operations shifted to include the Baltic front in an effort to match Russian activities there.

It was the successful Kharkiv counteroffensive by Ukraine that altered the trajectory of the crisis. Putin not only escalated nuclear use rhetoric but also conducted a dangerously provocative exercise that generated combat potential against Alaska and North America. NORAD and USSTRATCOM ran demonstrative exercises expressing the message that Russian provocations vis-à-vis North America would not be tolerated. It was at this point that high-level rhetoric increased with President Biden making explicit deterrence statements and Putin retorting with spurious justifications. Russian information operations were employed to generate "nuclear fear" in the West, supported again with significant nuclear force movements. U.S. nuclear forces were used to backstop high-level statements on the situation. The ultimate gray-zone attack took place in late September when the Nord Stream pipelines were blown up. This produced significant Russian activity in the information space which distracted observers from the main issue regarding nuclear moves. More gray-zone activities followed against Norway and Sweden. These were clearly designed to disrupt and distract from the issues at hand, that is, the Russian response to the Kharkiv counteroffensive. This was temporary, however, and by October attention shifted back to whether or not Russia would employ nuclear weapons in Ukraine.

CHAPTER 8

◆ ◆ ◆

Why Do These Things
Always Happen in October?

October–December 2022

T he success of the Kharkiv counteroffensive in September was in part due to the pressure Ukrainian forces kept on the Russian forces north of the Dnipro River. Ukrainian deep strikes conducted against bridges and cross-river shipping increased the logistics pressure on Russian occupation forces in Kherson, as did a lively Ukrainian resistance movement in the city itself. The outer Russian positions were struck on 2 October and started to erode everywhere, with units falling back to avoid encirclement. By 13 October, Russian authorities ordered the evacuation of Kherson. The global information space was charged with the same victory fever that it developed during the Kharkiv counteroffensive. Anticipation was everything and speculation grew as to just how humiliated the Russian forces were and what the possible effects of this might be, including Russian nuclear weapons use. Rumblings in the Russian power structure over the future of the Putin regime started, with names like Dugin, Strelkov, Prigozhin, and Kadyrov as possible alternatives echoing in the chamber. It is in this psychological environment that Putin and his *siloviki* had enmeshed themselves in throughout October.

Fear Factor: Early October

At this point the Russian information operations enhanced the nuclear fear line in Western media. On 1 October, footage of a train passing through Zagorsk depicting fifteen vehicles being moved by train was deployed on Telegram. Observers identified one as a security vehicle used by the 12th GUMO nuclear custodial organization. Another identified what they believed to be a 12th GUMO depot six kilometers east of the rail line and concluded the vehicles came from there. Another observer asserted that the train was heading west toward Ukraine. This information moved rapidly into Western media which then extrapolated that it was related to Putin's threats to use nuclear weapons in Ukraine.[1]

In reality, the footage originated from Rybar, a Telegram channel funded by Yevgeny Prigozhin, the infamous leader of Wagner and the Internet Research Agency.[2] Analysis of the footage shows the train headed northeast, not west. The "12th GUMO site" does not possess the security apparatus for nuclear weapons storage, nor does it appear on any list of 12th GUMO storage sites. Most of the vehicles on the train have no nuclear weapons handling or custodial function.[3] And what everybody missed is that the train passed by the Zagorsk-6 facility, a center for biological warfare development where smallpox and other nasties were weaponized.[4]

The same day, seventeen Iskander TELs were spotted moving by train into the Kherson region. Situating Iskanders in Kherson put them in range of Lviv and Kyiv. At this point, Ukraine mounted a violent assault on Russian positions east of Kherson city and started to roll up the Russian right flank and the vehicles were withdrawn. It was unclear if there were nuclear custodial vehicles accompanying the TELs.[5] This event's target audience was the intelligence space and likely designed to reinforce the arguments made in the public information environment with policymakers.

The next facet of the Russian nuclear fear operation was the reanimation of the existing *Belgorod* nuclear UUV story. A leftist Italian tabloid published an article claiming that NATO warned its members that the K-329 *Belgorod* equipped with Poseidon doomsday nuclear torpedoes had "deployed." Other media outlets enhanced it to "disappeared" and warned that a live test of the system with a nuclear warhead was in the offing. In this case, social media observers worked to discredit the story, but it still reached mainstream Western news outlets.[6]

Another nuclear fear line emerged the next day. Footage depicting the nighttime move of four Topol-M TELs on the northern Moscow ring road along with some support vehicles was disseminated but did not achieve mainstream media coverage.

Indeed, it was recycled footage from Topol-M TEL movements for the 9 May parade and there were no specialized security vehicles in the convoy.[7] Finally, NOTAMS also went up on 3 October and there was speculation of another RS-28 Sarmat test.[8]

The nuclear fear lines coincided with statements made by influential entrepreneur Elon Musk, which suggests he was being used by the Putin regime. On 3 October, Musk tweeted the Russian position in ongoing negotiations: conduct UN-supervised elections in "annexed" regions; recognize Crimea is formally part of Russia "as it has been since 1783"; assure water supply to Crimea; Ukraine remains neutral.[9] This touched off a firestorm in the information environment, including assertions that Putin was personally in contact with Musk.[10] Former presidential advisor Fiona Hill stated, "It is very clear Elon Musk is transmitting a message for Putin."[11] The Ukrainians had no desire to be "saved" by Elon Musk: Andrij Melnyk from the Ukrainian foreign ministry took to Twitter to respond with: "Fuck off is my very diplomatic reply to you."[12]

Also on 3 October, former Trump national security advisor John Bolton asserted that "we need to make it clear that if he did use a nuclear weapon, not only would there be severe consequences for the Russian military, there would also be severe consequences for him. We should say publicly that if Putin did authorize the use of a nuclear weapon, he would sign his own suicide note."[13] Bolton was commenting in his capacity of private citizen, but this would have been interpreted as a signal that Putin himself was targeted. The situation on 2 and 3 October may have stimulated an elevation of U.S. NC3I levels. The U.S. posture went from three E-6Bs aloft on 1 October to nine aircraft (E-6Bs and E-4Bs) on 2 October, and eight E-6Bs progressively aloft on 3 October, dropping to two on 4 October.[14]

By 4–5 October Ukraine continued to shove Russian forces back on the Kharkhiv and Kherson directions. Elsewhere Russian auxiliary vessels were in close proximity to sensitive Danish facilities, Norwegian submarine cables, and other facilities. Unidentified UAVs overflew these sites as well. The Russian command and control plot was a single Tver Loop. RCAF tankers refueled interceptors over the Beaufort Sea, and there was a single E-4B Nightwatch visible. A Cobra Ball lifted off in the evening over Alaska, indicating that there was information on an imminent Russian missile test.[15]

The next day Norwegian F-35s intercepted Russian aircraft of an unspecified type in the north. In the Baltic, four Russian fighters took off from Kaliningrad,

violated Polish airspace, and were intercepted by four Italian Typhoons. The
Russian aircraft turned into Swedish airspace before returning to Kalinin-
grad.[16] NATO moved to reinforce the air policing mission, and four Belgian
F-16s departed for Lithuania. One commentator, however, noted that two of the
F-16s came from Kleine Brogel, a NATO DCA base with an American custodial
detachment, and pointed out that weapons could be deployed forward rapidly.[17]
This just happened to be the day that Polish media released an interview with
President Andrzej Duda in which he noted that "Poland is open to hosting nuclear
weapons and has discussed the idea with the United States,"[18] remarks calculated
to get under the Putin regime's skin and add to Lukashenko's paranoia.

The situation shifted on 6 October with remarks made by Biden, including a
statement that Putin would use nuclear weapons if the Russian homeland was
threatened, a move that might trigger Armageddon.[19] The text of the remarks
suggests strongly that the United States was in possession of intelligence that the
Putin regime was considering battlefield nuclear weapons use and that Biden had
been in contact with Putin.

That afternoon, there were significant movements of the U.S. command
and control system. At 1400 hours EDT, two E-6B Mercuries were up and the
number of visible tankers was under twenty. Thirty minutes later, two E-4B
Nightwatches lifted off, one of which had the Order66 call sign. The visible tanker
count increased to forty-five. Twenty minutes after that a third E-6B took off,
and eight minutes later a fourth E-6B launched. The tanker count increased to
fifty-five. At 1600 hours, E-4B Order66, two E-6Bs and ten tankers were aloft.
There was also a Cobra Ball over Alaska.[20]

Earlier that day, the Russian command and control fleet was busy. A Tu-214PU
dispersed to Miniralnye Vody, while an Il-96–300PU took off and headed into
the Siberian Black Hole. At the same time, a Tu-214SR conducted a run to Yekaterinburg, and another Tu-214SR deployed to Sochi along with a Tu-214PU. Two
Il-96–300PUs dispersed to St. Petersburg.[21] This was a significant deployment
of high-value assets, especially the communications relay aircraft. There was no
public maneuvering of RVSN mobile ballistic missile forces, but these aircraft
movements had the signature of a significant communications exercise designed
to be seen. Alternatively, if the Russian communications satellite systems were

considered unreliable after the cyberattack earlier in the month, this could have been generated to confirm reliability.[22]

In the middle of the ongoing signaling maneuvers, Bolton asserted that Putin wanted to make the West believe that nuclear weapons use was "inevitable" and "wanted to see people nervous. . . . He's bluffing." It was crucial that "the West not to be deterred by Putin's use of this nuclear threat." When queried, Bolton said, "Well, you know, he's at the center of command and control of the Russian military. National Command Authority is what we call it. He's a legitimate military target. And I think while [there are] plenty of other things we can do as well, that he needs to know that he's on our target list at that point."[23] This may have not been an official statement, but there is no doubt that this had to be assessed by the Putin regime. Bolton had been read-in to American nuclear war planning when he served in the Trump administration, so his remarks had credibility in that regard. And linkage between U.S. nuclear forces, leadership signaling, and targeting Soviet leaders exists historically.[24]

The Russian leadership was preoccupied with dealing with the destruction of the Kerch Bridge on 8 October and retaliation efforts using LRA forces in a conventional role against Ukraine on 9 October. Between 9 and 22 October, four Russian ballistic missile submarines deployed from Gadzhiyevo joined one that was already at sea. This was interpreted by observers as preparations for the fall edition of Grom 2022, but as those exercises usually employed a single submarine firing a single missile, this nonpublic, unannounced surge of Russian strategic nuclear forces over a two-week period at the height of the crisis was significant.[25]

Putin confidant Dmitry Kiselyov opined that the United States would conduct a nuclear false flag operation in Ukraine and then use missiles against Russian troops fighting there.[26] Putin then went on television to announce "high-precision" missile strikes against Ukraine's "energy, military, and communications infrastructure."[27] The LRA strikes against Ukraine are believed to have been planned as early as 2 or 3 October, long before the Kerch Bridge was destroyed.[28] As some of the strikes were launched from Olenya, the repositioning of Tu-160 Blackjacks and TU-95MSs spotted at the beginning of October falls into place.

On the diplomatic front, Biden backed away from earlier statements and said that he did not think the Putin regime would use nuclear weapons, the same day

Lavrov indicated he was "open to dialogue with Western nations."[29] Meanwhile, Stoltenberg announced that nuclear exercise Steadfast Noon would continue as scheduled later in October. Later that day F-16s out of Aviano Air Base were conducting training runs over the Adriatic Sea.[30] NORAD's Noble Defender was also ongoing at this time, and, notably, footage of RCAF CF-18s fully armed with air-to-air missiles being refueled over the Beaufort Sea figured prominently in social media.[31] Also in the Arctic, the Russian Northern Fleet issued a press release claiming that two Tu-142 Bears and Su-24 Fencers conducted aerial refueling operations with Il-78 tankers while escorted by MiG-31s and Su-33s.[32] In terms of command and control movements, deployed Russian aircraft returned to bases in the Moscow area. On the U.S. side, there was only one aircraft visibly aloft during Biden's CNN appearance—an E-6B call sign Dirge65 that had a false track on public flight tracking systems that briefly flashed to its real location: Montana and Wyoming, home of the ICBM force.[33]

The next day, Biden appeared on CNN. "I am talking to Putin. He, in fact, cannot continue with impunity to talk about the use of a tactical nuclear weapon as if it is a rational thing to do." When asked about a response if Putin used nuclear weapons in Ukraine: "It would be irresponsible for me to talk about what we would or wouldn't do." Biden explained that Putin was a "rational actor," but his objectives "were not rational."[34] Disturbingly, Ukrainian information activities quoted a Russian Orthodox Church priest allegedly telling his flock, "If nuclear war starts you need to be close to the epicenter so everything will end faster."[35] Dmitry Peskov then excoriated Biden for his 6 October remarks, while the 14th Missile Division with its mobile missiles conducted an exercise. Secretary of Defense Austin told the media that he had not seen any indicators that Putin "might move ahead with a nuclear strike." One E-4B Nightwatch and two E-6Bs were up during the day and spelled by another E-4B in the early evening.[36]

On 13 October, there was significant mixed signaling in the Western European camp. The same day the NATO Nuclear Planning Group met, Macron tweeted, "We do not want a World War," urged a return to the negotiating table, and then stated in an interview, "Our doctrine rests on the fundamental interests of the nation. . . . They are defined clearly and wouldn't be directly affected at all if, for example, there was a ballistic nuclear attack in Ukraine, in the region."[37] This removal of ambiguity regarding the French independent deterrent was an extraordinary break from the front presented by the other members of NATO. Stoltenberg, coming out of the Nuclear Planning Group meeting, was asked about a NATO response to Russian nuclear weapons use and he deployed ambiguity: "We will not go into exactly how we will respond. . . . Even any use of a smaller nuclear weapon in Ukraine . . . would have consequences." The EU's Josep Borrell then remarked, "Any nuclear attack against Ukraine will bring a response, not a nuclear response, but such a powerful response from the military side that the Russian Army will be annihilated."[38]

The French nuclear flourish earlier in the year had been in response to threats directed at French national interests. Those threats were not seen to be in play this time and Macron confirmed this. France would not extend deterrence over Ukraine. There was no ambiguity. Borrell, on the other hand, believed that limited use of low-yield nuclear weapons in Ukraine by Russia would be met with a massive conventional response. Stoltenberg tried to keep NATO's response plans ambiguous: NATO had a series of nuclear and non-nuclear responses linked to each other to retain escalation dominance. Ukrainian Foreign Minister Dmytro Kuleba provided an interesting perspective: Canada and the United States were not afraid of Russia, "but some of Kyiv's allies are simply because they cannot imagine, in their boldest dreams and reflections, the defeat of Russia."[39]

Russian commentary was double-barreled. Medvedev stated that the American and European people did not want to die in a nuclear war on behalf of Ukraine. This echoed Cold War–era Soviet propaganda directed at NATO designed to generate fissures between the European and North American members of the alliance. Peskov remained "open" to negotiations. Supporting fire was provided by Alexander Venedikov of the Security Council: if Ukraine joined NATO, this would escalate to a third world war which would result in the destruction of Western Europe.[40]

The American leadership did not comment. Command and control movements consisted of three E-6B Mercuries, with two of them conducting trailing wire deployments over the Pacific and the Gulf of Mexico. On the Russian side, there were steady movements: one Tu-214PU and a Tu-214SR moved back and forth from Moscow to Sochi, while another Tu-214PU returned to Moscow from Kazakhstan. Unusually, a Tu-214PU-SBUS flew to Krasnoyarsk and back. Crucially, a Tu-154M flew into an airfield near Putin's palace on the Black Sea.

This has the appearance of keeping a web of communications available to Putin while he was on the move or in Sochi indicating some elevated concern over the situation.[41] USSTRATCOM helpfully tweeted that "B-52 long range bombers are heading to Europe for STEADFAST NOON." This was said to be a "routine event [that] helps ensure the Alliance's nuclear deterrent remains safe, secure, and effective."[42]

There were more threats issued by Medvedev from 14 to 16 October, but there were no official replies. Russian command and control aircraft, Tu-214PUs and Tu-214SRs, shuttled back and forth between Moscow and Sochi. Two U.S. Navy E-6Bs remained aloft for nearly a day and were replaced with an E-B and an E-4B Nightwatch code named Myers99, a reference to the fictional serial killer Michael Myers from the *Halloween* horror film series.[43] The launch of what was believed to be reconnaissance satellite Kosmos 2560 from Plesetsk coincided with the deployment of a Tu-214PU to Arkangelsk, and another that repositioned to Domodedovo, something not seen since early 2022. U.S. movements suggest an elevated posture: E-4B (Myers99 call sign), an E-6B over the Pacific, another over California, and a third over Omaha.[44] Again, there was a correlation between this posture and the Plesetsk launch, which went over the northern polar regions and North America.

Footage of Russian 2S7M self-propelled artillery being transported through Orsha, Belarus appeared on social media on 16 October. The 2S7M is a dual-capable system. It was not clear whether this was a Russian system being redeployed from Belarus or to Belarus from Russia. It is equally unclear whether this movement was meant to be seen and distributed on social media: it did not take off in social media or mainstream media venues as the "nuclear train" had.[45] U.S. NC3I movements included an E-4B and three E-6Bs up at the same time.[46]

The situation shifted significantly by 17 October to an emphasis on the northern flank. Russian gray-zone activity included substantial UAV movements around Norwegian oil and gas facilities, as well as airfields and military bases. Stavanger airport was shut down due to UAV interference, and two Russian intelligence operatives were apprehended by Norwegian security forces. Norway also visibly stepped up P-3 air and coast guard sea patrols.[47] Then Russia flooded the Barents and Kara Seas with NOTAMs, placing civilian ship movements at risk if these were ignored. This was interpreted by some observers as a response to the upcoming Steadfast Noon exercise or possibly another unannounced Russian Grom exercise.[48]

Dancing with the Devil: Mid- to End October

Russia also expressed its displeasure by sending two Tu-95MS Bears into the Alaskan ADIZ, where they were promptly intercepted by F-16s from Eielson AFB.[49] This activity had not been undertaken since spring 2022, but it was the day Steadfast Noon commenced. Unlike previous iterations that lasted one week, the 2022 exercise doubled this by combining the low-profile one-week "Cross servicing" or "X-servicing exercise whose goal was to test the ability of each partner to service other nation's [DCA] aircraft" while the exercise play took the second week with Tornadoes, F-16s, and F-35s practicing B61 deliveries in the North Sea area. Aviano-based F-16s conducted exercises over the Adriatic, while B-52Hs from Minot AFB were employed over the Pole and then Norway to the North Sea area before heading home.[50]

Other movements at this time included a Swedish Korpen surveillance aircraft watching the Russian LRA base at Olenya while the B-52Hs flew by Norway, and the departure of a *Vanguard*-class ballistic missile submarine from Faslane. Cobra Ball continued to deploy to North Dakota, while two E-6B Mercuries were up over the United States.[51]

Since the 29 September incident involving the RAF Rivet Joint surveillance aircraft, these aircraft, like their American cousins, were confined to Romanian air space and not allowed to fly over the Black Sea. Also the nearly daily MQ-4 Global Hawk coverage seen in August and September was dramatically scaled back. The RAF Rivet Joints flew Romanian missions on 10 and 13 October, but on 18 October, escorted by two Typhoon fighters, one RAF Rivet Joint suddenly conducted a patrol off occupied Crimea where it flew collection tracks. It did this again on 21 October.[52] On the 18 October mission, however, the RAF aircraft were met and shadowed by a Russian Tu-154AK signals collection and electronic warfare aircraft escorted by Sukhoi fighters. This particular aircraft had been employed over Belarus in February to collect intelligence on Ukraine.[53]

That day, British Secretary of State for Defence Ben Wallace flew to Washington, D.C. to confer with Austin and Sullivan on "shared security concerns."[54] A "security source" told British media that "the threat has increased recently" and this related to "speculation that Putin could detonate a nuclear bomb over the Black Sea."[55] Wallace said, "I went to the Pentagon, the State Department, the National Security Adviser and other meetings, and made sure that we are all understanding our planning processes about what we would do in the event of a whole range of things."[56]

A possible reason for this unusual visit is that the British intelligence apparatus acquired knowledge of the Putin regime's intended course of action in Ukraine and that likely involved the use of a nuclear weapon. If this occurred, then the specifics of the contingency options and what would trigger them had to be firmly locked in by both British and American leaders in face-to-face meetings instead of by electronic means that were susceptible to interception. These options probably included the progressive conventional targeting of Russian forces in Ukraine, and the nuclear contingencies of various intensities.

Russian command and control activity this day included an Il-96–300PU that was headed for Astrakhan and suddenly diverted to Voronezh. An hour later an Il-96–300PU was on a Tver Loop, followed by another in two hours' time. A Tu-214PU also dispersed to Sochi in the evening.[57] What went on in the United States that day was completely different in scale and scope. In the morning there were four E-6B Mercuries up, a possible shift change. In the afternoon KC-135 tankers scrambled from Topeka, Kansas, and B-52s were squawking over South Dakota. Other KC-135s orbited near former SAC bases like Griffiss in New York. An E-4B Nightwatch departed Offutt AFB, while an E-6B Mercury proceeded to Lake Superior. The E-4B and E-6B were rotated with new aircraft to include an E-4B with call sign Chucky11 (a reference to a child's doll serial killer from the eponymous horror film) and an E-6B. By early evening there were five E-6B Mercuries aloft: one of these deployed a trailing wire antenna over the Gulf of Mexico and another orbited the Colorado Springs–Denver area near NORAD headquarters (one aircraft may have been on a maintenance flight). Around this time nine KC-135s scrambled from Fairchild AFB. Interestingly, a lone KC-135 returned to base after a run out to St. Paul Island in the Bering Sea off Alaska.[58] Additionally, the presence of RCAF CF-18s at Thule Air Base, Greenland was posted on Twitter.[59] This suggests support to a NORAD-controlled intercept of Russian bombers in at least the Alaskan ADIZ and possibly the Canadian ADIZ.

Movements the next day are equally intriguing. One of two Tu-214PU-SBUS Ministry of Defence command planes departed Chkalovskiy Air Base and flew northwest toward Tver. Instead of conducting a Tver Loop, however, the aircraft proceeded to orbit over Vyshny Volochyok for three hours. The town is equidistant between the Novogorod-18 and the Tver-9 nuclear weapons storage depots. However, the 7th Missile Division complex with its Topol-M mobile launchers is located just northwest of the orbit track. It is the geographically closest RVSN unit to North America *and* the United Kingdom. This aircraft had never conducted these movements previously, nor did it afterward at any point in the crisis, suggesting that observers pay attention to the combat potential of the 7th Missile Division.

The U.S. command and control posture increased significantly that afternoon. In addition to a Cobra Ball up over North Dakota, five E-6Bs and an E-4B were up simultaneously, with at least one deployment over the Pacific. Over the course of the day, eleven U.S. NC3I aircraft were aloft. Over Russia a Tu-214SUS special command aircraft conducted a Tver Loop at the same time.[60]

Reports then came in that undersea cables from the southern coast of France had been cut, generating widespread internet outages. Reports also revealed that British undersea cables to the Shetland Islands and the Faroe Islands had been interfered with, severing communications.[61]

A Russian talk show discussion declared that Ukraine had a nuclear bomb primed and ready to detonate in Mykolaiv. This was allegedly a Ukrainian false flag operation designed to place the blame on Russia to trigger an American missile launch against Russia.[62] This was not specifically connected to assertions that appeared in an American pro-Russia conspiracist website on 10 October, but the similarity in the narrative suggests there was some preliminary information space laydown before commencement of an operation.[63]

An article in Russian technical media included a surprising amount of procedural detail. The Russian early warning system features satellites and OTH radar to ensure that RVSN "is ready for retaliatory measures long before the launch of ballistic missiles from the US military base in Grand Forks or from submarines." Given that Steadfast Noon was in play "through a special communications channel, the military will try to call the US and NATO headquarters. If they don't

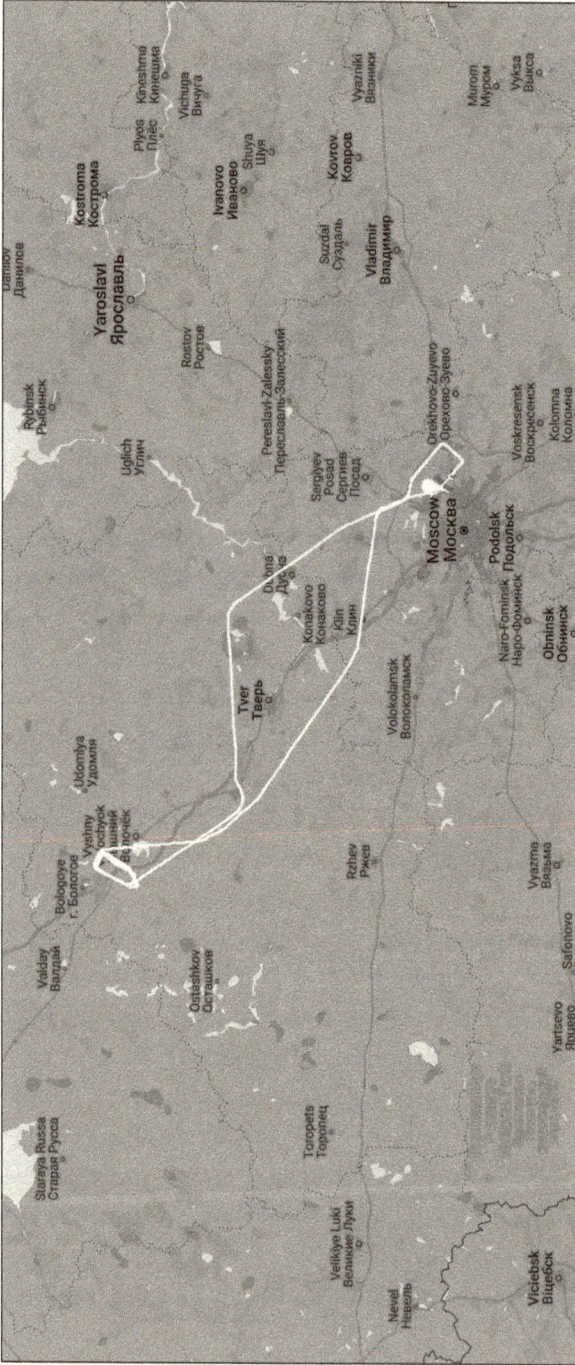

PHOTO 21 ◆ Signaling effort involving one of two Tu-214PU-SBUS Ministry of Defence command-post aircraft and the 7th Missile Division located near Bologoye northwest of Moscow, 19 October 2022, *flightradar24*

answer at the other end, this will be a sure signal for the beginning of the Third World War." Another key statement: "It is also important to understand something else—the control systems of the strategic nuclear arsenal are set in motion as soon as the stations from the SPRN (Missile Attack Warning System) begin to detect incomprehensible objects." Then "if all channels of communication with a potential adversary are cut off, then the strategic nuclear forces will receive a 'green light'" to fire. Notably, "it is important to understand that this happens approximately in the third, in extreme cases, the fifth minutes of the flight of American ballistic missiles." This apparently involves "half of the nuclear arsenal." The frightening conclusion: "The famous DEAD HAND [in Mount Kosvinsky Stone] analyses the situation and periodically checks the key subscribers who make decisions. If the President, the Minister of Defense, and Chief of the General Staff suddenly stop responding to calls, then [a program called] Perimeter will decide to launch missiles on its own." This was in effect a Russian ERCS in a silo.[64]

The possibility of Kahn-esque Russian Doomsday machinery did not deter the next American move. On the afternoon of 19 October, CENTCOM itself tweeted that the commander of USCENTCOM was visiting the USS *West Virginia* while it was surfaced in the northern Arabian Sea. The unmasking of a U.S. Navy ballistic missile submarine via USSTRATCOM's social media while it was on patrol at sea was unprecedented and had significant implications.

Viewed through the Russian lens of combat potential, the USS *West Virginia* and its location was extremely problematic. This ballistic missile submarine could carry twenty Trident II D5 submarine-launched ballistic missiles, each capable of carrying twelve reentry bodies with W76 warheads yielding 475 kt each. Those missiles could range out to seven thousand miles, placing the entire RVSN—from the command posts, silos, and mobile-based launchers of the 28th Missile Division at Kozelsk to the 29th Missile Division at Irkutsk—at risk. Using depressed trajectory launches, these weapons could strike their targets in Russia in under fifteen minutes if volley-fired from the submarine. Simultaneously, there were three E-6Bs up, with one conducting a trailing wire antenna deployment over the Atlantic.[65]

Russian SPRN consists of ground-based radar, satellites, and signals intelligence collection all funneled into a warning center in Moscow. In its ideal state, the SPRN consisted of Tundra early warning satellites of the EKS network as the first warning echelon, with the ground-based Voronezh phased array and

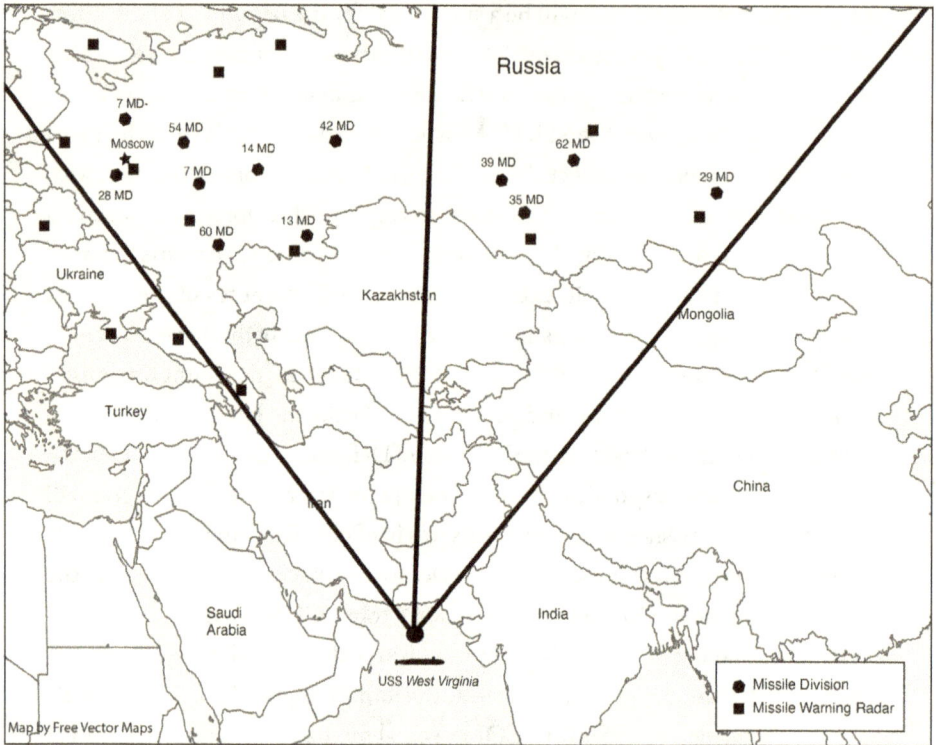

MAP 8 ◆ Range gates for Trident II SLBMs against Russian Strategic Missile Forces bases from the ballistic missile submarine USS *West Virginia* stationed in the Arabian Sea, October 2022 *Created by the author*

Konteyner OTH radars in the second warning echelon. The satellite system is designed to cover the U.S. ICBM fields and the oceans where ballistic missile submarines operate. In theory, the first echelon provides thirty minutes of warning and the second fifteen minutes.[66]

The reality of the situation in October 2022 is that SPRN was a lash-up of three Cold War legacy early warning radar types that were in the middle of a major replacement project with Voronezh. The prefabricated Voronezh radars were plagued with electronic component substitution corruption that affected their reliability.[67] Of the ten planned Tundra satellites, only five were in space, and this did not provide 24/7 coverage of U.S. ICBM and potential SLBM launch areas. Tundra, as well as GLONASS, GPS satellite construction was delayed due to the "lack of domestic space electronics," a result of the sanctions regime established in 2014–15.[68] Sanctions also interfered with launch systems adding further delays.[69] The USS *West Virginia* was geographically positioned to exploit the weaknesses in SPRN.

There were an estimated four or five Russian ballistic missile submarines deployed from the Northern Fleet. U.S. Navy and Royal Navy hunter-killer submarines were within reach of these particularly loud Russian submarines, at least in the Arctic and the Barents Sea, and capable of destroying them if required. This put any Russian second-strike capability at risk.[70]

On 20 October, Putin issued a decree increasing alert levels: Basic for the country and High Readiness for all oblasts adjacent to Ukraine.[71] There was no announcement of a corresponding increase in nuclear force alert levels. He then boarded a plane and took a flotilla of command and control aircraft with him to the Eurasian Intergovernmental Council Meeting in Yerevan. A Tu-215PU-SUS conducted a Tver Loop while an Il-96–300PUs, Tu-214PUs, and a Tu-214-SUS flew to Armenia.[72]

This was the equivalent of posting a "Gone fishing" sign on the Kremlin in what was supposedly the worst crisis since 1962. The United Kingdom and France were unraveling their respective undersea cable difficulties and finding evidence of nefarious activities, when news of the Mikoliev false flag narrative circulated widely on social media.[73] British media had headlines like "Putin's atomic plans are being disrupted from within with top brass sabotaging plans to use nuclear weapons" and breathless accounts of the deliberate sabotage of a Russian underground nuclear test near Archangelsk. The source of the information was Valery Solovey, former head of the Moscow State Institute of International Relations (MGIMO) public affairs and the man behind the SVR General Telegram channel that engaged in platforming conspiracies (on whose behalf remains unclear).[74] To ensure matters remained opaque, stories of a Ukrainian false flag attack plan against the Kakhovka Dam on the Dnipro made it into the wilds of Twitter and were immediately identified by the knowledgeable as the Russians "setting information conditions" to carry out this attack and blame Ukraine.[75]

In what has to be the strangest form of crisis communication since the Soviets used the Western Union office in Washington, D.C. during the Cuban Missile Crisis, an article appeared in the *Vzglyad* ("Look" in English) media outlet.

The piece started off with the presence of the USS *West Virginia* in the Arabian Sea, acknowledged that this was signaling, and entered into a lecture on what a "decapitation strike" was. The piece provided a detailed and accurate analysis that concluded, "The combination of these factors creates a technical possibility for the United States to launch a successful disarming nuclear strike on Russia without receiving a strike in return that is significant in terms of losses." The piece then addressed American analysts who wanted to believe that the submarine had Communist China in its sights and dismissed this argument. It then offered a technical discussion of depressed trajectory SLBM launches and why they are important.

This "article" reads like a briefing paper for Putin or one of the *siloviki* reworked into an information product. Indeed, *Vzglyad* has been controlled by the Expert Institute for Social Research since 2017. This entity is "associated with the Russian Presidential Administration," which in turn was run by Anton Vaino, the chief of staff of the Presidential Executive Office. In other words, this "article" was generated rapidly and meant to be found in the intelligence space. In effect it appears to have been a graceful but deniable climb down: "You got us." And everything that followed after that bears this out.

Aloft U.S. NC3I aircraft on 20 October consisted of three E-6B Mercuries with two of them conducting training wire operations over water, indicating a slightly heightened posture but nothing extraordinary.[76] The morning of Friday, 21 October an RAF Rivet Joint escorted by Typhoons was off occupied Crimea, while Steadfast Noon continued with B-52Hs and tankers over the North Sea. The Russian command and control plot was heightened. The aircraft accompanying Putin in Yerevan returned home, but one of two Tu-214-SUSs flew to Kaliningrad, landed, sat on the ground for three hours, and headed back, overflying Polish airspace instead of taking the long way over the Baltic. This suggests a high-level face-to-face meeting with Kaliningrad commanders, presumably on the status of the nuclear forces stationed in the enclave. One of the two Tu-214PU-SBUS Ministry of Defence command posts was on station northeast of Moscow for more than five hours, with a Tu-214SR communications relay aircraft flying out past Yekaterinburg. This has the signature of a high-level communications check with RVSN forces, though it is unclear if any of the missile divisions deployed their launchers. It may also have been a stopgap measure until Russian satellite communications could be sorted out.[77]

These moves were made while Shoigu finally permitted a phone call from Secretary of Defense Austin. This was the first contact between the two men in many months, despite Austin's repeated attempts.[78] Konstantin Gavrilov, a Russian delegate to the Vienna Negotiations on Military Security and Arms Control, was quoted by TASS and Rossiya1 as saying that the talks were intended to "clear up misunderstandings and eliminate risks" to prevent "incidents at sea, in midair over ships, and on a more global scale, the nuclear aspect."[79] This was also likely related to the launch of three vitally needed Russian GONETS-M communications satellites from Plesetsk that took place the next day in order to assure the United States that this was not part of an attack.[80] Another aspect of the conversation was to ensure that the upcoming Grom exercise was not mistaken for the same given the heightened tensions. Indeed the U.S. NC3I posture on 21 October included first an E-6B, then three E-6Bs and a Cobra Ball over North Dakota at 1611 Z, followed by an E-4B, two E-6Bs, and a Cobra Ball, for a total of nine NC3I aircraft active throughout the day.[81]

There was a flurry of contradictory activity on Sunday, 23 October. The day started with Ukrainian social media warning that Russian state-controlled media was preparing the ground for a "dirty bomb" false flag operation to be blamed on Ukraine. Four hours later, RIA Novosti claimed that Ukraine would use a "dirty bomb" or low-yield nuclear weapon to make accusations about Russia. This entered the Western information environment immediately.[82] A pro-Russian European news outlet made the same claim but added that Kyiv was threatening to blow up the Kakhovka Dam on the Dnipro in a false flag attack to blame Russia while simultaneously stopping the water supply to occupied Crimea.[83]

Three hours later, Shoigu called his opposite numbers in the United States, the United Kingdom, France, and Turkey and repeated the Ukrainian false flag "dirty bomb"/low-yield nuclear weapon claim and that it was going to be used as an excuse for the West to "escalate" the conflict. After his conversation with Shoigu, Wallace "refuted these claims and cautioned that such allegations should not be used as a pretext for greater escalation."[84] Austin had a nearly identical conversation with Shoigu, called Wallace, and then went on social media to note that their conversation was a "continuation" of topics covered during Wallace's visit.[85, 86]

These exchanges led to conversations between the foreign ministers of the four NATO countries. That led to a joint statement by the United States, the

United Kingdom, and France: "Our countries made clear that we all reject Russia's transparently false allegations that Ukraine is preparing to use a dirty bomb on its own territory. The world would see through any attempt to use this allegation as a pretext for escalation by Russia."[87] There was no Russian NC3I movement, but an E-6B and E-4B rotated with another pair of the same aircraft, and unusually a Cobra Ball over Lake Huron and an E-6B over the Pacific with its antenna deployed.[88]

The next day, Monday, 24 October, Chairman of the JCS General Mark Milley and Admiral Sir Tony Radakin, the British Chief of Defence Staff, had telephone conversations with Valery Gerasimov. This was the first time Milley and Gerasimov had conversed since May. The same conversations that took place at the ministerial level were essentially repeated. Pentagon spokesman John Kirby said, "The United States has not seen any evidence that Russia has decided on or is preparing for such an attack."[89] Despite all of this, the Putin regime continued with its information operations with the Russian Ministry of Defence providing details of the supposed weapons including radioactive material and the technological base that Ukraine allegedly had to make such weapons. Its spokesmen provided more spin, asserting that the dirty bomb would be used to simulate a Russian "low-capacity warhead." The foreign ministry provided supporting fire claiming that the objective of the exercise was to "generate anti-Russian sentiment" and "isolate Russia in the international arena."[90] Russia then demanded the proceedings be moved to the United Nations. The map used in the Russian briefings was the same one used on 18 August when Russia made allegations about the possibility of the Zaporizhzhia nuclear power plant being attacked by Ukraine in a false flag attack.[91] Interestingly Belarus pulled its troops back from the Ukrainian border near the Chernobyl exclusion zone on this day, as if they were no longer required.[92]

That day, the U.S. command and control plot consisted of an E-4B and two E-6Bs, then three E-6Bs.[93] The Russian plot was very different. Five An-148–100E regional jets belonging to the Special Flight Detachment departed Chkalovsky Air Base one after another; their departure was observed by ATC, but then they squawked off. Two went in the direction of Krasnodar, another Astrakhan, one to Sochi, and one went to a location near Saratov. This had the appearance of deploying high-ranking personnel to give or rescind orders in person.[94]

High Plateau: End of October and into November

The purported Russian stand-down was not fully accepted in the American camp. The first possibility was that the Putin regime was going to detonate a low-yield nuclear weapon as a warning shot to force negotiations and then freeze the conflict in place. This weapon could be used against Ukrainian forces or over the Black Sea. The second was a variant where a low-yield weapon or a radiological dispersion device (aka "dirty bomb") was detonated by Russian forces but blamed on Ukraine to reduce Western material and moral support. This may have involved the Zaporizhzhia nuclear power plant, the Chernobyl nuclear power plant, or the Kakhovka Dam. A third was a low-yield weapon or radiological dispersion device detonated by Russian forces as a false flag to justify the use of a small number of low-yield weapons to stop the Ukrainian advances. In all possibilities, there would be positive benefits in the Russian domestic information space, that is, this would offset the Kharkiv and Kherson failures with a strong blow that would appease the increasingly restive Russian power structure as well as the population.

The relationship of all of this to the latest iteration of Grom is significant. The fall 2022 Grom was apparently not announced using the deconfliction mechanism until the day before it started.[95] Ukrainian media reported on 25 October that Grom would take place the next day, and this rebounded in the information environment.[96] This was supported by increasingly bizarre Russian information lines: TASS "reported that Russia's Security Council pledged to 'desatanize' Ukraine" which was "urgent" because there were "hundreds of sects operating there."[97]

The role of the February Grom is instructive for comparative purposes. The original Russian plan had a chemical weapon false flag generated to provide justification for invading Ukraine. Grom was timed to demonstrate strategic nuclear capability as a signal to NATO to stay out of the arena. Then the coup de main was supposed to go in and take down the Ukrainian government in Kyiv. It was as if this plan was repackaged for the situation in October. A nuclear false flag generated justification for limited nuclear weapons use was supposed to be used to alter the balance on the battlefield in Russia's favor, while Grom was timed to demonstrate strategic nuclear capability as a signal to prevent a U.S./NATO conventional response and forestall any escalatory moves.

In effect, the United Kingdom and the United States came into possession of information on the trigger event and then mounted a demonstration to signal the Putin regime. The day of the exercise, RIA Novosti insisted that Ukraine was loading a captured Iskander missile or a Totchka-U missile modified to look like an Iskander, with radioactive material provided by Yuzhmash. This missile would be shot down over the Chernobyl exclusion zone and used to smear Russia and gain NATO intervention or other support.[98] The relationship of this accusation to the 24 October pullback of Belarusian troops from the Chernobyl exclusion zone is interesting. It suggests that they may have been the target of one of the false flag plans, which could have potentially activated Belarusian-Russian "defense" provisions involving nuclear weapons.

Fall Grom 2022 was not that dramatic. The usual imagery of Putin communicating orders was juxtaposed with imagery of a mobile YARS launch from Pletsetsk, while a Delta IV ballistic missile submarine, the *Tula*, fired a Sineva missile. A Tu-95MS launched a cluster of cruise missiles. Of note, none of the sub-strategic systems like Iskander were highlighted.[99] In terms of command and control, there was a communications exercise the day before Grom officially started. Two Tu-214PUs flew from Moscow to the Sochi and Stavropol areas, then returned to base. One of the two Tu-214PU-SBUS command planes orbited northeast of Moscow for three hours, then flew out to Ivanovo, home of the 27th Guards Missile Army, 54th Missile Division, and 7th Missile Division. A Tu-214SR communications relay plane flew from Moscow to Ulyanovsk, Saratov, and back.[100]

American moves included a Cobra Ball over North Dakota, single E-6Bs with one over the Gulf of Mexico; then three E-6Bs with one over the Gulf of Mexico, two single E-6Bs, and then three later in the day.[101] The United States also announced the accelerated deployment of B61–12 nuclear weapons for December 2022.[102] Blinken explained to the media that they had "communicated directly and very clearly about the consequences of using nuclear weapons" and especially in a false flag format.[103]

Putin attended the Valdai Discussion Club on 27 October and denied everything while continuing to flog the Ukraine "dirty bomb false flag" attack scenario. It was all, apparently, Liz Truss' fault for her comments back in February.[104] Clearly frustrated at the performance of Russian forces, the inability to gain traction with false flag scenarios, and outright deterrence signaling by the United States, the Putin regime shifted laterally. Foreign ministry representative Konstantin Vorontsov stated to the UN that unspecified commercial satellite support used by Ukraine "may be a legitimate target for a retaliatory strike."[105] Observers assumed that this threatened action was some kind of proportionate activity, like an ASAT attack, that could be directed at StarLink or Maxar constellations. They completely missed that the threat implied nuclear weapons use. There was an immediate White House response by Karine Jean-Pierre and John Kirby, which is indicative of the seriousness of the threat. If Russia went forward with such activity, it would be "met with a response at a time and manner of our choosing."[106]

The Putin regime pulled out of the Black Sea Grain Initiative on 29 October citing a Ukrainian strike against Russian naval facilities in Sevastopol. Russian ships from that base were flinging Kalibr cruise missiles at Ukraine from the Black Sea, so this was an obvious excuse. The next grain convoy consisting of fourteen ships out of Odesa continued without incident on 31 October.[107] It is evident that Putin was constrained, deterred, and he knew it. And to top it all off, the Finnish government announced that it would allow nuclear weapons on its soil if it were allowed to join NATO, prompting supportive commentary asserting that "Finland must be able to defend itself with tactical nuclear weapons." With stealthy F-35s coming on line for the Finnish air force, a future pathway to a DCA capability existed.[108] Instead of imaginary Ukrainian nuclear weapons, soon there could be real NATO-controlled nuclear weapons in a country with 1340 km of border with Russia.

Winter of Our Discontent: November and December

After the frenzy of activity in October, the next two months settled into a pattern. The United States maintained its deterrent posture and Russia tried to reconstitute theirs with a focus on command and control. The Putin regime scaled back some of its activities on the northern and the Baltic flanks. And Russian information activities got increasingly bizarre with their messaging.

Vladimir Solovoyov referred to the war in Ukraine as "a counterattack launched in response to eight years of Ukrainian genocide against those who won't accept LGBT transgender-Nazi values."[109] Medvedev implied the West was provoking a global nuclear war.[110] He followed up, "They are a bunch of insane Nazi drug addicts, a nation drugged and intimidated by them, and a large pack of barking dogs from the Western kennel."[111] Dugin weighed in as well: Russia "is fighting a holy war against the satanic West, a final apocalyptic, eschatological battle against the antichrist."[112] And the Russian Orthodox Church contributed its opinion: "The Lord is punishing Ukraine with the hands of Russia. Russians are just a weapon of God, it's like a belt with which one punishes a child."[113]

Simultaneously, British intelligence identified and then publicly called out the Putin regime for deploying MiG-31Ks and Kh-47s.[114] That day, the ballistic-missile submarine USS *Rhode Island* arrived in Gibraltar, announced via Twitter USSTRATCOM, which very pointedly stated that "the U.S. & U.K. share a strong history of cooperation, through exercises, operations, and cooperation activities such as this, that enhance or combined capabilities and partnership."[115] Cryptically, Command Submarine Forces Atlantic tweeted, "and we still hold the line."[116] The implication here is that there was some sort of nonpublic-specific threat directed at the United Kingdom by the Putin regime and this activity was a response to it.

Russian movements were unusual at this time. On 1 November, an Il-96–300PU entered the Siberian Black Hole, while a Tu-214PU-SBUS took a run out to Krasnoyarsk but came back along the northern Arctic coast. A Tu-214PU changed call signs while claiming to fly to Sochi but remained in the Moscow area. Three Tu-214PU command posts were on the ground at Sochi.[117] The only visible move was a Ministry of Defence announcement that the Tu-22M3 regiment at Belaya near Irkutsk were involved in an exercise.[118] There were flights of Tu-95MS in the Amur region at this time as well.[119] Of interest, a USAF Rivet Joint aircraft was in a collection pattern off Murmansk.

The Russian movements correspond to statements made by the Russian foreign ministry that emphasized, "Russian doctrine only mandated defensive goals" when it came to nuclear weapons use" and then Patrushev came out with a statement that the West was engaging in nuclear provocation to expand its anti-Russian coalition. They corresponded with the launch of a Tundra early warning satellite, bringing the constellation up to six satellites and thus significantly improving the capabilities of the SPRN early warning organization. Notably, this launch from Pletsetsk did not overfly North America.[120]

The next day, 3 November, the *Borei*-class ballistic missile submarine *Generalissimo Suvorov*, launched several Bulava ballistic missiles, which were tracked by a Cobra Ball orbiting over Alaska. Three E-6B Mercuries were up, with one conducting trailing wire operations over the Atlantic.[121] The Russian command and control plot was also different: an Il-96–300PU flew in an unusual square flight orbit northeast of Moscow, while another conducted a Tver Loop. A Tu-214PU performed a run to Saratov and back. The Putin pack (an Il-96–300PUs and Tu-214-SUS) headed for Sochi.[122]

Exercise Prairie Vigilance, with B-52H and tanker operations, was usually confined to North America. On the day USS *Rhode Island* departed Gibraltar, the 6th Air Refueling Wing at MacDill AFB, the 121st Air Refueling Wing at Rickenbacker ANGB, and the 155th Air Refueling Wing at Lincoln ANGB conducted nuclear ORIs. Tankers were observed operating over Nova Scotia, Canada and Duluth, Minnesota in what looked like Cold War–era Hostile Aircraft Evacuation Plan orbits.[123] This was followed by several tanker NORIs on 6 and 8 November from Altus AFB and the former SAC base at Bangor, Maine.[124] The U.S. Air Force also released pictures of ARRW hypersonic missile loading procedures undertaken at Barksdale AFB.[125] The bombers, of course, were not squawking.

Pressure was maintained on the Putin regime on the northern flank. Enhanced security measures included Norwegian P-8 Poseidon operations in and around oil and gas platforms, Swedish Korpen flights surveillance of the Kola Peninsula, U.S. Navy P-8A Poseidon flights from bases in Finland, U.S. Air Force Rivet Joint operations off Murmansk and ongoing spy and saboteur apprehensions in all of the Scandinavian countries.[126] The Royal Navy deployed the aircraft carrier *Queen Elizabeth* strike group to northern Norway.[127] Norway released a picture of the Heimal gas field platforms as seen through a submarine periscope, implying that their *Ula*-class submarines were keeping watch against any underwater or seabed activities.[128] A Norwegian naval commander stated, "I am sleeping better than I slept before the Ukraine war. I do so because the Kola Peninsula is emptied of ground forces. They have bled down in Ukraine."[129]

Another important signal from the northern flank was sent by STRATCOM twice on 13 November, which clearly was designed to underscore a previous post on 9 November by U.S. Special Operations Command, Europe.[130] This involved testing a system called Rapid Dragon "above the Arctic Circle." Rapid Dragon essentially drops pallets of Joint Air-to-Surface Standoff Missile (JASSM)–Extended Range out of the backs of MC-130 and C-17 aircraft, thus making transports cruise missile Pez dispensers. The JASSM variants can range out to 1,200 km. The Rapid Dragon test was launched from a base in Poland, and the weapon was deployed off the northern Norwegian coast by two MC-130J Commando II aircraft. This implied that the system was already in Poland and could be deployed elsewhere in the region.[131]

It was not a coincidence that there was very public coverage of a rapid deployment exercise of U.S. Army HIMARS from Germany to Latvia by C-130 Hercules in under three hours on 12 November.[132] This demonstrated that the Russian Iskander missile units south of St. Petersburg and in Kaliningrad could be brought into range rapidly with conventional missile systems. It is not clear if the Voronezh missile early-warning radar at Pionersky in Kaliningrad has 360-degree coverage. In theory HIMARS could destroy this system in the early stages of a conflict. Similarly, the Russian Volga radar in Belarus could also be attacked from HIMARS based in Latvia.

Shoigu announced Russian withdrawal from Kherson on 9 November and on 11 November the city was liberated. A Russian Ministry of Foreign Affairs spokesperson stated that Putin was "open to negotiations."[133] This may have reflected a purported split within the U.S. administration about whether Ukraine should negotiate to end the war. But Ukraine, despite the manipulations of Russian information operations, stood firm that the only acceptable negotiating point involved all Russian forces departing Ukrainian soil.

There were significant movements of Russian command and control aircraft on 10 November. Two Tu-214PUs rotated from Moscow to Sochi; a Tu-214SUS conducted a Tver Loop, as did an Il-96–300PU; one of the Tu-214PU-SBUSs flew to Krasnoyarsk and back to Moscow; an Il-96–300PU deployed in a wide loop, almost into the Siberian Black Hole before returning to Moscow; and a Tu-214SR communications plane dispersed to Kazakhstan.[134] This had all the markings of a major RVSN exercise, or at the very least a comprehensive communications check of the entire Russian nuclear command and control system. This suggests that there could have been something wrong with the GONETS satellite communication system or that the RVSN was conducting an exercise to test operating in a satellite-degraded environment.

The corresponding visible American movements were five E-6B Mercuries up singly or in pairs on a day-to-day basis, dropping to one on 11 November. However, an E-6B did arrive on 5 November in Rota, Spain and it proceeded to operate from there over the Atlantic until 10 November.[135] An E-6B working

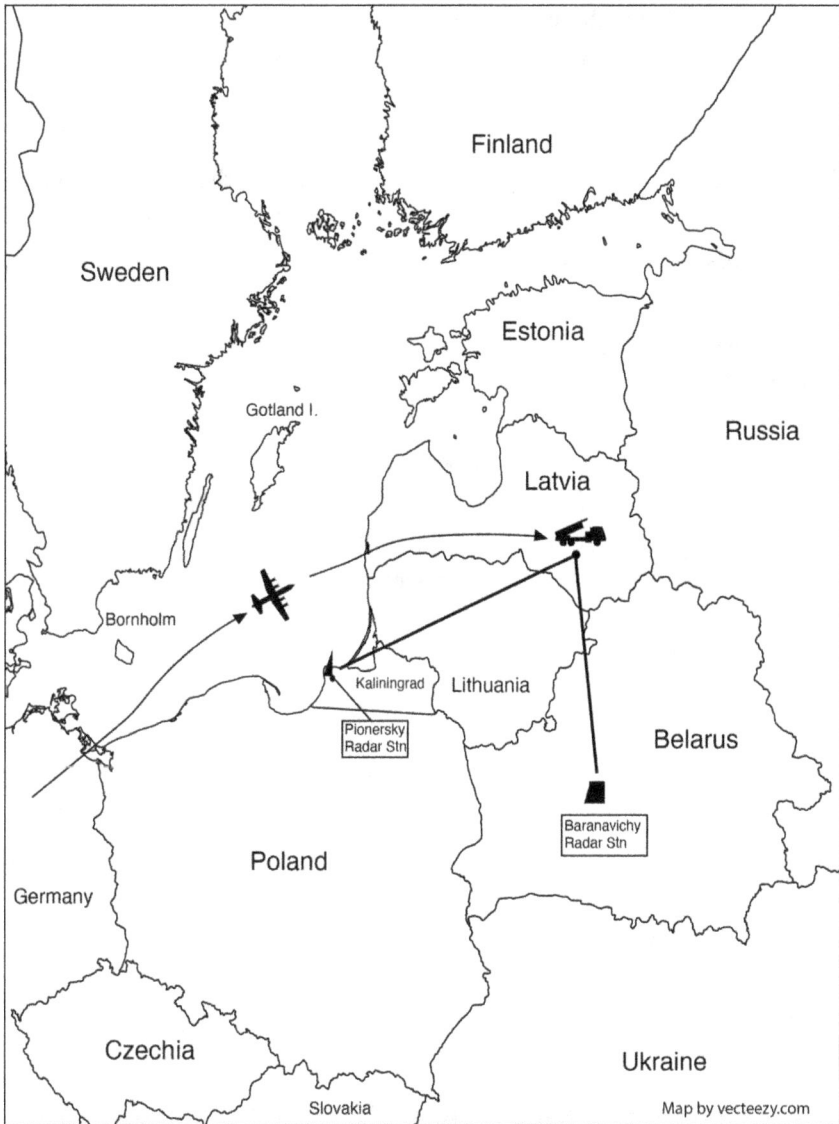

MAP 9 ◆ U.S. Army rapid HIMARS deployment from Germany to Latvia, 12 November 2022 with range gates in relationship to Russian ballistic missile early-warning radar sites. The ability to destroy nuclear command, control, and early-warning facilities using conventional weapons in the early phases of a nuclear conflict, thus blurring the nature of the attack and the war itself, is a factor planners must take into account. *Created by the author*

with the ballistic missile submarine could have been a test of the submarine's communications system, or it could have been specific signaling given events.

The U.S. command and control apparatus increased its posture on 12 and 13 November. The starting state on 12 November was a single E-6B out of Travis AFB throughout the night, replaced with two E-6Bs in the afternoon, which grew to five E-6Bs in the evening. Trailing wire deployments were conducted over the Atlantic and Gulf of Mexico. On 13 November, three E-6Bs were aloft, joined by an E-4B Nightwatch in the afternoon. The Russian movements these days consisted of a single Tver Loop and an Il-96–300PU went into the Siberian Black Hole.[136]

The relationship between the United States and Russian command and control apparatuses and a low-profile meeting between Director of Central Intelligence Bill Burns and SVR Chief Sergi Naryashkin is intriguing. Burns flew to Istanbul to meet with Naryashkin on 14 November. All parties denied that the discussion related to negotiations over Ukraine and the focus was, as Naryashkin explained cryptically later through RIA Novosti, on "strategic stability" and "nuclear security."[137] Official U.S. sources asserted, "He is not discussing settlement of the war in Ukraine. He is conveying a message on the consequences of the use of nuclear weapons by Russia and the risks of escalation to strategic stability."[138] The United States' visible command and control plot remained a baseline two aircraft. The Russian command and control plot, however, consisted of a Tu-214SR into the Siberian Black Hole, a Tu-214PU in a long loop out to Tyumen, and a Novosibirsk run by an Il-96–300PU.[139] The next day Russia unleashed a massive cruise-missile attack against Ukraine.

By 15 November the Russians had been pushed back on the Kharkhiv front, had retreated from Kherson, and had been deterred from using nuclear devices or using a false flag operation to alter conditions on the battlefield. It was being confronted on NATO's northern and Baltic flanks and its gray-zone activities deterred. It had replenished its early warning satellite constellation, and it had exercised its aircraft-based command and control architecture.

The missile strikes conducted on 15 November employed dual-capable bomber assets of the Russian LRA force. It is probable that the American intelligence apparatus saw preparations for this operation, which had to extend back several days, as well as the command and control aspects of it. This accounts for the

Burns-Naryashkin meeting. It possibly accounts for the reveal of the USS *Rhode Island* and the unusual deployment of the E-6B to Spain.

The same day, the SSGN USS *Florida* with its 240 Tomahawk cruise missiles was observed and photographed resupplying by helicopter in the Ionian Sea.[140] A B-52H call sign Chill23, operating from Minot in the clear, conducted a flight that was identical to a Cold War–era Large Charge weapons delivery profile using four nuclear gravity bombs in sequence.[141] The number of E-6Bs increased from one to three, with two of the Mercuries conducting ALCS checks over the Montana ICBM field.[142] On the other side there were three Tver Loops: a Tu-214SUS, a Tu-214PU, and an Il-96-300PU. A Tu-214SR flew into the Siberian Black Hole, while another Tu-214PU flew from Moscow to Tyumen to Ulyanovsk. These moves were definitely an elevated posture.[143]

Lavrov and the Russian delegation had been kicked out of the G-20 summit after declaring that the West had launched a hybrid war against Russia, but the reason actually related to a missile that struck Poland during the Russian cruise-missile attack on Ukraine. This is an example of how a near-random event can trigger catastrophic thinking when the event is contextualized by a nuclear crisis.

On the night of 14–15 November a missile struck a rural area of eastern Poland, killing two people. The strike appeared to be either a missed target in western Ukraine or a deliberate shot into Poland as a signal to deter further military support. This event led to wild speculation on social media and then mainstream media as to whether this could activate Article 5 of the North Atlantic Treaty. It then led to predictable accusations that the event was a NATO/Ukraine false flag attack or "provocation" to pave the way to open NATO military operations in Ukraine, Belarus, or both. These accusations were led by pro-Russian elements in social media and their confreres in mainstream media. The overall effect was to ramp up "nuclear fear" in the information space.[144]

In the intelligence space, it must have been something for Burns to have just delivered his message to Naryashkin about backing off on the nuclear rhetoric and the signaling, to be confronted by both a ramp-up in Russian nuclear command and control *and* a possible Article 5 situation. The reality was that the missile was a Russian-built, Ukrainian-owned S-300 air defense missile that went astray during the Russian cruise-missile attack. These revelations generated a whole new downward spiral on social media as pro-Russian elements tried to generate confusion and trouble with a "Zelensky was trying to start a nuclear war" information operation line.

The United States maintained an elevated signaling posture on 16 November. A B-52H visibly conducted Large Charge–type runs against mothballed industrial facilities in the Detroit area, before disappearing over the Upper Peninsula, former home to two major SAC bases. A KC-135 tanker orbited Lake Superior, while an E-6B Mercury was off Florida conducting trailing wire communications. Eventually four E-6Bs and an E-4B Nightwatch were aloft, with one E-6B conducting an ALCS test over the Montana ICBM fields. Comparable Russian moves included a Tu-214PU, a Tu-214SR, and an Il-96–300PU moving about central Russia communicating with RVSN forces.[145]

Both Russia and the United States maintained elevated command and control deployments from 16 to 18 November. In both cases this involved increased numbers of aircraft aloft above the baseline numbers, that is, nine E-4Bs and E-6Bs on the U.S. side and combinations of six Tu-214PUs variants on the Russian side, dropping to two aircraft each on 19 November, and up again 23–25 November.[146]

Two important events took place on 28 November. First, Russia replenished its GLONASS GPS constellation from Plesetsk, with the delivery system positioned to not fly over North America.[147] Second, STRATCOM tweeted that the USS *West Virginia* pulled into Diego Garcia for a full crew change. In effect, the reveal of the ballistic missile submarine again, and the fact it had visibly pulled back away from the Arabian Sea, signaled reduced Russian vulnerability while at the same time communicating that it could go back on station if required.[148]

On December 13, TASS announced the loading of a single YARS ICBM into a silo at Kozelsk.[149] The Kozelsk ICBM formation had a missile comparable to the American Emergency Rocket Communications System, an ICBM equipped with a transmitter that could send the "go code" to other units after launch. On 18 December, it was reported that the 76th Missile Regiment at Yurya "received the first Siren-M . . . a new mobile missile system with a command missile created on the basis of YARS."[150] Instead of an ERCS-like ICBM situated in a static missile silo between Ukraine and Moscow, there were now three such systems on mobile launchers based in the Russian interior. The combination of a replenished communications and early warning satellites constellations with

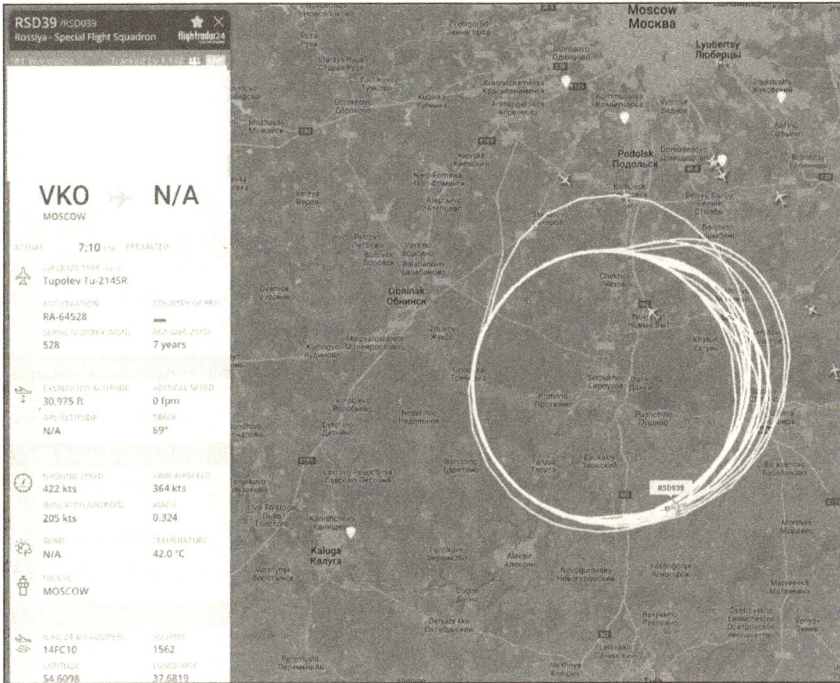

PHOTO 22 ♦ Tu-214R strategic communications aircraft orbit the Serpukhov-15 satellite control facility southwest of Moscow. *flightradar24, Google Earth*

a mobile ERCS-like communications capability covered off the vulnerabilities of the Russian command and control system exposed by the USS *West Virginia* flourish back in October.

The flights of two Tu-214SR communications relay aircraft that orbited the Serpukhov-15 facility southwest of Moscow were connected to this in some way. This was done in shifts throughout 23 December, movements not seen before or since. Serpukhov-15 is sixteen kilometers from the Chekhov bunker complex and twenty-three kilometers from the Chekhov ABM/early warning radar site. The large facility consists of eight huge domes for satellite communications and is the command center for the early warning satellite constellation. Both aircraft kept Serpukhov-15 dead center during their all-day orbits. The message could have been: "We're back." Or there could have been some technical fault, possibly a cyberattack, necessitating the temporary deployment of the aircraft to cover off a vital communications gap, and the Russian side wanted to ensure that hostile observers knew they were on watch while the fault was corrected.

Also of interest that day was a flight of a Tu-214PU-SBUS from Moscow to Izhevsk and then back to Moscow after four hours on the ground. The only facilities of strategic interest in that area are the chemical weapons depots at Kizner and Kambarka, which are equidistant from the airfield at Izhevsk.[151] Given that this aircraft is generally only used by either Shoigu or Gerasimov, it suggests a high-level visit to check readiness or to get a briefing.

Putin was signaling that strategic stability, from the Russian perspective, had returned to the pre-October status quo. USSTRATCOM's response was public affirmation that a successful ARRW hypersonic missile test had taken place on 9 December.[152] It was followed by a picture of an American ballistic missile submarine and a British ballistic missile submarine sailing together.[153] That coincided with a ceremony at RAF Fairford inaugurating a new USAF headquarters organization, the 501st Combat Support Wing, that would "strengthen its command-and-control capability from a strategic location"; as the media noted, "it is the only forward-operating base in Europe that can host and deploy American nuclear-capable heavy bombers like the B-52."[154]

What was not publicized was a USAF C-17 flight from Albuquerque to Aviano Air Base carrying the promised B61–12 upgraded nuclear gravity bombs for the DCA forces stationed in Italy.[155] Russian TU-95MSs probed the Alaskan ADIZ that day and were intercepted by F-22s, and the signaling cycle continued.[156] Shoigu also announced that the two Russian brigades stationed opposite Finland would be expanded into two divisions as soon as possible to offset the poor correlation of forces in northern Russia. And Russian naval operations started up again off Bornholm island.[157] The U.S. NC3I posture throughout December was significantly reduced from October and November, averaging five to six aircraft per day until 12 December when it dropped one to three per day.[158]

As 2022 came to an end, the Putin regime felt more secure than it had two months previously. Cruise-missile attacks on Ukraine escalated and new false flag accusations involving the Zaporizhzhia nuclear power plant as well as "American biolabs in Kazakhstan" ramped up.[159] This included Chechen prophet Boris

Malakhov who declared that "Russia will defeat Western bourgeois fascism and reptilian aliens."[160] Historian Timothy Snyder remarked, "Those of you talking about security guarantees for Russia should recall that Russia defines its security threat as Satan."[161] Italian Minister of Defense Guido Crosetto confirmed in the media that this drama was not over, that "the use of tactical nuclear weapons is planned by Russia. For us this is unthinkable, but for Moscow, yes, if the point of no return is passed, if they risk defeat."[162]

Conclusion

This book is called *The Cool War* for a reason: we are looking at a new conflict and several new means by which it manifests itself, even though it resembles the Cold War. The tempo of the Cold War involved military deterrent activities with the ideological and informational confrontation as a backdrop. The DEFCON system and associated movements of nuclear forces indicating how serious crises were became of decreasing utility later in the Cold War as entering into the escalatory logic of them became less useful for signaling and had an air of inevitability toward nuclear confrontation. There was nothing between day-to-day general deterrence and this escalatory path during a crisis. With the Cool War, ideological and informational confrontation became more and more integrated as information technologies and information spaces expanded dramatically. Nuclear force maneuvering became increasingly integral to informational confrontation itself. Thus in the Cool War most activity remained below the still-existing DEFCON threshold. There had to be ways of signaling that did not connote inevitability. There are now sub-DEFCON signaling subtleties. And that constitutes the bulk of the activity depicted in this narrative.

Indeed, there were and are more means of signaling. Deterrence and signaling is actually a combination of nuclear, conventional, space, and cyber maneuver. These moves directly interact with the information spaces, be they the intelligence space (between nations and alliances) the information space in the mainstream media, the overlapping information space of social media, and the interjection of coordinated information operations into all of these. The audience is not just national leadership: it includes the populations of the engaged states. Again, this is integrated to a much greater extent than it had been during the Cold War.

The rhythm of the Cool War is thus very different than the Cold War, even though some of the pieces on the board look (and like the B-52s, are) the same. There is near-constant maneuver of military forces with the objective of gaining exploitable advantage in the information spaces, to reduce or increase vulnerability. The Cool War is an ongoing series of complex and discrete moves and countermoves. Most are done with the understanding that they contribute to the greater whole. Most are not coincidental, some appear to be, and some are. Distinguishing between the three is a matter of judgment, context, and debate.

The tempo of these movements, and the protracted nature of them since early 2022, far exceeds the tempo of the Cold War, or any specific Cold War crisis, most of which tended to be limited to weeks. The acceleration of the speed of information is a given: we have all lived through these dramatic changes in how we are exposed to information in the past fifteen years. There was a definite shift in signaling methods and tempo in January 2022 and that shift was sustained throughout the year and continued into 2023. For example, the pattern and tempo of command and control, intelligence, and logistics aircraft movements are measurable changes. Others, like most bomber movements and ballistic-missile submarines movements, are not as easily measurable but in some cases are identifiable. Operations in the information spaces are subject to substantial interpretation requiring a robust framework with which to understand them, but they can be discerned and understood.

Ideally, analyzing the Cool War should be based on archival primary source documentation and interviews with key players in the drama. We can only reliably do this for parts of the Cold War because of the extreme secrecy involving nuclear weapons and their role in national security. Now with the explosion in information, it is possible to generate a preliminary narrative of the Cool War in ways we could never hope to for the Cold War. This book was written with the full understanding that it is not the whole picture. That said, the phenomenon of the Cool War exists and there are recognizable components of it. The ongoing confrontation with the People's Republic of China, North Korea, and, yes,

Russia demands that we have at least a preliminary understanding of how this all works. And as history is something that evolves, the Cool War was written with eyes wide open and humility, with the recognition that others will build on this foundation or create their own.

What are some of these early and preliminary findings? Let us start with who was deterred and when. The collective West was deterred from outright military intervention to support Ukraine. That was based on over seven years of "nuclear fear" built up by Russia's coordinated information operations. The West, however, was not deterred from providing certain types of military support to Ukraine. It was deterred in other areas, specifically, military systems that could appear to directly impact Russian forces on Russian territory, that is, aircraft and long-range missile systems. The West was not deterred from providing intelligence collected via systems deployed in close proximity to Russian forces and territory, or from space.

Russian moves in the 2022 phase of the Cool War have only succeeded in drawing Lukashenko's Belarus deeper into the embrace of the Putin regime and making it an accomplice to its abhorrent actions. Russian moves did slow down the provision of certain weapons systems. And ultimately the largest Russian success was intimidating all players from supporting Ukraine directly and effectively from 2014 to 2021 and deterring Russian attack in 2022.

Russia, on the other hand, was unable to accomplish its objectives in Ukraine mostly due to the resilience of the Ukrainian people and its leaders, but also because it failed to compel its Western opponents. Russia was unable to intimidate Norway, a key opponent on its Barents Sea flank, which was crucial for the protection and deployment of its ballistic missile submarine forces which was a sine qua non for attaining deterrence parity or dominance. Russia was unable to intimidate the Baltic neutrals, Finland and Sweden, with gray-zone activities and outright bullying with military forces. Indeed, Finland joined NATO, thus complicating the Russian national security domain permanently. Sweden joined NATO in 2024, again as a direct result of Russian military activities elsewhere. Germany, which Russia had worked on for nearly twenty years using a variety of political, economic, informational, and subversive means, was not deterred from supporting Ukraine and, most important, did not play the role of disruptive spoiler within NATO, as the Putin regime thought it might. Poland has rearmed and has the means to interfere with or even seize the Kaliningrad enclave. This would have been unthinkable before 2022. Importantly Russia was unable to deter Ukrainian conventional attack on Russian soil, nor was it able to deter Ukrainian conventional attack against Russian dual-capable strategic aircraft and their bases.

Of course, we cannot underestimate the role that the barbarity of Russian operations in Ukraine has played in tandem with the West's deterrent efforts and Ukrainian resistance. Nazi-like behavior against Ukrainian civilians and prisoners of war; the missile campaign against civilian infrastructure; the utter disregard for practically all international norms; and the open bullying of states not directly involved in the conflict have all played a role. But the key and central aspect of Russian bullying—the deliberate instillation of nuclear fear in Western populations and their leaders— backfired.

Russian declaratory policy and strategy has been shown to be utterly bankrupt in the face of Western deterrence signaling and maneuvering. The vaunted "escalate to de-escalate" option has not taken place, though the threat was there, repeatedly conveyed through coordinated information operations. The fact that the Putin regime felt the need to resort to plausibly deniable mechanisms, in this case chemical weapons, and low-yield nuclear weapons using a false flag as a screen underscores their insecurity with their own declaratory policy and strategy. The "nuclear threshold" that analysts were so worried about ten years before the conflict has been much higher than we were all led to believe, at least thus far in 2022.

Russian hybrid methods, or what some in the West categorized as hybrid warfare since 2014, have also been shown to be utterly bankrupt. Any actions that Russia takes in any space, information or otherwise, is automatically assessed with suspicion and assumed to be part and parcel of a larger malevolent objective, which they are. This facade was ripped away the second they attacked Ukraine in February 2022.

Now, why is this the case? And what role did signaling, including signaling with nuclear forces, play in limiting Russian ambitions? The most important aspect was the maintenance of credible forces that were trained, exercised, and equipped to undertake nuclear operations at the global level. Russia saw this to be true in the intelligence and information spaces. The fact that it was made explicit through the information spaces provided a salient into Russian thinking, as well as allied thinking. Assurance came with deterrence, and this was repeatedly demonstrated at the regional level in conjunction with other forces, most important, conventional forces and forces capable of countering gray-zone activities. Using older terminology, deterrence was successfully extended over Baltic neutrals as well as NATO allies at all three levels: nuclear, conventional, and gray zone. To a certain extent, we can argue that deterrence has been extended over Ukraine, though such an assertion should be massively qualified.

Signaling had specific roles to play when intelligence indicated that Russia was going to engage in some form of limited chemical or low-yield nuclear action. The decision to call out the Putin regime on its planned false flag action to trigger the 2022 phase of the war was backstopped with nuclear force signaling. This threw off the Russian plan and likely contributed, along with the provision of intelligence and Ukrainian valor, to the failure of the key operations in and around Kyiv in the first twenty-four hours of the war. This doomed the Russian effort and led to Russia's first open nuclear threat. That threat was matched with more nuclear force signaling. And as we know, Russia did not employ nuclear weapons in response.

A pattern emerged. When Russia failed to gain traction in Ukraine, it lashed out against NATO's northern flank, the Baltic flank, or the Black Sea flank using combinations of gray-zone activity, conventional harassment, and then nuclear flourishes. In each instance, these were met with conventional responses backed up with nuclear force signaling. None of this prevented or deterred Western support and resupply of Ukraine, nor did it interfere with the movement of Ukrainian grain to the world market.

In the fall of 2022 the Putin regime suffered its greatest setback to date—the Kharkhiv counteroffensive. Nuclear fear was ramped up in the information space, and there were intelligence indicators that a nuclear false flag operation would be used to justify low-yield nuclear weapons use to halt the Ukrainian counteroffensive. U.S. nuclear forces were employed to demonstrate Russian strategic vulnerability, and the Putin regime backed away from this course of action. However, once those vulnerabilities were mitigated, Russia started up a new cycle of activities intended to accomplish its objectives, in this case the practically pre-conventional and bloody Bakhmut offensive in 2023. This new cycle had its own rhythm, and how nuclear force signaling fit into it will be the subject of a future work.

At this point in history, we can tentatively state that the Putin regime's dreams of Eurasianist domination were put in check in 2022 by Ukrainian resistance and NATO's unwillingness to let its members be pushed around, which included maneuvering nuclear forces as deterrence signaling. I would, however, repeat that we are in the preliminary stages of our understanding of these events. It will be interesting to see how this narrative and its analysis holds up over the next five, ten, and twenty years. It will be exciting to find out what we don't know. And we are not beyond the use of nuclear forces as an integral part of how nations conduct and express themselves on the world stage.

NOTES

Chapter 1. Getting Here from There

1. The financial elements are discussed in detail in Dawisha, *Putin's Kleptocracy* and Belton, *Putin's People*. Michael Pullara's *The Spy Who Was Left Behind* examines the narcotics network. Charles Clover uses the phrase "yesterday forever" in relation to Kryuchkov and his activities in *Black Wind, White Snow*, 169. See also Andrei A. Kovalev, *Russia's Dead End: An Insider's Testimony* (Lincoln: Potomac Books, 2017), 208–210.

2. Dawisha, *Putin's Kleptocracy*, 34–35. Belton, *Putin's People*, 15–16, 74.

3. Walter Laqueur, *Black Hundred: The Rise of the Extreme Right in Russia* (New York: Harper Perennial, 1994); Anton Shekhovtsov, *Russia and the Western Far Right* (New York: Routledge, 2018).

4. Nina Tumarkin, *The Living and the Dead: The Rise and Fall of the Cult of World War II in Russia* (New York: Basic Books, 1994).

5. Philip Short, *Putin* (New York: Henry Holt and Co., 2022), 222–225, 231, 235, 239–240, 293–295.

6. Shekhovtsov, *Russia and the Western Far Right*, 71.

7. Marelene Laruelle, *Russian Eurasianism: An Ideology of Empire* (Baltimore: Johns Hopkins University Press, 2012), see ch. 2; Snyder, *The Road to Unfreedom*, 87–88.

8. Shekhovtsov, *Russia and the Western Far Right*, 52–54.

9. Clove, *Black Wind, White Snow*, 200–206.

10. John B. Dunlop, "Aleksadr Dugin's *Foundations of Geopolitics*," *Demokratizatsiya* Vol. 12, No. 1 (31 January 2004), https://demokratizatsiya.pub/archives/Geopolitics.pdf

11. Dunlop, "Aleksadr Dugin's *Foundations of Geopolitics*."

12. Dunlop, "Aleksadr Dugin's *Foundations of Geopolitics*."

13. Clover, "In Moscow, a New Eurasianism," StopFake.org, 10 Feb 15, https://www.stopfake.org/en/in-moscow-a-new-eurasianism/

14. Belton, *Putin's People*, 182; Dawisha, *Putin's Kleptocracy*, 71.

15. Clove, *Black Wind*, 295.

16. Short, *Putin*, 443–445, Snyder, *The Road to Unfreedom*, 58–59.

17. Laqueur, *Black Hundred*, 84–85, 283.

18. Paul Hanebrink, *A Specter Haunting Europe: The Myth of Judeo-Bolshevism* (Cambridge, MA: Harvard University Press, 2018), 51.

19. Snyder, *The Road to Unfreedom*, 16.

20. Clove, *Black Wind*, 295.

21. McFaul, *From Cold War to Hot Peace*, 122.

22. Shekhovtsov, *Russia and the Western Far Right*, 78.

23. Kovalev, *Russia's Dead End*, 198.

24. Snyder, *The Road to Unfreedom*, 91–93.

25. Snyder, *The Road to Unfreedom*, 99–100.

26. Alexander Cooley and John Heathershaw, *Dictators without Borders: Power and Money in Central Asia* (New Haven: Yale University Press, 2017), 139.

27. Daria Sobakina, "Nikolai Patrushev against 'Velvet Revolutions,'" PolitiCom, 16 May 05, https://politcom.ru/274.html

28. "On the economic consequences of color revolutions,'" Pravda.ru, 9 Nov 05 https://www.pravda.ru/economics/66764-ob_ekonomicheskikh_posledstvijakh_cvetnykh_revoljucii/; Alexy Arbatov, "Color Revolutions and Greater Europe," RIA Novosti, 31 May 05, https://www.yabloko.ru/Publ/2005/2005_06/050622_arbatov_rian.html

29. "Gennaday Seleznev is convinced that 'orange revolutions' will not happen in either Belarus or in Russia," Union State website, https://xn—c1anggbdpdf.xn—p1ai/news/various/169978/

30. "Belarus: 'Revolution' is not expected," https://xn—c1anggbdpdf.xn—p1ai/news/various/170089/

31. "Color revolutions in the CIA as a consequence of the disintegration of the post-Soviet space: New geopolitical realities of the Union State," Union State website, 27 Apr 05, https://xn—c1anggbdpdf.xn—p1ai/news/various/170014/

32. "Color revolutions as a consequence of the network war: New geopolitical realities of the Union State," Union State website, https://xn—c1anggbdpdf.xn—p1ai/news/various/169905/. On Messner's theories and their relationship to so-called hybrid warfare, see Ofer Fridman, *Russian Hybrid Warfare: Resurgence and Politicisation* (London: Hurst, 2018), ch. 3.

33. "Union state-a cure for the 'orange revolution?,'" Union State website, 17 Jun 05, https://xn—c1anggbdpdf.xn—p1ai/news/various/169761/

34. V.D. Nightingale, "Color Revolutions and Russian," *Comparative Politics* 1/2011, https://mgimo.ru/files2/z03_2013/solovei.pdf; D.A. Iglin, "Color revolutions in post-Soviet states as a socio-political phenomenon: Definition and basic views," *Problems of the Post-Soviet Space* No. 3 (2016), https://www.postsovietarea.com/jour/article/view/94; V.I. Batyuk, "Russia, USA, and the 'Color Revolutions,'" *Bulletin of the Peoples' Friendship University of Russia*, 2006, https://cyberleninka.ru/article/n/rossiya-ssha-i-tsvetnye-revolyutsii

35. Mark B. Schneider, "Russian Nuclear Strategy," *Journal of Strategy and Politics* Vol. 2, No. 1 (2017), 121–140; S.J. Main, "Russia's Military Doctrine," Conflict Studies Research Centre Occasional Brief 77, Royal Military Academy Sandhurst, April 2000.

36. "Russia's National Security Concept" as translated by FBIS at https://www.arms control.org/act/2000–01/features/russias-national-security-concept

37. "Russia's National Security Concept."

38. Samuel Charap, "Strategic Sderzhivanie: Understanding Contemporary Russian Approaches to Deterrence," George C. Marshal Center for European Security, Security Insights, September 2020, No. 062, https://www.marshallcenter.org/en /publications/security-insights/strategic-sderzhivanie-understanding-conteporary -russian-approaches-deterrence-0

39. Samual Charap, "Strategic Sderzhivanie."

40. "Russia's Military Doctrine," https://www.armscontrol.org/act/2000–05/russias -military-doctrine

41. "Russia's Military Doctrine."

42. S.J. Main, "Russia's Military Doctrine," Conflict Studies Research Centre Occasional Brief 77, Royal Military Academy Sandhurst, April 2000.

43. Milton Leitenberg, "False allegations of biological-weapons use from Putin's Russia," *The Nonproliferation Review* 12 Oct 21, https://cissm.umd.edu/sites/default/files /2021–10/NonProliferationReview_False%20allegations%20of%20biological%20 weapons%20use%20from%20Putin%20s%20Russia.pdf

44. "The Foreign Policy Concept of the Russian Federation," 28 June 2000, https:// nuke.fas.org/guide/russia/doctrine/econcept.htm

45. "The Foreign Policy Concept of the Russian Federation," 12 July 2008, https:// russiaeu.ru/userfiles/file/foreign_policy_concept_english.pdf

46. Mark B. Schneider, "Russian Nuclear Strategy," *Journal of Strategy and Politics* Vol. 2, No. 1 (2017), 121–140.

47. Schneider.

48. Schneider.

49. Schneider.

50. "The Foreign Policy Concept of the Russian Federation 2013," https://www.vol tairenet.org/article202037.html

51. "The Foreign Policy Concept of the Russian Federation 2013."

52. "The Foreign Policy Concept of the Russian Federation 2013."

53. "The Military Doctrine of the Russian Federation, 25 December 2014," https://www. researchgate.net/publication/330109811_The_Military_Doctrine_of_the_Russian _Federation

54. Schneider, "Russian Nuclear Strategy," 121–140.

55. Michael Pietkiewicz, "The Military Doctrine of the Russian Federation," *Polish Political Science Yearbook*, No. 3 (2018), 505–520.

56. "Russian National Security Strategy, 31 December 2015," https://www.russiamatters .org/node/21421

57. "Russian National Security."

58. "The Foreign Policy Concept of the Russian Federation 2016," https://www.vol tairenet.org/article202038.html

59. Oscar Jonsson, "Myths and Misconceptions around Russian Military Intent," Chatham House, 14 Jul 22, https://www.chathamhouse.org/2022 /06/myths-and-misconceptions-around-russian-military-intent/myth-1-russia -waging-grey-zone

60. Kevin N. McCauley, *Russian Influence Campaigns against the West: From the Cold War to Putin* (North Charleston: Create Space, 2016); Brian D. Dailey and Patrick J. Parker, *Soviet Strategic Deception* (Toronto: Lexington Books, 1987).

61. Per Hogselius, *Red Gas: Russia and the Origins of European Energy Dependence* (New York: Palgrave-Macmillan, 2013); Jeronim Perovic (ed.), *Cold War Energy: A Transnational History of Soviet Oil and Gas* (New York: Palgrave-Macmillan, 2016).

62. Peter Schweizer, *Victory: The Reagan Administration's Secret Strategy that Hastened the Collapse of the Soviet Union* (New York: Basic Books, 1994); Peter Schweizer, *Reagan's War: The Epic Story of His Forty-Year Struggle and Final Triumph over Communism* (New York: Doubleday, 2002); Douglas E. Streusand et al., *The Grand Strategy that Won the Cold War: Architecture of Triumph* (London: Lexington Books, 2016).

63. Andrei Soldatov, "Kremlin.com," *Index on Censorship*, Vol. 39, No. 1 (2010), 71–78.

64. Antonio Missiroli et al. "Issue Report No. 30: Strategic Communications East and South," European Union Institute for Security Studies, https://ethz.ch/content /dam/ethz/special-interest/gess/cis/center-for-securities-studies/resources /docs/EUISS%20Strategic%20Communications%20East%20and%20South.pdf; Guillaume Lasconjarias and Jeffrey A. Larsen (eds.), *NATO's Response to Hybrid Threats* (Rome: NATO Defense College, 2015), 127.

65. NATO Strategic Communications Centre of Excellence, "2007 Cyber Attacks on Estonia, https://stratcomcoe.org/cuploads/pfiles/cyber_attacks_estonia.pdf

66. Lasconjarias and Larsen, *NATO's Response to Hybrid Threats*, 126; Clint Watts, *Messing with the Enemy: Surviving in a Social Media World of Hackers, Terrorists, Russians, and Fake News* (New York: Harper Collins, 2018), 139.

67. Pomerantsev, *This Is Not Propaganda*, 65.

68. Victor Litovkin, "CSTO set up against color revolutions," *Independent Military Review*, 9 Sep 11, https://nvo.ng.ru/realty/2011–09–09/1_odkb.html

69. Ivan Safronov Jr., "The General Staff is preparing for war," *Kommersant*, 18 Nov 11, https://www.komersant.ru/doc/1818296

70. When the Interregional Public Organization's "Joint Edition of the Cossack Mass Media Cossack Information and Analytical Center" publishes Makarov's remarks in detail on the same day they were made, you know deliberate official widespread distribution is in progress. "The Russian General Staff puzzled America," KIAC, 18 Nov 11, https://kazak-center.ru/news/rossijskij _genshtab_ozadachil_ameriku/2011–11–18–1687

71. "The Army announced the call for contract soldiers," MKRU, 28 Mar 11, https:// www.mk.ru/politics/2011/03/28/576306-armiya-obyavila-prizyiv-kontraktnikov .html; Victor Litovkin, "The Chief of the General Staff Diagnosed," *Independent Military Review*, 1 Apr 11. See also Roger McDermott, "The Brain of the Russian Army Degenerates," *Eurasia Daily Monitor*, 6 Apr 11, https://jamestown.org/program/the -brain-of-the-russian-army-degenerates/

72. Timothy D. Conley, "Protecting the Status Quo: The Defense against a Russian Color Revolution," Naval Post Graduate School Thesis, December 2017, 16.

73. Pomerantsev, *This Is Not Propaganda*, 89–91.

74. Watts, *Messing with the Enemy*, 92.

75. Martin Moore, *Democracy Hacked: Political Turmoil and Information Warfare in the Digital Age* (New York: OneWorld Publications, 2018), xiii.

76. Bill Gertz, *iWar: War and Peace in the Information Age* (New York: Threshold Editions, 2017), 18–19.

77. "Band of Brothers: The Wagner Group and the Russian State," Center for Strategic and International Studies, 21 Sep 20, https://www.csis.org/blogs/post-soviet-post/band -brothers-wagner-group-and-russian-state

78. Moore, *Democracy Hacked*, 94.

79. Valery Gerasimov, "The Value of Science in Foresight," *Military Industrial Courier*, 27 Feb 13, https://vpk.name/news/85159_cennost_nauki_v_predvidenii.html

80. Fridman, *Russian Hybrid Warfare*, 77–81, 83–86, 95.

81. Fridman, 163.

82. Fridman, 143.

83. Fridman, 144.

84. Marek Menkiszak, *Russia's Long War on Ukraine* (Washington, D.C.: Transatlantic Academy, 2016), https://www.gmfus.org/sites/default/files/Menkiszak_Russias LongWar_Feb16_web.pdf

85. Michael Korman et al., *Lessons from Russia's Operations in Crimea and Eastern Ukraine* (Santa Monica: RAND Corporation, 2017), 30; Oleksandr Gavrylyuk, "Russian Invasion: Improvisation or Long-Term Planning?," Eurasian Monitor, 12 Dec 14, https://jamestown.org/program/russian-invasion-improvisation-or-long-term-planning/. Andrei Illarionov, Putin's economic advisor, says contingency planning started in 2003 because Putin thought it was unacceptable that two key places of the Russian Orthodox Church were in Ukraine and then the plan was

adopted in 2004 after the Orange Revolution. This smacks of post hoc justification and is reminiscent of Serb excuses for repression in Kosovo in 1999.

86. Taras Kuzio and Paul D'Anieri, "Annexation and Hybrid Warfare in Crimea and Eastern Ukraine," E-International Relations, 25 Jun 18, https://www.e-ir.info/2018/06/25/annexation-and-hybrid-warfare-in-crimea-and-eastern-ukraine/

87. "Hearing before the Subcommittee on Europe and Regional Security Cooperation of the Committee on Foreign Relations United States Senate, March 4, 2015," https://www.govinfo.gov/content/pkg/CHRG-114shrg97882/html/CHRG-114shrg97882.htm

88. "Koval: Russia began preparations for a hybrid war in Ukraine at least 8 years ago," ZN.UA, 14 Jul 14, https://zn.ua/UKRAINE/koval-rossiya-nachala-podgotovku-k-gibridnoy-voyne-v-ukraine-kak-minimum-let-8-nazad-148949_.html

89. Cathy Young, "The Sci-Fi Writers War: They predicted and possibly inspired the conflict in the Ukraine and now they're fighting it," *Slate*, 11 Jul 14, https://slate.com/news-and-politics/2014/07/science-fiction-writers-predicted-ukraine-conflict-now-theyre-fighting-it.html

90. Menkiszak, *Russia's Long War on Ukraine*, 1–7.

91. Menkiszak, 1–7.

92. "Turchynov: Yanukovych knew about Russia's preparations for aggression as early as 2013," LB.UA., 27 Jun 14, https://lb.ua/news/2014/06/27/271207_turchinov_yanukovich_znal.html

93. *"Little Green Men": A Primer on Modern Russian Unconventional Warfare, Ukraine 2013–2014*, (Ft. Bragg: U.S. Army Special Operations Command, n.d.), 53.

94. Michael Korman et al., *Lessons from Russia's Operations in Crimea and Eastern Ukraine*, 30.

95. Menkiszak, *Russia's Long War on Ukraine*.

96. Korman et al., *Lessons*, 8.

97. "Little Green Men," 54.

98. Halya Coynash, "Odesa Smoking Gun Leads Directly to Moscow," Kharkiv Human Rights Protection Group, 20 Sep 19, https://khpg.org/en/1473972066?fbclid=IwAR308l4OrCpai9yBd2TRZLWoqocBx77Psrfi5WY5W4t68Yx42RsgiYftRTM

99. "Little Green Men," 44, 58.

100. "Little Green Men," 58; Korman et al., *Lessons*, 34.

101. Catalin Alin Costea, "Russia's Hybrid War in Ukraine (2014–2018)," SETA Foundation for Political, Economic, and Social Research, Istanbul, 2019, https://setav.org/en/assets/uploads/2019/12/R147En.pdf, 11.

102. Coynash, "Odesa Smoking Gun,"

103. Costea, "Russia's Hybrid War," 51.

104. Korman et al., *Lessons*, 34, 38, 42, 43, 60, 61; "Little Green Men," 31–32.

105. Confidential correspondence.

Chapter 2. Red Forces

1. Peter Conradi, *Who Lost Russia? How the World Entered a New Cold War* (London: Oneworld Publications, 2017), 177.

2. Markus Ekstrom, *Rysk operative-strategisk ovningsverksamhet under 2009 och 2010* (Stockholm: Swedish Defense Research Agency, 2010), https://www.foi.se/rappo rtsammanfattning?reportNo=FOI-R—3022—SE, 4; Mark Galeotti, *Putin's Wars: From Chechnya to Ukraine* (Oxford: Osprey, 2022), 88–89.

3. Ekstrom, *Rysk operative-strategisk ovningsverksamhet under 2009 och 2010*, 30–39, 42–48.

4. Ekstrom, 49–56, 61.

5. Johan Norberg, *Training to Fight: Russia's Military Exercises 2011–2014* (Stockholm: Swedish Defense Research Agency, 2015), 7.

6. Norberg, 27–30.

7. Norberg, 33.

8. Norberg, 34–35.

9. Norberg, 38.

10. Norberg, 42.

11. Norberg, 42–43.

12. Ian Brzezinski and Nicholas Varangis, "The NATO-Russia exercise Gap . . . Then, now, & 2017," Atlantic Council, https://www.atlanticcouncil.org/blogs/natosource /the-nato-russia-exercise-gap-then-now-2017/

13. "Combat Potential," *Encyclopedia of the Strategic Missile Forces,* https://rvsn.aca demic.ru/1683/%D0%91%D0%BE%D0%B5%D0%B2%D0%BE%D0%B9_%D0% BF%D0%BE%D1%82%D0%B5%D0%BD%D1%86%D0%B8%D0%B0%D0%BB

14. Jody Edmonstone, "Canada's Expanded ADIZ," Canadian Forces College paper, 2018.

15. Frédéric Lasserre and Pierre-Louis Têtu, "Russian Air Patrols in the Arctic: Are Long-Range Bomber Patrols a Challenge to Canadian Security and Sovereignty?," *Arctic Yearbook 2016,* https://arcticyearbook.com/images/yearbook/2016/Scholarly _Papers/11.Lasserre-and-Tetu.pdf

16. Nigel Walker, *Ukraine Crisis: A Timeline (2014-Present)* (London: House of Commons Library, 2022), https://commonslibrary.parliament.uk/research-briefings/cbp -9476/, 1–11; Korman et al., *Lessons,* 7.

17. Norberg, *Training to Fight,* 49.

18. Norberg, *Training to Fight.*

19. Norberg, 50.

20. Jacek Durkalec, *Nuclear-Backed "Little Green Men": Nuclear Messaging in the Ukraine Crisis* (Warsaw: Polish Institute of International Affairs, 2015), https://www.files. ethz.ch/isn/193514/Nuclear%20Backed%20%E2%80%9CLittle%20Green%20

Men%E2%80%9D%20Nuclear%20Messaging%20in%20the%20Ukraine%20Crisis.
pdf 9.

21. Walker, *Ukraine Crisis: A Timeline (2014–Present)*, 13–14; Korman et al., *Lessons*, ch. 3.

22. Zachery Keck, "Russia's Military Begins Massive Nuclear War Drill," *The Diplomat*, 29 Mar 14, https://thediplomat.com/2014/03/russias-military-begins-massive-nuclear-war-drill/; Oleg Vladykin, "Readiness for a massive nuclear strike," *Nezavisimaya Gazeta*, 26 Mar 14, https://www.ng.ru/armies/2014–03–26/6_strike.html

23. Durkalec, *Nuclear-Backed*, 10.

24. "UPDATE-1: Russian jet's passes near U.S. ship in Black Sea 'provocative'-Pentagon," 14 Apr 14, https://www.reuters.com/article/usa-russia-blacksea-idUSL2N0N6OV 520140414#TmlqFb14qYOCTCUy.

25. Thomas Frear, Lukasz Kulesa, and Ian Kearns, *Dangerous Brinksmanship: Close Military Encounters between Russia and the West in 2014*, European Leadership Network Policy Brief, November 2014, https://www.europeanleadershipnetwork. org/wp-content/uploads/2017/10/Dangerous-Brinkmanship.pdf

26. "Putin oversees Russian nuclear exercise amid Ukraine tensions," CBS News, 8 May 14, https://www.cbsnews.com/news/putin-oversees-russian-nuclear -exercise-amid-ukraine-tensions/

27. CBS News; "Russia hold military drills to repel nuclear strike," Russia Today (RT), 8 May 14, https://www.rt.com/news/157644-putin-drills-rocket-launch/; Pavel Podvig, "Multiple missile launches during a command and control exercise," http://russianforces.or/blog/2014–05/multiple_missile_launches_duri.shtml

28. Durkalec, *Nuclear-Backed*, 10–11.

29. Norberg, *Training to Fight*, 53.

30. Durkalec, *Nuclear-Backed*, 7.

31. Durkalec, 7.

32. Frear, Kulesa, and Kearns, *Dangerous Brinksmanship*.

33. Frear, Kulesa, and Kearns.

34. Durkalec, *Nuclear-Backed*.

35. Bill Gertz, "Russian Strategic Bombers Near Canada Practice Cruise Missile Strikes on US," *Washington Free Beacon*, 8 Sep 14, https://freebeacon.com/national-security/ russian-strategic-bombers-near-canada-practice-cruise-missile-strikes-on-us/

36. Alexi Anishchuck, "Don't mess with nuclear Russia, Putin says," *Reuters*, 29 Aug 14, https://www.reuters.com/article/us-russia-putin-conflict-idUSKBN0GT1D 420140829

37. Paula Meija, "U.S. and Canada Intercept Eight Russian Military Jets off North American Coastline," *Newsweek*, 20 Sep 14, https://www.newsweek.com/us-and -canada-intercept-eight-russian-military-jets-north-american-coastline-272042

38. Norberg, *Training to Fight*, 47–49.

39. Michael Birnbaum, "NATO says Russian jets, bombers circle Europe in unusual incidents," *Washington Post*, 29 Oct 14; John Vandiver, "NATO cites 'unusual' Russian air activity as intercepts rise," *Stars and Stripes*, 30 Oct 14, https://www.stripes.com/news/nato-cites-unusual-russian-air-activity-as-intercepts-rise-1.311104

40. Foxall, *Close Encounters*, 12–13.

41. "Russian military planes buzzed HMCS Toronto in Black Sea," CBC, 8 Sep 14, https://web.archive.org/web/20180217155109/http://www.cbc.ca/news/world/russian-military-planes-buzzed-hmcs-toronto-in-black-sea-1.2759843; Frear, Kulesa, and Kearns, *Dangerous Brinksmanship*.

42. Korman et al., *Lessons*, 44–45, 47, 48, 52, 70.

43. Lucy Pawle, "UK summons Russian envoy after bombers fly over English Channel," CNN, 30 Jan 15, https://www.cnn.com/2015/01/29/world/uk-russia-bombers-intercepted; "Russia says patrols near UK airspace were 'routine,'" BBC, 30 Jan 15, https://www.bbc.com/news/uk-31053371; "RAF jets scramble after Russian bombers seen off Cornwall," BBC, 20 Feb 15, https://www.bbc.com/news/uk-31530840

44. "RAF jets intercept Russian planes near UK airspace," BBC, 14 Apr 15, https://www.bbc.com/news/uk-32308307; regarding the disruptive nature of Russian activities on air traffic control, there is a useful discussion of the technical matters at hand at "The Professional Pilots Rumour Network" (PPRUNE) thread, "TU-95 Intercept," located at https://www.pprune.org/archive/index.php/t-556792.html, discussion points from 18 Feb 15 to 7 Mar 15.

45. Heather A. Conley and Caroline Rohlof, *The New Ice Curtain: Russia's Strategic Reach into the Arctic* (London: Rowman and Littlefield, 2013), 2–6, 87.

46. "Japan Scrambles Jets to Intercept Russian Bombers," SOFREP, https://sofrep.com/fightersweep/japan-scrambles-jets-intercept-russian-bombers/; "Why the Russian TU-95 bombers that intruded U.S. air space weren't intercepted," *Business Insider*, 4 May 15, https://www.businessinsider.com/why-the-russian-tu-95-bombers-that-intruded-us-air-space-werent-intercepted-2015-5

47. Adrian A. Mojica, "Two Pairs of Russian Bombers Intercepted by US Fighter Jets," FOX17 News, 6 Jul 15, https://fox17.com/news/local/gallery/report-two-pairs-of-russian-bombers-intercepted-by-us-fighter-jets; Vasudevan Sridharan, "Washington jets intercept two-nuclear capable Russian bombers on Independence Day," *International Business Times*, 7 Jul 15, https://www.ibtimes.co.uk/washington-jets-intercept-two-nuclear-capable-russian-bombers-us-independence-day-1509707; Brian Todd and Jethro Mullen, "July Fourth message not the first from Russian bombers," CNN, 23 Jul 15, https://www.cnn.com/2015/07/23/politics/us-russian-bombers-july-4-intercept

48. Matthew Bodner, "6th Russian Air Force Crash Raises Concerns over Aircraft Safety," *Moscow Times*, 13 Jul 15, https://www.themoscowtimes.com/2015/07/13

/putin-orders-cuts-to-interior-ministry-payroll-a48102; Juliusz Sabak, "Tu-95 Crashes in Russia. Sixth Airframe Lost This Summer," *Defense* 24, 15 Jul 15, https://defense24 .com/armed-forces/tu-95-crashes-in-russia-sixth-airframe-lost-this-summer

49. "Photos Show RAF Typhoons Intercepting Russian Tu-160 Bombers," *Military .com*, 11 Sep 15, https://www.military.com/defensetech/2015/09/11/photos-show-raf-typhoons-intercepting-russian-tu-160-bombers; "RAF fighter jets intercept Russian planes over North Sea," BBC, 11 Sep 15, https://www.bbc.com/news/uk-scotland -north-east-orkney-shetland-34227249; "RAF Lossiemouth fighter jets scrambles over Russian planes," BBC, 20 Nov 15, https://www.bbc.com/news/uk-scotland -north-east-orkney-shetland-34880523; "US fighter jets intercept Russian bombers over the Sea of Japan," *The Guardian*, 20 Oct 15, https://www.theguardian.com /us-news/2015/oct/30/us-fighter-jets-intercept-russian-bombers-over-sea-of-japan

50. Tom Cooper, *Moscow's Game of Poker: Russian Military Intervention in Syria, 2015–2017* (Warwick: Helion, 2018), 34.

51. Peter Konradi, *Who Lost Russia? How the World Entered a New Cold War* (London: Oneworld Books, 2017), xi.

52. "British fighter jets scrambles to intercept 2 Russian strategic bombers over Atlantic," RT, 20 Nov 15, https://www.rt.com/news/322857-uk-fighters-russian-bombers/

53. Duncan Allen and Ian Bond, *A new Russian policy for post-Brexit Britain* (London: Chatham House, 2022), https://www.chathamhouse.org/sites/default/files/2022 –01/2022–01–27-russia-policy-post-Brexit-britain-allan-bond.pdf, 9; Rachel Elle-huus, "Did Russia Influence BREXIT?," CSIS, 21 Jul 20, Brexit.

54. "Britain scrambles fighters to intercept Russian bombers," Reuters, 17 Feb 16, https://www.cnbc.com/2016/02/17/britain-scrambles-fighters-to-intercept-russian -bombers.html; "Jets from four countries intercept supersonic Russian bomb-ers known as 'White Swans' over the Atlantic," *National Post*, 5 Oct 16, https:// nationalpost.com/news/world/jets-from-four-countries-intercept-two -russian-bombers-known-as-white-swans-over-the-atlantic; "UK fighters scrambled to intercept Russian bombers," RT, 23 Sep 22, https://www.rt.com/uk/360344 -uk-fighters-russian-bombers/

55. "Panic in Western Media over 'Russian Bombers' Intercepted by French Jets," Sputnik News, 5 Oct 16, https://sputniknews.com/20161005/panic-media-russian -bombers-1046037076.html

56. "Video of TU-95MS flight off the coast of Alaska accompanied by Su-35, Top-War.ru, https://en.topwar.ru/115071-video-poleta-tu-95ms-u-beregov-alyaski-v -soprovozhdenii-su-35.html

57. Sam LaGrone, "Pentagon: Russian Fighter Conducted 'Unsafe' Intercept of U.S. Recon Plane over Black Sea," USNI News, 28 Jan 16, https://news.usni.org/2016/01/28/penta-gon-russian-fighter-conducted-unsafe-intercept-of-u-s-recon-plane-over-black-sea;

Sophie Tatum and Barbara Starr, "Russia denies wrongdoing after jet barrel-rolls over U.S. aircraft," CNN, 17 Apr 16, https://www.cnn.com/2016/04/16/politics/russian-jet-barrel-rolled-us-aircraft/; Sam LaGrone, "Updated: Russian Fighter Came within 10 Feet of Navy Surveillance Plane over Black Sea," USNI News, 7 Sep 16, https://news.usni.org/2016/09/07/russian-fighter-came-within-10-ft-navy-surveillance-plane-black-sea

58. "The Strategic Missile Forces conducted large-scale exercises throughout the country," TopWar.Ru, 31 Oct 16, https://en.topwar.ru/102909-rvsn-proveli-krupnye-ucheniya-po-vsey-strane.html; "Attack 'saboteurs' on the missile system Topol-M repulsed in the suburbs," TopWar.ru, 14 Dec 16, https://en.topwar.ru/105580-ataku-diversantov-na-raketnyy-kompleks-topol-m-otbili-v-podmoskove.html; "Loading the YARS complex rocket into the mine," TopWar.ru, 15 Dec 16, https://en.topwar.ru/105621-zagruzka-rakety-kompleksa-yars-v-shahtu.html; "Yars entered combat duty in the Kozelsky division," TopWar.Ru, 16 Dec 16, https://en.topwar.ru/105692-yarsy-postupili-na-boevoe-dezhurstvo-v-kozelskuyu-diviziyu.html

59. Joby Warrick, *Red Line: The Unraveling of Syria and America's Race to Destroy the Most Dangerous Arsenal in the World* (New York: Doubleday, 2021), ch. 19.

60. Chris Klint, "Air Force fighters scramble from Alaska to intercept Russian bombers," *Alaska Dispatch News*, 18 Apr 17, https://www.arctictoday.com/air-force-fighters-scramble-from-alaska-to-intercept-russian-bombers/; Luis Martinez, "Russian aircraft fly close to Alaska for 4th time in 4 days," ABC7 News, 21 Apr 17, https://abc7.com/news/russian-aircraft-fly-close-to-alaska-for-4th-time-in-4-days/1900780/

61. Martinez, "Russian aircraft."

62. Lee Berthiuame, "Canadian jets intercept Russian bombers, 1st time since 2014," CBC, 21 Apr 17, https://www.cbc.ca/news/canada/north/russia-jets-intercepted-canada-north-coast-1.4081048

63. Luis Martinez, "US intercepts Russian bombers and fighters near Alaska," ABC7 News, 4 May 17, https://abc7ny.com/1954849/; Steven Thorne, "Testing North American Air Space," *Legion Magazine*, 10 May 17, https://legionmagazine.com/en/testing-north-american-air-space/

64. Jody Edmonstone, "Canada's Expanded ADIZ," Canadian Forces College paper, 2018.

65. David Axe, "Russia and America Are Taking Turns Probing Each Other's Air Space," *The National Interest*, https://nationalinterest.org/blog/buzz/russia-and-america-are-taking-turns-probing-each-other%E2%80%99s-air-space-59027

66. "Russia Sends Nuclear-Capable Bombers to Base Near the Alaskan Coast in Exercise," RFE/RL, 14 Aug 19, https://www.rferl.org/a/russia-sends-nuclear-capable-bombers-to-base-near-u-s-alaskan-coast-in-exercise/30110004.html

67. Pavel Podvig, "Annual exercise of the strategic forces, this time without ICBMs," 11 Oct 18, www.russianforces.or/blog/2018/10/annual_exercise_of_the_strateg .shtml

68. Thomas Nilsen, "Salvo of four Bulava missiles hit test range on Kanin Peninsula," *Barents Observer*, 12 Dec 20, https://thebarentsobserver.com/en/security/2020/12 /salvo-four-bulava-missiles-hit-test-range-cape-kanin

69. Maloney, *Deconstructing Dr. Strangelove*, 296.

70. Maxim Starchak, "Key features of Russia's GROM 2019 Nuclear Exercise," *Eurasia Daily Monitor* Vol. 16, Issue 150 (29 Oct 19), https://jamestown.org/program/key -features-of-russias-grom-2019-nuclear-exercise/

71. Pavel Baev, "The GROM-2019 Exercise Illuminated the Risks of Nuclear Renaissance in Russian Strategic Culture," Marshal Center, January 2020, https://www .marshallcenter.org/en/publications/security-insights/grom-2019; Thomas Nilsen, "Russia announces massive trans-Arctic nuclear war games," *Barents Observer*, 14 Oct 19, https://thebarentsobserver.com/en/security/2019/10/russia-announces -massive-nuclear-war-games; Franz-Stefan Gady, "Russian Sub Aborts Test Firing of Ballistic Missile during Exercise," *The Diplomat*, 22 Oct 19, https://thediplomat .com/2019/10/russian-sub-aborts-test-firing-of-ballistic-missile-during-exercise/

72. "Submarine Launched Ballistic Missiles," Russian Space Web, http://www.russian spaceweb.com/rockets_slbm.html

73. Baev, "GROM-2019."

74. Thomas Nilsen, "Salvo of four Bulava missiles hit test range on Kanin Peninsula," *Barents Observer*, 12 Dec 20, https://thebarentsobserver.com/en/security/2020/12 /salvo-four-bulava-missiles-hit-test-range-cape-kanin

75. Matthew Stein, *Compendium of Central Asian Military and Security Activity* (Ft. Leavenworth: Foreign Military Studies Office, 2021), 61.

76. Alexander Golts and Michael Korman, *Russia's Military: Assessment, Strategy, and Threat* (Washington, D.C.: Center on Global Interests, June 2016), 13.

77. Norberg, *Training for War*, 36, 42, 53.

78. Anna Maria Dyner, "KAVKAZ 2016: The Next Test of Russia's Armed Forces," Polish Institute of International Affairs, https://pism.pl/publications/_Kavkaz _2016___The_Next_Test_of_Russia_s_Armed_Forces

79. Pavel Felgenhauer, "Massive Russian Troop Deployments and Exercises Held Close to Ukraine," *Eurasian Daily Monitor* Vol. 13, No. 44, https://jamestown.org/program /massive-russian-troop-deployments-and-exercises-held-close-to-ukraine/

80. Norberg, *Training for War*, 39.

81. "ZAPAD-2017: Lessons Learned," Warsaw Institute Foundation, 2017, www .warsawinstitute.org

82. "ZAPAD-2017: Lessons Learned."

83. Dave Johnson, "ZAPAD 2017 and Euro-Atlantic Security," *NATO Review*, 14 Dec 17, https://www.nato.int/docu/review/articles/2017/12/14/zapad-2017-and-euro -atlantic-security/index.html

84. Dominik P. Janowski, "The Eastern Flank after ZAPAD-2017," Barcelona Institute for International Affairs, https://www.cidob.org/en/content/download/68077/2066573 /version/6/file/09–16_DOMINIK%20P.%20JANKOWSKI_ANG.pdf

85. "ZAPAD-2017: Lessons Learned."

86. Keir Giles, "Russia Hit Multiple Targets with ZAPAD-2017," Carnegie Endowment for International Peace, January 2018, https://carnegieendowment.org/2018/01/25 /russia-hit-multiple-targets-with-zapad-2017-pub-75278

87. Dave Johnson, "ZAPAD 2017 and Euro-Atlantic Security," *NATO Review*, 14 Dec 17, https://www.nato.int/docu/review/articles/2017/12/14/zapad-2017-and-euro -atlantic-security/index.html

88. Janowski, "The Eastern Flank after ZAPAD-2017."

89. Johnson, "ZAPAD 2017."

90. *Russia's Strategic Exercises: Messages and Implications*, NATO Strategic Communications Centre of Excellence, July 2020, https://stratcomcoe.org/cuploads/pfiles /ru_strat_ex_29–07-e147a.pdf

91. Dave Johnson, "VOSTOK 2018: Ten years of Russian strategic exercises and warfare preparation," *NATO Review*, 20 Dec 18, https://www.nato.int/docu/review /articles/2018/12/20/vostok-2018-ten-years-of-russian-strategic-exercises-and -warfare-preparation/index.html

92. *Russia's Strategic Exercises: Messages and Implications*.

93. *Exercise KAVKAZ 2020-A Final Test of Russian Military Reform?*, NATO Strategic Communications Centre of Excellence, https://stratcomcoe.org/cuploads/pfiles /Kavkaz-StratCom_30–03–2021.pdf

94. FAA, "Notice to Air Missions (NOTAMS)," https://www.faa.gov/about/initiatives /notam/what_is_a_notam#:~:text=Notice%20to%20Air%20Missions%20(%20 NOTAMs%20)&text=A%20NOTAM%20is%20a%20notice,)%20E2%80%93%20 not%20the%20normal%20status.

95. Kristian Atland, Thomas Nilsen, and Torbjorn Pedersen, "Military Muscle-Flexing as Interstate Communication: Russian NOTAM Warnings off the Coast of Norway, 2015–2021," *Scandinavian Journal of Military Studies* Vol. 5, No. 1 (2022), 63–78.

96. "Iskander short-range ballistic missiles," http://www.military-today.com/missiles /iskander.htm#:~:text=The%20Iskander%20(Western%20reporting%20name ,the%20SS%2DX%2D26.

97. "Iskander short-range ballistic missiles."

98. "Iskander short-range ballistic missile," *Russian Space Web*, 5 May 19, http://www.russian-spaceweb.com/iskander.html; Piotr Zochowski, "Iskander missiles in Kaliningrad:

A constant element of Russia's policy of intimidation," Centre for Eastern Studies, 11 Oct 16, https://www.osw.waw.pl/en/publikacje/analyses/2016–10–11/iskander -missiles-kaliningrad-a-constant-element-russias-policy

99. "Russia transporting Iskander missile system to Kaliningrad," *Lithuanian Tribune*, 7 Oct 16, https://lithuaniatribune.com/russia-transporting-iskander -missile-system-to-kaliningrad/

100. "Interview with CGS Nikolai Makarov," RG.RU, 23 Mar 10, https://rg.ru/2010/03/23 /makarov.html

101. Zochowski, "Iskander missiles."

102. Wales Summit Declaration, 5 Sep 14, https://www.nato.int/cps/en/natohq/official _texts_112964.htm

103. Congressional Research Service, *Russian Compliance with the Intermediate Range Nuclear Forces (INF) Treaty: Background and Issues for Congress*, 2 Aug 19, https:// sgp.fas.org/crs/nuke/R43832.pdf, 2.

104. Adarsh Vijay, "Iskander-M in Kaliningrad: The Changing Equations of Deter- rence," Institute of Peace and Conflict Studies, 9 Nov 16, http://www.ipcs.org /comm_select.php?articleNo=5172; "Russia: New missiles in Kaliningrad answer US Shield," Al-Jazeera, 21 Nov 16, https://www.aljazeera.com/news/2016/11/21 /russia-new-missiles-in-kaliningrad-answer-us-shield; "Russia Moves Nuclear Capable Missiles into Kaliningrad," *Politico*, 8 Oct 16, https://www.politico.eu/article /russia-moving-nuclear-capable-missiles-into-kaliningrad/; "Troubled Waters," *Medium*, 7 Oct 16, https://medium.com/dfrlab/troubled-waters-7a1319917ef7# .n59iaupa7

105. Sergey Sukhankin, "Kaliningrad: From boomtown to battlestation," European Council on Foreign Relations, 27 Mar 17, https://ecfr.eu/article/commentary _kaliningrad_from_boomtown_to_battle_station_7256/

106. "Media reports Russian troops deploying Iskander systems to Kaliningrad," *Kyiv Post*, 4 Dec 17, https://www.kyivpost.com/eastern-europe/media-reports-russian -troops-deploying-iskander-systems-kaliningrad-region.html

107. Julian Borger, "Kaliningrad photos appear to show Russia upgrading nuclear weap- ons bunker," *Guardian*, 18 Jun 18, https://www.theguardian.com/world/2018/jun/18 /kaliningrad-nuclear-bunker-russia-satellite-photos-report; Hans Kristensen, "Russia Upgrades Nuclear Weapons Storage Site in Kaliningrad," Federation of American Scientists, 18 Jun 18, https://fas.org/blogs/security/2018/06/kaliningrad/; "Russia deploys Iskander nuclear-capable missiles to Kaliningrad: RIA," Reuters, 5 Feb 18, https://www.reuters.com/article/us-russia-nato-missiles-idUSKBN1FP21Y; "Russia May Send More Iskander Missiles to Kaliningrad," Radio Free Europe /Radio Liberty, 13 Oct 17, https://www.rferl.org/a/russia-reported-may-send iskander-missiles-kanliningrad-us-armored-divisions-poland/28790998.html;

"Russian Baltic Fleet simulates Iskander missile launches," Xinhua, 26 Aug 19, http://www.xinhuanet.com/english/2019–08/26/c_138337663.htm

Chapter 3. Blue Forces

1. William W. Mendel and Graham H. Turbiville Jr., *The CINCs' Strategies: The Combatant Command Process* (Carlisle: U.S. Army War College Press, 1997), 4, https://press.armywarcollege.edu/cgi/viewcontent.cgi?article=1174&context=monographs
2. Mendel and Turbiville, 63.
3. Mendel and Turbiville, 15.
4. Mendel and Turbiville, 16.
5. "Chronology of US-North Korean Missile Diplomacy," Arms Control Association, https://www.armscontrol.org/factsheets/dprkchron specifically 1998–99 entries.
6. *Strategic Force Planning Support Program*, Defense Threat Reduction Agency, September 2000, https://apps.dtic.mil/sti/pdfs/ADA388677.pdf, 2.
7. *Strategic Force Planning Support Program*, 7.
8. *Strategic Force Planning Support Program*, 9.
9. *Strategic Force Planning Support Program*, A-10.
10. *Strategic Force Planning Support Program*, A-15, A-27.
11. *Strategic Force Planning Support Program*, A-30.
12. *Strategic Force Planning Support Program*, A-34.
13. *Strategic Force Planning Support Program*, A-39–A40.
14. USSTRATCOM, "Annex C to USCINCSTRAT OPLAN 8044–98: Operations," https://www.governmentattic.org/38docs/USSTRATCOMannexcOPLAN 8044_2001.pdf, C-2.
15. For the details, see Maloney, *Emergency War Plan*.
16. USSTRATCOM, "Annex C," C-3-C-2.
17. FOIA USSTRATCOM, "Annex C to USCINCSTRAT OPLAN 8044–98: Operations," https://www.governmentattic.org/38docs/USSTRATCOMannexc OPLAN8044_2001.pdf, C-3-D-5, C-3-D_6
18. FOIA USSTRATCOM, "Annex C to USCINCSTRAT OPLAN 8044–98: Operations," https://www.governmentattic.org/38docs/USSTRATCOMannexc OPLAN8044_2001.pdf, C-1-B-4, C-1-C2, C-7-7.
19. Nuclear Information Project, "US Changes Name of War Plan," 21 Dec 04, https://www.nukestrat.com/us/stratcom/siopname.htm
20. GAO, "Military Transformation: Additional Actions Needed by U.S. Strategic Command to Strengthen Implementation of Its Many Missions and New Organization," 8 Sep 2006, https://www.govinfo.gov/content/pkg/GAOREPORTS-GAO-06-847/html/GAOREPORTS-GAO-06-847.htm

21. The Emergency War Plans of the 1950s had prehostilities aspects to them, specifically, altering and generating aircraft during times of tensions to various degrees in this preDEFCON environment.

22. Hans M. Kristensen, "Global Strike: A Chronology of the Pentagon's New Offensive Strike Plan," Federation of American Scientists, 15 Mar 2006, https://www.nukestrat.com/pubs/GlobalStrikeReport.pdf, 4.

23. Kristensen, 5.

24. Kristensen, 7.

25. Kristensen, 13–14,

26. Kristensen, 47.

27. Kristensen, 40.

28. "Trident D-5," http://www.astronautix.com/t/tridentd-5.html

29. Kristensen, "Global Strike," 24.

30. Kristensen, 9.

31. Kristensen, 9.

32. GAO, "Military Transformation"; "108th ORI, Best Seen to Date, *Guardlife*, Vol. 32, No. 4, https://www.nj.gov/military/publications/guardlife/volume32no4/108th_ori.html

33. "Air Force Programs: Integrated Strategic Planning and Analysis Network," https://www.globalsecurity.org/military/library/budget/fy2007/dot-e/af/2007ispan.pdf; Rich Bungaden, "Exercises important to McGuire readiness," 27 Apr 2007, https://www.jbmdl.jb.mil/News/Article-Display/Article/246542/exercises-important-to-mcguire-readiness/

34. Robert M. Gates, *Duty: Memoirs of a Secretary at War* (New York: Alfred A. Knopf, 2014), ch. 5.

35. U.S. Strategic Command, "CDRUSSTRATCOM OPLAN 8010–08 Global Deterrence and Strike," 1 Feb 08, https://www.governmentattic.org/38docs/USSTRATCOM oplans8010–08_8010–12.pdf, v–vi.

36. U.S. Strategic Command, viii.

37. U.S. Strategic Command, xv–xxi.

38. U.S. Strategic Command, vi–x.

39. U.S. Strategic Command, xvii–xviii.

40. Maloney, *Deconstructing Dr. Strangelove*, ch. 4.

41. U.S. Strategic Command, "USSTRATCOM OPLAN 8010–12," xviii.

42. U.S. Strategic Command, xx–xxii.

43. Lynn Kuok, *The U.S. FON program in the South China Sea: A lawful and necessary response to China's strategic ambiguity* (Washington, D.C.: Brookings Institution, 2016), 2–4.

44. W. E. Butler, "Innocent Passage and the 1982 Convention: The Influence of Soviet Law and Policy," *American Journal of International Law*, April 1987, at https:// www.iilj.org/wp-content/uploads/2016/08/Butler-Innocent-Passage-and-the -1982-Convention-1987.pdf; Richard Halloran, "2 U.S. Ships Enter Soviet Waters off Crimea to Gather Intelligence," *New York Times*, 19 Mar 86.

45. "The Presidential Nuclear Weapons Initiative at a Glance," Arms Control Association, July 2017, https://www.armscontrol.org/factsheets/pniglance

46. DoD Annual Freedom of Navigation Reports 1991–2021 archived at https://policy .defense.gov/OUSDP-Offices/FON/

47. Gates, *Duty*, 180–181; Mark T. Esper, *A Sacred Oath: Memoirs of a Secretary of Defense during Extraordinary Times* (New York: William Morrow, 2022), 191, 201.

48. U.S. Navy briefing, "Ice Exercise 2018 (ICEX 18)," https://media.defense.gov/2018 /Jul/19/2001944483/-1/-1/1/ICEX%2018%20INFORMATION.PDF

49. Parliamentary Business, Written Answers to Questions, 2 November, 2010, https:// publications.parliament.uk/pa/cm201011/cmhansrd/cm101102/text/101102w0001 .htm#10110298000032; CinCFLEET "Board of Inquiry into the Collision of HMS *Tireless* on 13 May 03," https://web.archive.org/web/20081210074748/http:// www.mod.uk/NR/rdonlyres/C762975E-6E2E-43A0-9B6D-901125D84878/0 /summary_tireless_boi.pdf

50. David K. Stumpf, *Minuteman: A Technical History of the Missile that Defined American Nuclear Warfare* (Fayetteville: University of Arkansas Press, 2020), 37, 395–396; General Accounting Office National Security and International Affairs Division, Report to the Chairman, Committee on Armed Services, House of Representatives, *Strategic Forces: Minuteman Weapons System Status and Current Issues*, September 1990; "MIRV-SRV," Minuteman Missile, at https://minutemanmissile. com/mirvsrv.html; see also the Minuteman III historical timeline available at Association of Air Force Missileers, https://www.afmissileers.org/MMIII-ICBM

51. "A Moment of Glory," Los Alamos National Laboratory, 20 Apr 20, http://www .lanl.gov/discover/publications/national-security-science/2020-spring/glory-trip. php

52. These data were distilled from a Minuteman III historical timeline available at Association of Air Force Missileers, https://www.afmissileers.org/MMIII-ICBM

53. "Minuteman III Launches," *Peterson Space Observer*, 11 Aug 17, https://csmng .com/2017/08/11/minuteman-iii-launches/

54. George Koutsoumpos, "Submarine Launched ICBM Trident II D5 and Conventional Trident Modification," *Journal of Computations & Modelling*, Vol. 4, No. 1 (2014), 245–265. See also Hans Kristensen, "Alert Status of Nuclear Weapons (version 2)," 21 Apr 17, https://uploads.fas.org/2014/05/Brief2017_GWU_2s.pdf

55. Compiled at "Trident D-5," http://www.astronautix.com/t/tridentd-5.htmlv

56. "Trident D-5."

57. Alexander Levakov, "Trident II D5 night flight mystery," Oct 19, https://rstudio-pubs-static.s3.amazonaws.com/536630_32d74f784e7c43c7b8f4b5d75c9ee6d5.html; Tyler Rogway, "Amazing cockpit video of unusual Trident ballistic missile test may point to New Warhead," *The War Zone*, 5 Oct 19, https://www.thedrive.com/the-war-zone/30170/amazing-cockpit-video-of-unusual-trident-ballistic-missile-test-may-point-to-new-warhead; Keir A. Lieber and Daryl G. Press, "The New Era of Counterforce: Technological Change and the Future of Nuclear Deterrence," *International Security*, Vol. 41, No. 4 (Spring 2017), 9–49.

58. "US Navy Tests Updated Trident II Missile System," *The Watch*, 6 Jul 22, https://thewatch-magazine.com/2022/07/06/u-s-navy-tests-updated-trident-ii-missile-system/

59. "SSBN/SSGN Ohio Class Submarine," *Naval Technology*, https://www.naval-technology.com/projects/ohio-class-submarine/

60. Office of the Undersecretary of State for Arms Control and International Security, *Strengthening Deterrence and Reducing Nuclear Risks, Part II: The Sea-Launched Cruise Missile-Nuclear (SLCM-N)*, Vol. 1, No. 11, 23 Jul 20, 2; "The Presidential Nuclear Weapons Initiative at a Glance," Arms Control Association, July 2017, https://www.armscontrol.org/factsheets/pniglance

61. Hans Kristensen, "US Navy Instruction Confirms Retirement of Nuclear Tomahawk Cruise Missile," Federation of American Scientists, 18 Mar 13, https://fas.org/blogs/security/2013/03/tomahawk/

62. Loren Thompson, "The Four Biggest Hurdles to Rebuilding the Nuclear Bomber Force," *Forbes*, 8 Jan 21, https://www.forbes.com/sites/lorenthompson/2021/01/08/the-four-biggest-hurdles-to-rebuilding-the-nuclear-bomber-force/?sh=7adb632b3ea4; John A. Tirpak, "Last B-1B Bombers Retire until B-21 Comes Online," *Air & Space Forces Magazine*, 24 Sep 21, https://www.airandspaceforces.com/last-b-1b-bombers-retire-until-b-21-comes-online/; David Axe, "The Steward of America's Nukes Is Sending Mixed Messages about B-1 Bombers-And That's Dangerous," *Forbes*, 10 Dec 20, https://www.forbes.com/sites/davidaxe/2020/12/10/the-steward-of-americas-nukes-is-sending-mixed-messages-and-thats-dangerous/?sh=40362006dc90

63. The White House Office of the Press Secretary, "Fact Sheet: European Reassurance Initiative and Other U.S. Efforts in Support of NATO Allies and Partners," 3 Jun 14, https://obamawhitehouse.archives.gov/the-press-office/2014/06/03/fact-sheet-european-reassurance-initiative-and-other-us-efforts-support-

64. "Operation REASSURANCE," Fact Sheet, 1 Nov 22, https://www.canada.ca/en/department-national-defence/services/operations/military-operations/current

-operations/operation-reassurance.html; "HMCS *Regina* starts voyage home after successful two-mission deployment," 14 Aug 14, https://www.canada.ca/en/news/archive/2014/08/hmcs-regina-starts-voyage-home-after-successful-two-mission-deployment.html

65. "Russia's jets pass near U.S. ship in Black Sea 'provocative'-Pentagon," Reuters, 14 Apr 14, https://www.reuters.com/article/usa-russia-blacksea-idUSL2N0N60V52 0140414#TmlqFb14qYOCTCUy.97

66. U.S. European Command, Operation ATLANTIC RESOLVE fact sheet, 26 Jun 14.

67. Tony Roper, "Global Lightning," Planes and Stuff Blog, 17 May 14, https://planesandstuff.wordpress.com/2014/06/

68. Oriana Pawlyk, "B-2, B-52 bombers deploy to Europe for military exercises," *Military Times*, 6 Jun 15, https://www.militarytimes.com/news/your-military/2015/06/06/b-2-b-52-bombers-deploy-to-europe-for-military-exercises/; "B-1's Conduct South China Sea Mission, Demonstrates Global Presence," PACAF public affairs, 30 Apr 20, https://www.pacom.mil/Media/News/News-Article-View/Article/2171834/b-1s-conduct-south-china-sea-mission-demonstrates-global-presence/

69. Jonathan King, "Bomber Assurance and Deterrence Missions: Effect on North Korean Sicourse," USAF Center for Unconventional Weapons Studies, CUWS Trinity Site Papers, July 2017, https://www.airuniversity.af.edu/Portals/10/CSDS/assets/trinity_site_paper8.pdf

70. Eric C. Paulson, "Strategic Assurance and Signaling in the Baltics," Air War College, Air University, 6 Apr 17, https://apps.dtic.mil/sti/pdfs/AD1038066.pdf

71. Paulson, "Strategic Assurance and Signaling in the Baltics."

72. NATO, "Wales Summit Declaration," 5 Sep 14, https://www.nato.int/cps/en/natohq/official_texts_112964.htm

73. Sean M. Maloney, "Fire Brigade or Tocsin: NATO's ACE Mobile Force, Flexible Response, and the Cold War," *Journal of Strategic Studies*, Vol. 27, No. 4 (2004).

74. NATO, "Wales Summit Declaration," 5 Sep 14, https://www.nato.int/cps/en/natohq/official_texts_112964.htm

75. "Russian military planes buzzed HMCS *Toronto* in the Black Sea," CBC News, 8 Sep 14, https://www.cbc.ca/news/world/russian-military-planes-buzzed-hmcs-toronto-in-black-sea-1.2759843; Steven Chase, "Russia denies military jets approached Canadian frigate near Ukraine," *The Globe and Mail*, 8 Sep 14, https://www.theglobeandmail.com/news/politics/russian-jets-buzz-canadian-frigate-near-ukraine/article20478538/

76. "Russian planes circle HMCS *Toronto* in 'provocative' move," CTV News, 8 Sep 14, https://www.ctvnews.ca/world/russian-planes-circle-hmcs-toronto-in-provocative-move-on-black-sea-1.1997347

77. Ukraine Ministry of Foreign Affairs, "Ukrainians Worldwide," https://mfa.gov
.ua/en/about-ukraine/ukrainians-worldwide
78. "US Pacific Command Forces Come Together for Valiant Shield 2014," PACOM
news release, 9 Sep 14, https://www.pacom.mil/Media/News/News-Article-View
/Article/564579/us-pacific-command-forces-come-together-for-valiant-shield
-2014/; Brok McCarthy, "Valiant Shield 2014 Comes to a Successful End," PACAF
news release, 23 Sep 14, https://www.pacaf.af.mil/News/Article-Display/Article
/591144/valiant-shield-2014-comes-to-successful-end/
79. STRATCOM released photos to Flickr: https://www.flickr.com/photos/usstrat
com/albums/72157648601044899/; DVIDs pictures: https://www.dvidshub.net
/image/1651086/global-thunder-15
80. "USSTRATCOM conducts key command and control exercise," U.S. Strategic Command Public Affairs, 28 Oct 14, https://www.stratcom.mil/Media
/News/News-Article-View/Article/983763/usstratcom-conducts-key-command
-and-control-exercise/
81. Carter, *Inside the Five-Sided Box*, 263.
82. Carter, 274, 276.
83. David B. Larter, "In challenging China's claims in the South China Sea, the U.S.
Navy is getting more assertive," Defense News, 5 Feb 20, https://www.defensenews
.com/naval/2020/02/05/in-challenging-chinas-claims-in-the-south-china-sea
-the-us-navy-is-getting-more-assertive/
84. Greg Hadley, "B-2's Deploy to Australia for Bomber Task Force Mission," *Air &
Space Forces Magazine*, 12 Jul 22, https://www.airandspaceforces.com/b-2s-deploy
-to-australia-for-bomber-task-force-mission/; "Chronology of US-North Korean
Missile Diplomacy," Arms Control Association, https://www.armscontrol.org
/factsheets/dprkchron specifically 2015 entries.
85. "B-52 Stratofortress's Participate in 'POLAR GROWL' Training," *Tactical Life*, 11 Jan
16, https://www.tactical-life.com/firearm-news/b-52-stratofortresses-polar-growl/
86. "USAF's B-52 Stratofortress bombers participate in Exercise POLAR GROWL,"
Airforce Technology, 7 Apr 15, https://www.airforce-technology.com/uncategorized
/newsusafs-b-52-stratofortress-bombers-participate-in-exercise-polar-growl
-4548986/
87. Hans Kristensen, "Increasing Nuclear Bomber Operations," Federation of
American Scientists, 25 Sep 16, https://fas.org/blogs/security/2016/09/nuclear
-bomber-operations/
88. "CF-18 Fighters Intercept B-52 Bombers in Training Exercise," *Canadian Military
Family Magazine*, 22 Apr 15, https://www.cmfmag.ca/duty_calls/cf-18-fighters
-intercept-b-52-bombers-in-training-exercise/
89. "B-52s Demonstrate Long-Range Strategic Capabilities in Europe," USSTRAT-
COM Public Affairs, 5 Jun 15, https://www.afgsc.af.mil/News/Article-Display

/Article/629232/b-52s-to-demonstrate-long-range-strategic-capabilities-in-europe
/; "Key NATO & Allied Exercises," NATO Fact Sheet, October 2015.

90. Oriana Pawlyk, "B-2, B-52 bombers deploy to Europe for military exercises," *Military Times*, 6 Jun 15, https://www.militarytimes.com/news/your-military/2015/06/06/b-2-b-52-bombers-deploy-to-europe-for-military-exercises/

91. "B-52s demonstrate Long-Range Strategic Capabilities in Europe," USSTRATCOM Public Affairs, 5 Jun 15, https://www.afgsc.af.mil/News/Article-Display/Article/629232/b-52s-to-demonstrate-long-range-strategic-capabilities-in-europe/

92. Maloney, *Emergency War Plan*, 234–241.

93. Joseph Raatz, "Bombers receive warm welcome for COLD RESPONSE," 2nd Expeditionary Bomb Group Public Affairs, 29 Feb 16, https://www.af.mil/News/Article-Display/Article/681968/bombers-receive-warm-welcome-for-cold-response/; Joseph Raatz, "Mighty Shadow over the Fjords: a B-52's Cold Response," 20th Expeditionary Bomb Squadron Public Affairs, 7 Mar 16, https://www.usafe.af.mil/News/Features/Display/Article/749439/mighty-shadow-over-the-fjords-a-b-52s-cold-response/; "B-52s deploy in support of exercise Cold Response," USSTRATCOM Public Affairs, 26 Feb 16, https://www.stratcom.mil/Media/News/News-Article-View/Article/983856/b-52s-deploy-in-support-of-exercise-cold-response/

94. Kristensen, "Increasing Nuclear."

95. "Tankers enable long-range bomber capability during Polar Roar," Mobility Command Public Affairs, https://www.afgsc.af.mil/News/Article-Display/Article/908687/tankers-enable-long-range-bomber-capability-during-polar-roar/; "Strategic Bomber Force Showcases Allied Interoperability during POLAR ROAR," USSTRATCOM Public Affairs, 3 Aug 16, https://www.stratcom.mil/Media/News/News-Article-View/Article/983671/strategic-bomber-force-showcases-allied-interoperability-during-polar-roar/; "NATO Air Policing jets train with US bomber," NATO News, 1 Aug 16, https://www.nato.int/cps/en/natohq/news_134318.htm; USAF video which includes a briefing is available at https://militaryaviationreview.com/video-b-52-on-board-footage-from-polar-roar

96. Hans Kristensen, "Increasing Nuclear."

97. "Risk of nuclear attack rises," CBS News, 25 Sep 16, https://www.cbsnews.com/news/60-minutes-risk-of-nuclear-attack-rises/

98. "Dyess B-1s deploy to Europe for Ample Strike," USAFE Public Affairs, 3 Sep 16, https://www.usafe.af.mil/News/Article-Display/Article/932479/dyess-b-1s-deploy-to-europe-for-ample-strike/

99. "Exercise GLOBAL THUNDER 17 concludes," USSTRATCOM Public Affairs, 2 Nov 16, https://www.afgsc.af.mil/News/Article-Display/Article/993846/exercise-global-thunder-17-concludes/

100. "Global Thunder 17 exercise concludes," DVIDS, 1 Nov 16, https://www.dvidshub.net/feature/GlobalThunder

101. Carter, *Inside the Five-Sided Box*, 275.

102. Derived from the memoirs of Jim Mattis, *Call Sign Chaos*; John Bolton's *The Room Where It Happened*; and Mark Esper's *A Sacred Oath*. It is important to note that Mattis has not released a memoir of his time as secretary of defense.

103. David B. Later, "In challenging China's claims in the South China Sea, the US Navy is getting more assertive," *DefenseNews*, 5 Feb 20, https://www.defensenews.com/naval/2020/02/05/in-challenging-chinas-claims-in-the-south-china-sea-the-us-navy-is-getting-more-assertive/

104. "Chronology of US-North Korean Missile Diplomacy," Arms Control Association, https://www.armscontrol.org/factsheets/dprkchron specifically 2017 entries

105. "Around the Air Force," Air Force TV, 14 Jun 17, https://www.af.mil/News/Air-Force-TV/videoid/531705/About-Us/AF-75th-Anniversary/

106. "Russian Back at It Again!," SOFREP, 7 Jun 17, https://sofrep.com/fightersweep/russia-back-b-52-intercepted-su-27-fighter-baltic-sea/

107. "Canadian NORAD Region aircraft practice intercept and escort procedures with United States Air Force bombers," 19 Jun 17, https://www.norad.mil/Newsroom/Article/1218304/canadian-norad-region-aircraft-practice-intercept-and-escort-procedures-with-un/

108. USAF photography of RED FLAG ALASKA 2017 showed nuclear-capable B-52Hs employed in the exercise. See Nellis AFB public affairs, https://www.nellis.af.mil/News/Photos/igphoto/2001792072/; on REFLEX ACTION deployments, see Maloney, *Emergency War Plan*, 364–367.

109. "US Bombers Arrive in Europe for Exercise to Deter Russia," Military.com, 23 Aug 17, https://www.military.com/daily-news/2017/08/23/us-bombers-arrive-europe-exercise-deter-russia.html

110. Dario Leone, "Cool Photos Show B-2 Spirit Strategic Bombers Taking Part in USSTRATCOM Exercise GLOBAL THUNDER 18," 9 Nov 17, https://theaviationgeekclub.com/cool-photos-show-b-2-spirit-strategic-bombers-taking-part-usstratcom-exercise-global-thunder-18/

111. "World is a dangerous place: US prepares to put B-52 nuclear bombers on high alert," RT, 23 Oct 17, https://www.rt.com/usa/407494-nuclear-b52-bombers-alert/

112. "Air Force denies report that it is preparing to put B-52 bombers on high alert," NBC News, 23 Oct 17, https://www.cnbc.com/2017/10/23/us-nuclear-bombers-on-high-alert-as-north-korean-threat-grows.html

113. Justin Armstrong, USAFE Public Affairs, 15 Jan 18, https://www.usafe.af.mil/News/Article-Display/Article/1415510/minot-b-52s-airmen-deploy-to-raf-fairford/

114. J.T. Armstrong, "Minot Airmen, B-52s complete UK deployment," USAFE Public Affairs, 31 Jan 18, https://www.usafe.af.mil/News/Article-Display/Article/1428395/minot-airmen-b-52s-complete-uk-deployment/

115. Dan Kaszeta, *Toxic: A History of Nerve Agents from Nazi Germany to Putin's Russia* (Oxford: Oxford University Press, 2021), 227–229; Tyler Rogoway, "This Awesome Chart Shows All the Assets Used in the Trilateral Missile Strikes on Syria," *The Warzone,* 29 Apr 18, https://www.thedrive.com/the-war-zone/20509/this-awesome-chart-shows-all-the-assets-used-in-the-trilateral-missile-strikes-on-syria

116. *The Guardian* consolidated its coverage on the matter here: https://www.theguardian.com/world/live/2018/apr/14/syria-donald-trump-announcement-chemical-attack-live

117. Mark Hookham and Tim Ripley, "British submarine in duel with Kremlin's Black Hole hunter killer," *The Times,* 15 Apr 18, https://www.thetimes.co.uk/article/british-submarine-in-underwater-duel-with-kremlins-black-hole-hunter-killer-dhxhlpwc9

118. "USAF B-52 Stratofortress conducts Arctic flight," blogbeforeflight, 17 Sep 18, https://www.blogbeforeflight.net/2018/09/usaf-b-52-stratofortress-conducts-artctic-flight.html

119. Thomas Nilson, "US and Russian bombers test airspace over European Arctic," *Independent Barents Observer,* 24 Sep 18, https://www.rcinet.ca/eye-on-the-arctic/2018/09/24/russia-usa-bombers-barents-norway-sea-airspace-military-mission/

120. "2018 Bomber Task Force Squadron Photos," Whiteman AFB Public Affairs, https://www.flickr.com/photos/whitemanafb/albums/72157695988537300; "Bomber Task Force 2018 Crew Chiefs," USAF Public Affairs, https://www.adr.af.mil/News/Photos/igphoto/2002045943/

121. Ben Werner, "U.S. Destroyer Conducts FONOP Near Russian Pacific Fleet Headquarters," USNI News, 5 Dec 18, https://news.usni.org/2018/12/05/39351

122. Bolton, *The Room Where It Happened,* 165.

123. "GLOBAL THUNDER 19," USAF Public Affairs, https://www.flickr.com/photos/usstratcom/albums/72157675616562628/

124. "B-1 Lancers return to Indo-Pacific for bomber task force deployment," Air Force News Agency, 5 May 20, https://www.defencetalk.com/b-1b-lancers-return-to-indo-pacific-for-bomber-task-force-deployment-75380/

125. "American B-52s worked out the bombing of the northern part of Russia," Avia-pro, 23 Feb 19, https://avia.pro/news/amerikanskie-b-52-otrabotali-bombardirovku-severnoy-chasti-rossii

126. "U.S. Air Force B-52s deploy to Europe," USAFE New Release, 14 Mar 19, https://www.usafe.af.mil/News/Press-Releases/Article/1784804/us-air-force-b-52s-deploy-to-europe/; Tessa Corrick, "B-52s in England: BTF 2019 begins," 2nd Bomb Wing Public Affairs, 21 Mar 19, https://www.usafe.af.mil/News/Press-Releases/Article/1784804/us-air-force-b-52s-deploy-to-europe/

127. William Echols, "Updated: Moscow Claims US B-52 'Forced' from Russian Border, No-Says US Air Force," Polygraph.info, 23 Mar 19, https://www.polygraph.info/a /b52-intercepted-russian-jets-fact-check/6742096.html

128. Alexey Degtyarev, "The United States told its version of the interaction of the Su-27 and B-52 over the Baltic," View Business Paper, 21 Mar 19, https://vz.ru/news /2019/3/21/969478.html; "Two Sukhoi jets chase US B-52 bomber away from Russian border," TASS, 21 Mar 19, https://tass.com/defense/1049759; Dario Leone, "USAF B-52 flying over the Baltic Sea turned around after being tracked by Russian Air Defense System, Russian MoD says," The Aviation Geek Club, 17 Mar 19.

129. "Named the main danger of the American B-52H, which the military does not talk about," Avia.pro, 31 Feb 19, https://avia.pro/news/nazvana-glavnaya-opasnost amerikanskih-b-52h-o-kotoroy-ne-rasskazyvayut-voennye

130. Stephen Losey, "A glimpse of deployments to come? Largest bomber rotation since the start of the Iraq War comes home," Air Force Times, 16 Apr 19, https://www .airforcetimes.com/news/your-air-force/2019/04/16/a-glimpse-of-deployments -to-come-largest-bomber-rotation-since-start-of-iraq-war-comes-home/

131. David Cenciotti, "Russian fighter jets intercepted US B-52 bombers during an unusual flight over eastern Europe," The Aviationist, 18 Jun 19, https://www.insider .com/russian-fighter-jets-intercept-b52-bombers-over-black-baltic-seas-2019-6

132. "American B-52s simulated a missile attack on Kaliningrad," Avia.pro, 11 Jul 19, https:// avia.pro/news/amerikanskie-b-52-symitirovali-raketnyy-udar-po-kaliningradu

133. "The targets of American B-52 strikes on Russia are shown, there are more than a hundred of them," Avia.pro, 2 Apr 19, https://avia.pro/news/pokazany -celi-udarov-amerikanskih-b-52-po-rossii-ih-bolee-sotni

134. "American B-52 bombers unexpectedly flew near the Russian islands of Shikotan and Habomai," Avia.pro, 23 Apr 19, https://avia.pro/news/amerikanskie-bombard- irovshchiki-b-52-neozhidanno-proleteli-ryadom-s-rossiyskimi-ostrovami

135. Esper, Sacred Oath, 73.

136. Esper, 76.

137. Esper, 436.

138. "B-2 Bomber Task Force Europe 2019," Jet Wash Aviation Photos, http://www.jet washaviationphotos.com/useucom-b-2-deployment-2019.html

139. Stefano D'Urso, "Four U.S. B-52s of Bomber Task Force Europe 20–1 Have Deployed to RAF Fairford UK," The Aviationist, 12 Oct 19, https://theaviationist .com/2019/10/12/four-u-s-b-52s-of-bomber-task-force-europe-20–1-have-deployed- to-raf-fairford-uk/

140. Stephen Losey, "B-52s train in tense Black Sea region, in signal to Russia," Air Force Times, 23 Oct 19, https://www.airforcetimes.com/news/your-air-force/2019/10/23 /b-52s-train-in-tense-black-sea-region-in-signal-to-russia/

141. "Russian Aerospace Forces failed to drive two American B-52 bombers away from Crimea and Sochi," Avia.pro, 22 Oct 19, https://avia.pro/news/vks-rossii-ne-smogli -otognat-dva-amerikanskih-bombardirovshchika-b-52-ot-kryma-i-sochi

142. Thomas Nilsen, "B-52 flights close to homeport and patrol areas for Russia's ballistic missile subs," *Barents Observer*, 8 Nov 19, https://thebarentsobserver .com/en/security/2019/11/us-b-52-strategic-bombers-and-norwegian-f-16s-flying- wing-wing-over-barents-sea. See also "Three American B-52s 'attacked' the Russian border from the north at once," Avia.pro, 7 Nov 19, https://avia.pro/news/srazu -tri-amerikanskih-b-52-atakovali-rossiyskuyu-granicu-s-severa

143. Karen Singer, "U.S. Strategic Command Conducts Exercise GLOBAL THUNDER linked with VIGILANT SHIELD," USSTRATCOM Public Affairs, 18 Oct 19, https://www.stratcom.mil/Media/News/News-Article-View/Article/1992213/; Tory Patterson, "Dyess AFB Airmen prove readiness in Global Thunder 20," 7th Bomb Wing Public Affairs, 4 Nov 19, https://www.dyess.af.mil/News/Features /Article/2007682/dyess-afb-airmen-prove-readiness-in-global-thunder-20/; U.S. DoD, "GLOBAL THUNDER," https://www.defense.gov/Multimedia/Photos /igphoto/2002205315/

144. "Two American B-5s immediately bombed along the eastern Russian border," Avia.pro, 31 Oct 19, https://avia.pro/news/srazu-dva-amerikanskih-b-52-otbombilis -po-vostochnoy-rossiyskoy-granice

145. Esper, *Sacred Oath*, 566.

146. Diana Stacey Correll, "The Air Force has stopped its Continuous Bomber Pres- ence mission in Guam," *Air Force Times*, 21 Apr 20, https://www.airforcetimes .com/news/your-air-force/2020/04/21/the-air-force-has-stopped-its-continuous -bomber-presence-mission-in-guam/; Diana Stacey Correll, "B-1 Bombers are back in Guam for bomber task force deployment," *Air Force Times*, 4 May 20, https:// www.airforcetimes.com/news/your-air-force/2020/05/04/b-1-bombers-are-back -in-guam-for-bomber-task-force-deployment/

147. Tyler Rogoway, "B-1B Bomber Made Bold Flight into the Sea of Okhotsk," *The War Zone*, 22 May 20, https://www.thedrive.com/the-war-zone/33637/b-1b-bomber -made-bold-flight-into-the-sea-of-okhotsk-that-is-surrounded-by-russian-territory

148. Lee Berthiaume, "2 Russian bombers approached Canadian airspace in Arctic, NORAD say," The Canadian Press, 31 Jan 20, https://globalnews.ca/news/6489830 /russian-bombers-canada-norad/

149. Stephen Losey, "B-2 bomber task force deploys to Portugal," *Air Force Times*, 9 Mar 20, https://www.airforcetimes.com/news/your-air-force/2020/03/09/b-2-bomber -task-force-deploys-to-portugal/

150. David Cenciotti, "Norwegian F-35As on QRA at Orland Air Station Carry Out Type's First Intercept of Russian Aircraft off Norway," *The Aviationist*, Mar 20, https://

theaviationist.com/2020/03/07/norwegian-f-35as-on-qra-at-orland-air-station
-carry-out-types-first-intercept-of-russian-aircraft-off-norway/

151. "Royal Air Force Intercepts Russian Aircraft Three Times in Six Days," Euro-Atlantic News, 13 Mar 20, https://t-intell.com/2020/03/13/royal-air-force
-intercepts-russian-aircraft-three-times-in-six-days/

152. "Royal Air Force Intercepts Russian Aircraft Three Times in Six Days."

153. NORAD Twitter post, 10 Mar 20; Amy Hudson, "USAF, Canadian Fighters Intercept Russian Aircraft North of Alaska," Air & Space Forces, 10 Mar 20, https://www
.airandspaceforces.com/usaf-canadian-fighters-intercept-russian-aircraft-north-of
-alaska/; "NORAD intercepts Russian aircraft entering Air Defense Identification Zone," Peterson AFB Public Affairs, 10 Mar 20, https://www.dvidshub.net/news
/press-release/?iframe=true&width=90%2525&height=90%2525&page=1481

154. Thomas Newdick, "U.S. Navy Denies Russian Warships Forced Its Destroyer Out of Disputed Pacific Waters," The Drive, 24 Nov 20, https://www.thedrive.com
the-war-zone/37805/navy-denies-a-russian-warship-forced-one-of-its-destroyers
out-of-disputed-pacific-waters

155. Esper, A Sacred Oath, 435; Marc Lanteigne, "Cold Games? The U.S. Navy (Re-) Enters the Barents Sea," Over the Circle: Arctic News and Analysis, 11 May 20, https://
overthecircle.com/2020/05/11/a-frigid-fonop-the-us-navy-re-enters-the
-barents-sea/; "U.S., Norway, UK Carry Out FONOP Mission off Russia's Arctic Coast," Maritime Executive, 14 Sep 20, https://maritime-executive.com
/article/u-s-norway-uk-carry-out-fonop-mission-off-russia-s-arctic-coast

156. The usual panicked commentary by those who fear escalation with Russia is exemplified in "U.S. Freedom of Navigation Operation in the Arctic: 'Would Be a High-Risk Gesture with Unpredictable Consequences,'" High North News, 11 Sep 20, https://www.highnorthnews.com/en/us-freedom-navigation-operation
-arctic-would-be-high-risk-gesture-unpredictable-consequences. A counter to this is Troy J. Boufford, "A Developing Operational Maritime Environment: Forward Presence and Freedom of Navigation in the Arctic," Strategic Perspectives, 12 Jan 21, https://www.naadsn.ca/wp-content/uploads/2021/01/Strategic-Perspectives-A
-Developing-Maritime-Operational-Environment-Bouffard.pdf; "Into the Bear's Backyard: The Royal Navy in the Barents Sea," Navy Lookout, 6 May 20, https://www
.navylookout.com/into-the-bears-backyard-the-royal-navy-in-the-barents-sea/

157. Jonathan G. Odom, "The Rules-Based Order, Maritime Freedom and Recent Naval Operations in the Barents Sea," Modern Diplomacy, 14 May 20, https://
moderndiplomacy.eu/2020/05/14/the-rules-based-order-maritime-freedom-and
-recent-naval-operations-in-the-barents-sea/

158. Diana Stancy Correll, "Despite pandemic, Air Force continues long-range Bomber Task Force missions in Europe," Air Force Times, 30 Jun 20, https://www.airforcetimes

.com/news/your-air-force/2020/06/30/despite-pandemic-air-force-continues
-long-range-bomber-task-force-missions-in-europe/

159. "Two U.S. bomber fly over Ukraine in integration with Ukrainian fighter jets,"
Ukraine Information Agency, 29 May 20, https://www.unian.info/politics/bomber
-task-force-mission-u-s-bombers-fly-over-ukraine-with-ukrainian-fighter-jets
-11017442.html; Stephen Losey, "USAF B-1B Lancers practiced anti-ship missile strikes in Black Sea," *Air Force Times*, 1 Jun 20, https://www.airforcetimes
.com/news/your-air-force/2020/06/01/usaf-b-1b-lancers-practiced-anti-ship
-missile-strikes-in-black-sea/

160. Dario Leone, "Russia's MoD releases video of Su-27 jets intercepting USAF B-1B
bomber over the Black Sea," The Aviation Geek Club, 30 May 20, https://theaviation
geekclub.com/russias-mod-releases-video-of-su-27-fighter-jets-intercepting-usaf
-b-1b-bomber-over-the-black-sea/

161. "Turkish F-16s intercepted Russian Tu-22M3 bombers right off the coast of Crimea,"
Avia.pro, 22 May 20, https://avia.pro/news/tureckie-f-16-perehvatili-rossiyskie
-bombardirovshchiki-tu-22m3-pryamo-u-beregov-kryma

162. Stephen Losey, "B-52s train with Norwegian fighters on long-range flight to
Arctic," *Air Force Times*, 3 Jun 20, https://www.airforcetimes.com/news/your
-air-force/2020/06/03/b-52s-train-with-norwegian-fighters-on-long-range-flight
-to-arctic/

163. "US Intercepts Nuclear-Capable Russian Bombers and Fighter Jets Near
Alaska," *Statecraft*, 11 Jun 20, https://www.statecraft.co.in/article/us-intercepts
-nuclear-capable-russian-bombers-and-fighter-jets-near-alaska

164. Stephen Losely, "B-52s fly from Minot to Baltics to support naval exercise," *Air Force Times*,
17 Jun 20, https://www.airforcetimes.com/news/your-air-force/2020/06/17/b-52s
-fly-from-minot-to-baltics-to-support-naval-exercise/

165. "B-52 bombers worked out new nuclear strikes on St. Petersburg and the Far East,"
Avia.pro, 17 Jun 20, https://avia.pro/news/bombardirovshchiki-b-52-otrabotali
-novye-yadernye-udary-po-sankt-peterburgu-i-dalnemu-vostoku

166. Wyatt Olson, "B-52 bombers return to Alaska following three-year hiatus,"
Anchorage Daily News, 17 Jun 20, https://www.adn.com/alaska-news/military
/2020/06/17/b-52-bombers-return-to-alaska-following-three-year-hiatus/

167. Oriana Pawlyk, "For 8th Time This Year, US Fighters Intercept Russian Bombers
off Alaska," *Military.com*, 17 Jun 20, https://www.military.com/daily
-news/2020/06/17/8th-time-year-us-fighters-intercept-russian-bombers-off-alaska
.html

168. USTRANSCOM Twitter post, 18 Jun 20; Military Monitoring World Twitter post,
18 Jun 20.

169. "Russian Tu-95 nuclear bombers enchantingly 'congratulated' the Americans on Independence Day heading for the US borders," Avia.pro, 3 Jul 20, https://avia.pro/news/rossiyskie-yadernye-bombardirovshchiki-tu-95-feerichno-pozdravili-amerikancev-s-dnyom

170. "American B-52s staged a provocation near Kamchatka on Independence Day," Avia.pro, 4 Jul 20, https://avia.pro/news/amerikanskie-b-52-ustroili-provokaciyu-ryadom-s-kamchatkoy-na-den-nezavisimosti-ssha

171. "Russian Tu-160 strategic bombers were deployed closer to the United States and for a reason," Avia-pro, 13 Aug 20, https://avia.pro/news/rossiyskie-strategicheskie-bombardirovshchiki-tu-160-perebrosili-poblizhe-k-ssha-i-ne-prosto

172. "Nuclear performance: Tu-95MS bombers of the Russian Aerospace Forces responded to a serious provocation by Japan with demonstrative strikes," Avia.pro, 19 Aug 20, https://avia.pro/news/yadernyy-performans-bombardirovshchiki-tu-95ms-vks-rf-otvetili-na-seryoznuyu-provokaciyu

173. Thomas Newdick, "U.S. Navy Denies Russian Warships Forced Its Destroyer out of Disputed Pacific Waters," The Drive, 24 Nov 20, https://www.thedrive.com/the-war-zone/37805/navy-denies-a-russian-warship-forced-one-of-its-destroyers-out-of-disputed-pacific-waters

174. A Belarusian joke says it all. A guy driving in Minsk is stopped by the police, who immediately start beating him. The guy says, "Hey, wait, I voted for Lukashenko!" The policeman says, "Cut the bullshit, nobody voted for Lukashenko."

175. "Russia offer Belarus military assistance as 200,000 gather in largest demonstration yet," CNBC, 16 Aug 20, https://www.cnbc.com/2020/08/16/russia-offers-belarus-military-assistance-as-200000-gather-in-largest-demonstration-yet.html

176. "Exercise KAVKAZ 2020-A Final Test of Russian Military Reform? NATO Strategic Communications Centre of Excellence, 2021, https://stratcomcoe.org/publication/exercise-kavkaz-2020-a-final-test-of-russian-military-reform/4

177. Confidential correspondence.

178. Joseph Trevithick, "Six B-52 Strategic Bombers Fly alongside Norwegian Fighters in a Clear Signal to Russia," The Drive, 24 Aug 20, https://www.thedrive.com/the-war-zone/35962/six-b-52-strategic-bombers-fly-alongside-norwegian-fighters-in-clear-signal-to-russia

179. "U.S. Bombers & Allied Aircraft Integrate to Fly over All 30 NATO Nations in a Day," U.S. European Command Public Affairs, 28 Aug 20, https://www.usafe.af.mil/News/Press-Releases/Article/2328257/us-bombers-allied-aircraft-integrate-to-fly-over-all-30-nato-nations-in-a-day/

180. "Not Cool: Watch a Russian Fighter Dangerously Buzz a U.S. B-52 Bomber," Popular Mechanics, https://www.popularmechanics.com/military/aviation/a33852860

/russian-su-27-fighter-intercepts-american-b-52-bomber/; "Unsafe, unprofessional intercept of US bomber by Russian aircraft over the Black Sea," USAFE press release, 29 Aug 20, https://www.usafe.af.mil/News/Press-Releases/Article/2329819/unsafe-unprofessional-intercept-of-us-bomber-by-russian-aircraft-over-the-black/

181. Joseph Trevithick, "B-52 Bombers Fly Unprecedented Patrol along Edge of Russian-Controlled Territory in Ukraine (Updated)," *The Drive*, 4 Sep 20, https://www.thedrive.com/the-war-zone/36190/b-52-bombers-fly-unprecedented-patrol-along-edge-of-russian-controlled-territory-in-ukraine

182. Mike Lintott-Danks, "Bomber Task Force 20–5 'BUFFs are back,'" Air Speed Media, 17 Oct 20, https://www.airspeedmedia.net/post/bomber-task-force-20-4-buffs-are-back

183. "Black Sea Bombers," Key.Aero, 11 Dec 20, https://www.key.aero/article/black-sea-bombers

184. "U.S., Norway, UL Carry out FONOP Mission off Russia's Arctic Coast," *The Maritime Executive*, 14 Sep 20, https://maritime-executive.com/article/u-s-norway-uk-carry-out-fonop-mission-off-russia-s-arctic-coast

185. Navy Lookout Twitter post, 20 Sep 20.

186. Michael "Mac" Sims, "Freedom of Navigation in the Arctic," MA thesis, May 2022.

187. "Black Sea Bombers," Key.Aero, 11 Dec 20, https://www.key.aero/article/black-sea-bombers

188. "While the British Sentinel R1 interfered with the Russian S-400s, the Russian TU-160s 'destroyed' half of the UK," Avia.pro, 15 Sep 20, https://avia.pro/news/poka-britanskiy-sentinel-r1-meshal-rossiyskim-s-400-rossiyskie-tu-160-unichtozhili-polovinu

189. Nolan Peterson, "Russia scrambles fighters as US Air Force B-52 bombers fly over Ukraine," *Coffee or Die Magazine*, 15 Sep 20, https://www.coffeeordie.com/b52s-over-ukraine

190. "Black Sea Bombers."

191. "Operation NOBLE DEFENDER," NORAD press release, 24 Sep 20, https://www.norad.mil/Newsroom/Tag/176495/dynamic-force-employment/

192. "Operation NOBLE DEFENDER," NORAD fact sheet, 8 Feb 20, https://www.norad.mil/Newsroom/Fact-Sheets/Article-View/Article/2928028/operation-noble-defender/

193. Michah "Zeus" Fesler, "Agile Combat Employment for Continental Defense," RUSI briefing, 19 May 21, https://rusi-ns.ca/arctic_air_operations/; John Rohrer, "Colorado ANG builds readiness with Canadian partners," *National Guard*, 13 Nov 20, https://www.nationalguard.mil/News/Article/2413756/colorado-ang-builds-readiness-with-canadian-partners/

194. "Black Sea Bombers."

195. "Russian Su-3s for the first time staged a raid on US air defence," Avia.pro, 2 Sep 20, https://avia.pro/news/rossiyskie-su-35-vpervye-ustroili-nalyot-na-amerikanskie-pvo

196. "Royal Canadian Air Force CF-18 Hornet on Operation REASSURANCE intercept Russian aircraft near Romanian airspace," News Release, 25 Sep 20, https://www.canada.ca/en/department-national-defence/news/2020/09/royal-canadian-air-force-cf-18-hornet-on-operation-reassurance-intercept-russian-aircraft-near-romanian-airspace.html

197. "Two Russian TU-160s simultaneously worked out the destruction of Ukraine, Poland, Lithuania and Latvis from the airspace of Belarus," Avia.pro, 23 Sep 20, https://avia.pro/news/dva-rossiyskih-tu-160-odnovremenno-otrabotali-unich tozhenie-ukrainy-polshi-litvy-i-latvii-iz

198. Mike Lintott-Danks, "Bomber Task Force 20–5 'BUFFs are back,'" Air Speed Media, 17 Oct 20, https://www.airspeedmedia.net/post/bomber-task-force-20 -4-buffs-are-back

199. Screenshot, ADS-B Exchange, 23 Sep 20.

200. "The Americans, in response to the flights of the Russian TU-160s over Belarus, left Russian submarines without communications," Avia.pro, 25 Sep 20, https://avia.pro/news/amerikancy-v-otvet-na-polyoty-rossiyskih-tu-160-nad-belorussiey -razbombili-sotni-rossiyskih

201. Military Monitoring World Twitter post, 29 Sep 20.

202. "Russia deployed 8 bombers carrying 23 megatons of nuclear weapons to Belarus," Avia.pro, 4 Oct 20, https://avia.pro/news/rossiya-perebrosila-v-belorussiyu-8 -bombardirovshchikov-nesushchih-23-megatonny-yadernogo

203. "Russian Tu-160 mysteriously circled over the Barents Sea for several hours," Avia.pro, 15 Oct 20, https://avia.pro/news/rossiyskiy-tu-160-neskolko-chasov-zagadochno -kruzhil-nad-barencevym-morem

204. John A. Tirpak, "F-22s Intercept Bears, Flankers, and Russian AWACS off Alaska," Air & Space Forces, 20 Oct 20, https://www.airandspaceforces.com /f-22s-intercept-bears-flankers-and-russian-awacs-off-alaska/

205. Military Monitoring World Twitter post, 28 Nov 20.

206. Military Monitoring World Twitter posts, 14 and 15 Oct 20.

207. Thomas Nilsen, "Salvo of four Bulava missile hit test range on Kanin Peninsula," Barents Observer, 12 Dec 20, https://thebarentsobserver.com/en/security/2020/12 /salvo-four-bulava-missiles-hit-test-range-cape-kanin

Chapter 4. Buildup

1. Mark Galeotti, "Controlling chaos: How Russia manages its political war in Europe," European Council on Foreign Relations, 1 Sep 17, https://ecfr.eu/publication /controlling_chaos_how_russia_manages_its_political_war_in_europe/

2. Galeotti, "Controlling chaos."

3. Digital Threat Analysis Center, "Russia's Propaganda and Disinformation Eco-system-2022 Update and New Disclosures," 15 Feb 22, https://miburo.substack.com/p/russias-propaganda-and-disinformation

4. Andrei Soldatov, "Kremlin.com," *Index on Censorship*, Vol. 39, No. 1 (2010), 71–78.

5. "Commentary: The fog of mystery and silence around Russian cyber troops," NCI, 30 Dec 20, https://nci.org.pl/commentary-the-fog-of-mystery-and-silence-around-russian-cyber-troops/

6. John A. Farinelli, "Cyber Espionage and Information Warfare in Russia," *Small Wars Journal*, 10 Feb 22, https://smallwarsjournal.com/jrnl/art/cyber-espionage-and-information-warfare-russia

7. Bob Woodward and Robert Costa, *Peril* (New York: Simon and Schuster, 2021), xviii.

8. Woodward and Costa, *Peril*, xvi.

9. Herman Kahn, *Thinking about the Unthinkable* (New York: Discus Books, 1962), 47; H.R. Haldeman, *The Ends of Power* (New York: Dell, 1978), 122.

10. "Successful color revolution: Molotov's grandson assessed the events in the USA," TASS, 12 Jan 21, https://crimea.ria.ru/20210112/Udavshayasya-tsvetnaya-revolyutsiya-vnuk-Molotova-otsenil-sobytiya-v-SShA-1119141091.html; "Medvedev called the riots in the Capitol reckoning for the color revolutions," Regnum.ru, 1 Feb 21, https://regnum.ru/news/3178312.html; Vladimir Mozhegov, "America on the brink of civil war or liberal dictatorship," View, 22 Sep 22, https://regnum.ru/news/3178312.html; "Biden won the US presidential election, held according to the Ukrainian scenario," ukraina.ru, 7 Feb 21, https://ukraina.ru/20210107/1030179165.html

11. Military Monitoring World Twitter post, 6 Jan 21. The graphic shows departure from Minot, refueling tracks over Canada and the Atlantic for three B-52Hs call signs Grim.

12. "Four Russian SU-30M intercepted American B-52s off the coast of Crimea," Avia.pro, 25 Jan 21, https://avia.pro/news/chetyre-rossiyskih-su-30sm-perehvatili-amerikanskie-b-52-u-beregov-kryma

13. Diana Stacey Correll, "Air Force B-52s conduct third Bomber Task Force mission in Middle East this year," *Air Force Times*, 28 Jan 21, https://www.airforcetimes.com/news/your-military/2021/01/28/air-force-b-52s-conduct-third-bomber-task-force-mission-in-middle-east-this-year/

14. "F-16 fighters and B-52 bombers practises strikes on the Black Sea coast of Russia," Avia.pro, 1 Feb 21, https://avia.pro/news/istrebiteli-f-16-i-bombardirovshchiki-b-52-otrabotali-udary-po-chernomorskomu-poberezhyu-rossii

15. See the Wikipedia timeline for 2020–2021 Belarusian protests, https://en.wikipedia.org/wiki/Timeline_of_the_2020%E2%80%932021_Belarusian_protests

16. Simon Shuster, "The Untold Story of the Ukraine Crisis," *Time*, 2 Feb 22, https://time.com/6144109/russia-ukraine-vladimir-putin-viktor-medvedchuk/

17. "Initiation of Russian TU-22M3 strikes on NATO ships in the Black Sea will be the finale of large-scale exercises," Avia.pro, 4 Feb 21, https://avia.pro/news/imitaciya-udarov-rossiyskih-tu-22m3-po-korablyam-nato-v-chyornom-more-stanet-finalom

18. "TU-22M3 bomber launched multiple strikes with kh-32 cruise missiles," Avia.pro, 6 Feb 21, https://avia.pro/news/bombardirovshchik-tu-22m3-nanyos-mnozhestvennye-udary-krylatymi-raketami-h-32

19. "Russian TU-22M3 bombers escorted two American missile destroyers out of the Black Sea," Avia.pro, 10 Feb 21, https://avia.pro/news/rossiyskie-bombardirovshchiki-raketonoscy-tu-22m3-vyprovodili-dva-amerikanskih-raketnyh-esminca

20. "Russian strategic bombers practise a nuclear strike on the largest bases with US Air Force bombers," Avia.pro, 9 Feb 21, https://avia.pro/news/strategicheskie-bombardirovshchiki-rossii-otrabotali-yadernyy-udar-po-krupneyshim-bazam-s; "America media reported on an unexpected imitation of an attack by Russian TU-160 on US air bases in Europe," Avia.pro, 14 Feb 21, https://avia.pro/news/amerikanskie-smi-soobshchili-o-neozhidannoy-imitacii-ataki-rossiyskih-tu-160-na-aviabazy-ssha-v

21. Shuster, "The Untold Story of the Ukraine Crisis."

22. Diana Stacey Correll, "B-1B Lancers conduct first Bomber Task Force mission to Norway," *Air Force Times*, 26 Feb 21, https://www.airforcetimes.com/news/your-air-force/2021/02/26/b-1b-lancers-conduct-first-bomber-task-force-mission-from-norway/

23. "The Joint Expeditionary Force (JEF) explained," Medium.com, 30 Jun 21, https://medium.com/voices-of-the-armed-forces/the-joint-expeditionary-force-jef-explained-35fb42169d62

24. Muzyka, *Defending the Union*, 3.

25. Mykhaylo Zabrodskyi et al., *Preliminary Lessons in Conventional Warfighting from Russia's Invasion of Ukraine: February–July 2022* (London: RUSI, 2022), 7.

26. "Russian TU-95s simulated the bombing of Japan for 9 hours," Avia.pro, 12 Mar 21, https://avia.pro/news/rossiyskie-tu-95-9-chasov-imitirovali-bombardirovku-yaponii

27. Jamming Twitter post, 25 Mar 21, which was in part based on the now-defunct Hunter's Note Twitter page.

28. Jason Cutshaw, "SMDC participates in Global Lightning 21," US Army.mil, 15 Mar 21, https://www.army.mil/article/244285/smdc_participates_in_global_lightning_21

29. David Axe, "Four American Bombers Met up in the Arctic-It's a More Impressive Feat than You Might Think," *Forbes*, 18 Mar 21, https://www.forbes.com/sites

/davidaxe/2021/03/18/four-american-bombers-met-up-in-the-arctic-its-a-more
-impressive-feat-than-you-might-think/?sh=60ed89b97a70

30. Daio Leone, "For the first time ever, a B-1 Lancer strategic bomber landed in the Arctic Circle," The Aviation Geek Club, 11 Mar 21, https://theaviationgeekclub.com/for-the-first-time-ever-a-b-1-lancer-strategic-bomber-landed-in-the-arctic-circle/

31. Hilde-Gunn Bye, "The U.S. B-1 Lancers currently deployed in Norway conducted their first mission of the deployment in and around the Barents Sea Friday," High North News, 1 Mar 21, https://www.highnorthnews.com/en/us-b-1s-conducted-first-mission-norway

32. "Nuclear submarines with ballistic missiles: Project 667-BRDM 'Dolphin' (Delta-IV class)," TopWar.ru, 18 Sep 12, https://topwar.ru/18942-atomnye-podvodnye-lodki-s-ballisticheskimi-raketami-proekt-667-bdrm-delfin-delta-iv-class.html; "Project 667BRDM Dolphin-DELTA IV," Military Russia, 16 Dec 12, http://militaryrussia.ru/blog/topic-703.html; "Project 667BRDM (NATO-Delta IV)," DeepStorm.ru, http://www.deepstorm.ru/DeepStorm.files/45–92/nbrs/667BDRM/list.htm

33. "Projects 955 'Borey,'" Deepstorm.ru, http://www.deepstorm.ru/DeepStorm.files/on_1992/955/list.htm

34. "Arctic Exercise UMKA-2021 Shows Russian SSBN Can Deliver Massive Nuclear Strikes," Naval News, https://www.navalnews.com based on TASS release; Ryan White, "3 Russian Submarines Surface from under Arctic Ice for the First Time," Naval Post, 26 Mar 21, https://navalpost.com/russian-submarines-nuclear-arctic-exercise-umka-2021/; "3 Russian nuclear subs emerge from under Arctic ice for 1st time," Xinhua News, 27 Mar 21, http://www.xinhuanet.com/english/2021–03/27/c_139839125.htm

35. "Media: Not three, but five submarines surfaced in the Arctic, one of which launched an ICBM," TopCor.ru, 30 Mar 21, https://en.topcor.ru/19264-smi-v-arktike-vsplyli-ne-tri-a-pjat-podlodok-odna-iz-kotoryh-zapustila-mbr.html

36. "NATO intercepts Russian planes 10 times in a day," BBC, 20 Mar 21, https://www.bbc.com/news/world-europe-56577865

37. Flight Radar 24 screenshots of Il-96–300PU activity on 18, 20, and 27 March 2021.

38. "Russia draws troops to the border with Ukraine-Khomchak," Ukrainian Pravda, 30 Mar 21, https://www.pravda.com.ua/news/2021/03/30/7288381/

39. Mykola Bielieskov, "The Russian and Ukrainian Spring 2021 War Scare," CSIS, 21 Sep 21, https://www.csis.org/analysis/russian-and-ukrainian-spring-2021-war-scare; and Matthew P. Funaiole et al., "Unpacking the Russian Troop Buildup along Ukraine's Border," CSIS, 22 Apr 21, https://www.csis.org/analysis/unpacking-russian-troop-buildup-along-ukraines-border

40. See the Wikipedia timeline for 2020–2021 Belarusian protests, https://en.wikipedia.org/wiki/Timeline_of_the_2020%E2%80%932021_Belarusian_protests

41. "Russia Puts Eurofighter Typhoons, F-16s to 'Stern Test' as NATO Jets Chase Russian Bombers," Eurasian Times, 1 Apr 21, https://eurasiantimes.com/russia -puts-eurofighter-typhoons-f-16s-to-stern-test-as-nato-jets-chase-russia-bombers/

42. "Flight of 'alligators': How the Russian Armed Forces practically implement information operations against Ukraine during military exercises," InformNapalm, 2 Apr 21, https://informnapalm.org/ua/prolot-alihatoriv/

43. Bielieskov, "War Scare."

44. NATO, "Key NATO and Allied Exercises in 2021," https://www.nato.int/nato_static _fl2014/assets/pdf/2021/3/pdf/2103-factsheet_exercises.pdf

45. Bielieskov, "War Scare."

46. Pavel Litvinenko, "The third plane from the special flight squad 'Russia' is approaching the Saratov region-this time the 'presidential' Il-96," Versia.ru, 12 Apr 21, https:// nversia.ru/news/k-saratovskoy-oblasti-priblizhaetsya-uzhe-tretiy-za-den-samolet -iz-specialnogo-letnogo-otryada-rossiya-na-etot-raz-prezidentskiy-il-96/; Irina Filippova, "TU-214 plane of the special flight squad 'Russia' landed at the airbase in Engels," Versia.ru, 12 Apr 21, https://nversia.ru/news/na-aviabaze-v-engelse -prizemlilsya-samolet-tu-214-specialnogo-letnogo-otryada-rossiya/

47. Status-6 Twitter post, 15 Apr 21.

48. Woodward and Costa, Peril, 401–402.

49. "Something Was Badly Wrong: When Washington Realized Russia Was Actually Invading Ukraine," Politico, 24 Feb 23, https://www.politico.com/news /magazine/2023/02/24/russia-ukraine-war-oral-history-00083757

50. Muzyka, Defending the Union, 3; Matthew P. Funaiole et al., "Unpacking the Russian Troop Buildup along Ukraine's Border," CSIS, 22 Apr 21, https://www.csis .org/analysis/unpacking-russian-troop-buildup-along-ukraines-border

51. Bielieskov, "War Scare."

52. "Russian TU-160 broke through two Norwegian F-35s that squeezed it, turning on supersonic speed," Avia.pro, 25 Apr 21, https://avia.pro/news/rossiyskiy-tu -160-prorvalsya-cherez-dva-zazhavshih-ego-norvezhskih-f-35-vklyuchiv; "The Russian Tu-22M3 completed strikes against all NATO naval bases in the Mediterranean in one flight," Avia.pro, 26 Apr 21, https://avia.pro/news/rossiyskiy-tu -22m3-za-odin-polyot-otrabotal-udary-po-vsem-voenno-morskim-bazam-nato-v

53. Pavel Podvig, "Russian Strategic Nuclear Forces: Strategic Rocket Forces," https:// russianforces.org/missiles/

54. "Perimeter System," Military Review, 27 Jun 13, at https://topwar.ru/29887-sistema -perimetr.html

55. This is based on an analysis of Russian NC3I aircraft movements derived from the Flight Radar 24 website that includes movements of all Il-96–300PU and TU-214PU and TU-214SR aircraft.

56. ADS-B Exchange screenshots, January, February, and March 2021.

57. ADS-B Exchange screenshots, 11–12 Apr 21.

58. ADS-B Exchange screenshots, 14, 20, 23 27, and 29 Apr 21.

59. "Something Was Badly Wrong: When Washington Realized Russia Was Actually Invading Ukraine," *Politico*, 24 Feb 23, https://www.politico.com/news/magazine /2023/02/24/russia-ukraine-war-oral-history-00083757

60. Peter Suchiu, "B-52 Bombers Are Training Near the Arctic Circle to Send Russia a Message," 19fortyfive.com, 26 May 21, https://www.19fortyfive.com/2021/05/b-52 -bombers-are-training-near-the-arctic-circle-to-send-russia-a-message/

61. "Two TU-160s raided NATO air defense systems in Norway," Avia.pro, 19 May 21, https://avia.pro/news/dva-rossiyskih-tu-160-ustroili-nalyot-na -sredsvta-pvo-nato-v-novregii

62. "Russia successfully worked out the destruction of the NATO fleet off the coast of Romania and Ukraine," Avia.pro, 20 May 21, https://avia.pro/news/rossiyskie- bombardirovshchiki-krasivo-otrabotali-unichtozhenie-baz-vms-nato-i-ukrainy-v

63. "Russian Air Force Su-27 fighter intercepted an American B-52 'surfaced' near the Russian border," Avia.pro, 20 May 21, https://avia.pro/news/istrebitel-vks-rf-su -27-perehvatil-amerikanskiy-b-52-vynyrnuvshiy-u-rossiyskoy-granicy

64. Tony Wesolowsky, "Hope and Horror: How Belarus Has Changed since an Election Ignited a Crisis One Year Ago," RFE/RL, 8 Aug 21, https://www.rferl.org/a /belarus-2020-crisis-anniversary/31399497.html

65. Alesia Rudnik, "Belarus dictator turns hybrid war into a humanitarian crisis," Atlantic Council, 15 Nov 21, https://www.atlanticcouncil.org/blogs/belarusalert /belarus-dictator-turns-hybrid-war-into-humanitarian-crisis/

66. "Die Here or Go to Poland," Human Rights Watch, 24 Nov 21, https://www. hrw.org/report/2021/11/24/die-here-or-go-poland/belarus-and-polands-shared -responsibility-border-abuses

67. "Russian TU-22M3 bomber worked out a strike against NATO ships with Kh-22 cruise missile carrying a thermonuclear warhead," Avia.pro, 28 May 21, https:// avia.pro/news/rossiyskiy-bombardirovshchik-tu-22m3-otrabotal-udar-po -korablyam-nato-krylatoy-raketoy-h-22-s

68. "US Air Force B-52 bomber staged a provocation in front of Russian S-400s in Crimea," Avia.pro, 26 May 21, https://avia.pro/news/bombardirovshchik-b-52-vvs -ssha-ustroil-provokaciyu-pered-rossiyskimi-s-400-v-krymu

69. "Remarks by President Biden in Press Conference," 16 Jun 21, https://www.white house.gov/briefing-room/speeches-remarks/2021/06/16/remarks-by-president -biden-in-press-conference-4/

70. "From the Beaufort Sea to the Eastern Atlantic: AMALGAM DART Provides Valuable Training in the Arctic," Canadian NORAD Region, 23 Jun 21, https://

www.norad.mil/Newsroom/Article/2666915/from-the-beaufort-sea-to-the-eastern
-atlantic-norad-exercise-amalgam-dart-provi/

71. Wyatt Olson, "B-52 bombers traverse Arctic in 27-hour Europe-to Pacific mission,
    Stars and Stripes, 22 Jun 21, https://www.stripes.com/branches/air_force/2021–06
    –21/b-52-bombers-traverse-arctic-in-27-hour-europe-to-pacific-mission
    -1763438html

72. For example, Ed Adamczyk, "B-52H bombers fly over the Arctic from Spain to
    Louisiana," UPI, 18 Jun 21, https://www.upi.com/Defense-News/2021/06/18
    /b52h-bomber-stratofortress-barkdale/5511624040241/

73. Brendan Cole, "Russia Sends Fighter Jets to Intercept U.S. Strategic Bombers,"
    Newsweek, 15 Jul 21, https://www.newsweek.com/russia-mig-31-b-52-bering
    -sea-military-flight-1609890

74. Davis Cenciotii, "Two Russian Tu-16s and four Flankers intercepted by Italian
    F-35s, Danish F-16s and Swedish Gripens over the Baltic," The Aviationist, 15 Jun
    21, https://theaviationist.com/2021/06/15/tu-160-italian-f-35/

75. Dmitry Boltenkov and Alexey Ramm, "Island maneuver: What did the ships of the
    Navy do near the Hawaiian archipelago," Izvestia, 20 Jun 21, https://iz.ru/1180721/
    dmitrii-boltenkov-aleksei-ramm/ostrovnoi-manevr-chto-delali-korabli-vmf-vblizi
    -gavaiskogo-arkhipelaga; "We approached Pearl Harbor: What the Russian fleet
    did in the Pacific Ocean," RIA Novosti, 24 Jun 21, https://ria.ru/20210624/flot
    -1738306098.html

76. "To distant shores: The Russian fleet conducted exercises near Pearl Harbor for
    the first time," TopWar, 30 Jun 21, https://topwar.ru/184506-k-dalnim-beregam
    -rossijskij-flot-vpervye-provel-uchenija-okolo-perl-harbora.html

77. "Russian aviation and navy practises a large-scale attack on the US Navy base in
    Hawaii and American submarines," Avia.pro, 16 Jun 21, https://avia.pro/news
    /rossiyskaya-aviaciya-i-flot-otrabotal-i-krupnomasshtabnyy-udar-po-baze-vms-
    ssha-na-gavayyah-i

78. Newdick, "A wake-up call."

79. Edward Lundquist, "Exercise Sea Breeze 2021 comes to a close in the Black
    Sea," *Seapower Magazine*, 13 Jul 21, https://seapowermagazine.org/exercise-sea
    -breeze-2021-comes-to-a-close-in-black-sea/

80. "To distant shores: The Russian fleet conducted exercises near Pearl Harbor for the
    first time," TopWar, 30 Jun 21, https://topwar.ru/184506-k-dalnim-beregam-rossijskij
    -flot-vpervye-provel-uchenija-okolo-perl-harbora.html

81. ADS-B Exchange screenshots, May 2021.

82. ADS-B Exchange screenshots, June 2021.

83. Muzyka, "Defending the Union: ZAPAD 2021."

84. "Kursk Nuclear Power Plant," Wikipedia, https://en.wikipedia.org/wiki/Kursk
    _Nuclear_Power_Plant

85. Muzyka, "Defending the Union: ZAPAD 2021."

86. Mykhaylo Zabrodskyi et al., *Preliminary Lessons in Conventional Warfighting from Russia's Invasion of Ukraine: February–July 2022* (London: RUSI, 2022), 7.

87. See the timeline in "2021–2022 Belarus-European Union Border Crisis," https://en.wikipedia.org/wiki/2021%E2%80%932022_Belarus%E2%80%93European _Union_border_crisis

88. "Belarus: Speech on behalf of High Representative/Vice-President Josep Borrell at the EP Plenary," Strasbourg, 5 Oct 21, https://www.eeas.europa.eu/eeas/belarus -speech-behalf-high-representativevice-president-josep-borrell-ep-plenary _en; Janko Bekic, "Coercive Engineered Migrations as a Tool of Hybrid Warfare: A Binary Comparison of Two Cases on the External EU Border," *Croatian Political Science Review*, Vol. 59, No. 2 (2022), 141–169.

89. Bekic, "Coercive Engineered Migrations"; Alla Leukavets, "Crisis in Belarus: Main Phases and the Role of Russia, the European Union, and the United States," *Kennan Cable*, No. 74, 1 Jan 22, https://www.wilsoncenter.org/publication/kennan -cable-no-74-crisis-belarus-main-phases-and-role-russia-european-union-and -united?collection=32543; Grigol Julukhidze, *The Weaponization of Migration—New Tool of Russian Hybrid War: Crisis on the Polish-Belarusian Border*, Expert Opinion 172: The Georgian Foundation for Strategic and International Studies, 2021, https://gfsis.org.ge/publications/view-opinion-paper/172; Andrzej Szabaciuk, "The crisis on the Polish-Belarusian border in the context of rising tensions in Eastern Europe (part 2)," *Komentarze IES*, No. 471 (168/2021), 23 Dec 21, https://ies.lublin.pl/wp -content/uploads/2021/12/ies-commentaries-471.pdf

90. Borrell speech.

91. Bekic, "Coercive Engineered Migrations."

92. "Migrant Crisis on the Belarus-Poland Border," Congressional Research Service, 13 Dec 21, https://crsreports.congress.gov/product/pdf/IF/IF11983

93. Samatha de Bendern, "Belarus is new weapon in Putin's hybrid warfare arsenal," Chatham House, 18 Aug 21, https://www.chathamhouse.org/2021/08/ belarus-new-weapon-putins-hybrid-warfare-arsenal

94. Anton Shekhovtsov, *Russia and the Western Far Right* (New York: Routledge, 2018) for a detailed analysis of this issue.

95. Viktor Denisenko, "Elements of Information Warfare during Migration Crisis on Belarus-EU Border," Civic Resilience Initiative, 2021, https://cri.lt/wp-content/ uploads/2022/03/CRI_tyrimas.pdf

96. ADS-B Exchange screenshots, 26 Jul 21.

97. Examples include "Biden revealed the reason for the refusal of the war in Afghanistan," Lenta.Ru, 31 Agu 21, https://lenta.ru/news/2021/08/31/bidenzyv/; "Biden explained the flight of American troops from Afghanistan," RIA Novosti, 16 Aug 21, https://ria ru/20210816/bayden-1746009167.html; "On the geopolitical implications of Joseph

Biden's decision on Afghanistan," Rossiyskaya Gazeta, 17 Aug 21, https://www
.ng.ru/editorial/2021–08–17/2_8227_editorial.html

98. "Biden falls into the abyss," Vesti.ru, 22 Aug 21, https://www.vesti.ru/article/2603769;
"NYT: Biden will go down in history with crushing US defeat in Afghanistan," RT,
17 Aug 21, https://russian.rt.com/inotv/2021–08–17/NYT-Bajden-vojdet-v-istoriyu;
"Biden's rating collapsed after the events in Afghanistan," TVC.ru, 18 Aug 21, https://
www.tvc.ru/news/show/id/218285

99. Muzyka, "ZAPAD 2021."

100. Muzyka.

101. Muzyka.

102. Muzyka.

103. Muzyka.

104. "Global Storm highlights interoperability of U.S. strategic deterrence," Defense
Forum, 17 Sep 21, https://ipdefenseforum.com/2021/09/global-storm-highlights
-interoperability-of-u-s-strategic-deterrence/

105. David Axe, "U.S. Air Force B-2s arrive in Iceland to practice stealth bombing," Forbes,
25 Aug 21, https://www.forbes.com/sites/davidaxe/2021/08/25/arctic-stealth-us-air
-force-b-2s-arrive-in-iceland-to-practice-cold-weather-bombing/?sh=33c4acb282bf

106. ADS-B Exchange screenshots, 12–14 Aug 21.

107. FR24 screenshots for 8 and 9 Sep 21.

108. RivetJoint Twitter post, 13 Sep 21; FR 24 screenshots, 13 Sep 21.

109. "An unexpected raid of Russian Tu-160s lifted combat aircraft of four European
countries into the sky," Avia.pro, 22 Sep 21, https://avia.pro/news/neozhidannyy
-reyd-rossiyskih-tu-160-podnyal-v-nebo-boevuyu-aviaciyu-chetyryoh-stran-evropy

110. RivetJoint Twitter post, 6 Sep 21.

111. U.S. Air Force, "RC-135U fact sheet," https://www.af.mil/About-Us/Fact-Sheets
/Display/Article/104495/rc-135u-combat-sent/#:~:text=The%20Combat%20
Sent%20deploys%20worldwide,and%20ultra%20high%20frequency%20radios.

112. RivetJoint Twitter posts, 10 Sep 21; 13 Sep 21; 15 Sep 21.

113. RivetJoint Twitter posts, 14 Sep 21; 14 Sep 21; 16 Sep 21; 22 Sep 21.

114. RivetJoint Twitter posts, 24 Sep 21.

115. This is said with the full understanding that information on this may become
available in the future to contradict this statement.

116. RivetJoint Twitter post, 29 Sep 21.

117. "Something Was Badly Wrong: When Washington Realized Russia Was Actually
Invading Ukraine," Politico, 24 Feb 23, https://www.politico.com/news/magazine
/2023/02/24/russia-ukraine-war-oral-history-00083757

118. ADS-B Exchange screenshots, 13 Sep 21.

119. ADS-B Exchange screenshots, Sep 21.

120. "Something Was Badly Wrong: When Washington Realized Russia Was Actually Invading Ukraine," *Politico*, 24 Feb 23, https://www.politico.com/news/magazine /2023/02/24/russia-ukraine-war-oral-history-00083757

121. FR 24 screenshot, 7 Oct 21.

122. RivetJoint Twitter posts, 5 Oct 21; 13 Oct 21; 15 Oct 21.

123. "Electronic warfare systems again attacked an American aircraft in the Kerch Strait area," Avia.pro, 19 Oct 21, https://avia.pro/news/kompleksy-reb-vnov-atakovali -amerikanskiy-samolyot-v-rayone-kerchenskogo-proliva

124. "The YARSES of the Bernaul RVSA compound have entered the combat patrol routes," VPK, 13 Oct 21, https://vpk.name/en/548930_the-yarses-of-the-barnaul -rvsn-compound-have-entered-the-combat-patrol-routes.html?fbclid=IwARotlty -jyQcENfQ9_WY1EHzCDMiKdd4ZXpbJpbmT3GMMy3Wa_PUX8KXluo

125. Podvig, "Strategic Rocket Forces," https://russianforces.org/missiles/

126. "CSTO: Thunder-2021 special exercise to be held in Armenia this year," Dialog News, 3 Jun 21, https://www.dialogorg.ru/news/03.06.2021-chcghjhccgcjgckc gckhckiccc/; "The exercises of the collective forces of the CSTO 'GROM-2021' were held in Armenia," NewsRambler, 29 Sep 21, https://news.rambler.ru /troops/47295451-ucheniya-kollektivnyh-sil-odkb-grom-2021-proshli-v-armenii/

127. "Russian Armed Forces to hold strategic command and staff exercises 'Vostok' and 'Thunder,'" TASS, 21 Dec 22, https://tass.ru/armiya-i-opk/13258711

128. Department of Defense, "Secretary of Defense Lloyd J. Austin III's visit to Ukraine," 19 Oct 21, https://www.defense.gov/News/Releases/Release/Article/2815096 /secretary-of-defense-lloyd-j-austin-iiis-visit-to-ukraine/

129. "Two Russian Jets Intercept American Bombers over Black Sea," Global Times, 21 Oct 21, https://www.globaltimes.cn/page/202110/1236903.shtml

130. "Vladimir Putin Meets with Members of the Valdai Discussion Club. Transcript of the Plenary Session of the 18th Annual Meeting," 22 Oct 21, https://valdaiclub .com/events/posts/articles/vladimir-putin-meets-with-members-of-the-valdai -discussion-club-transcript-of-the-18th-plenary-session/#masha_0=251:5,251:82

131. RivetJoint Twitter post, 20 Oct 21.

132. U.S. Air Force, "RC-135S COBRA BALL fact sheet," https://www.af.mil/About -Us/Fact-Sheets/Display/Article/104498/rc-135s-cobra-ball/#:~:text=The%20 RC%2D135S%20Cobra%20Ball,electronic%20data%20on%20ballistic%20targets.

133. RivetJoint Twitter post, 28 Oct 21.

134. "Russia has worked out a nuclear strike on the largest US base in Alaska," Avia. pro, 23 Oct 21, https://avia.pro/news/rossiya-otrabotala-yadernyy-udar-po -samoy-krupnoy-baze-ssha-na-alyaske

135. FR 24 screenshot, 26 Oct 21.

136. Bart Noeth, "Royal Norwegian Air Force intercepts Russian bombers and fighters," Aviation24, 28 Oct 21, https://www.aviation24.be/organisations/nato/royal-norwegian-air-force-intercepts-russian-bombers-and-fighters/

137. "Russian Tu-22M3 bombers worked out a strike on Europe after threats from the German Ministry of Defence," Avia.pro, https://avia.pro/news/rossiyskie-bomb ardirovshchiki-tu-22m3-otrabotali-udar-po-evrope-posle-ugroz-minoboron y-germanii

138. "Team Minot Concludes Exercise Global Thunder 22," Minot AFB Public Affairs, 17 Nov 21, https://www.minot.af.mil/News/Article-Display/Article/2847019/team-minot-concludes-exercise-global-thunder-22/

139. Flight Radar 24 screenshots, 4, 8, 10, and 17 Nov 21.

140. USAF, "EC 135 Communications Center Electronic: Inflight Operations and Maintenance Manual," 10 Oct 80, 1–161 to 1–162.

141. Flight Radar 24 screenshots, 12 Apr, 14 May, 19 May, and 28 Sep 21; Assessment of ADS-B Exchange historical data, Jan–Dec 2021.

142. Betsy Woodruff Swan and Paul McLeary, "Satellite images show new Russian military buildup near Ukraine," *Politico*, 1 Nov 21, https://www.politico.com/news/2021/11/01/satellite-russia-ukraine-military-518337

143. "ACAPS Briefing Note: Belarus/Poland: Migration Crisis on the Belarus-Poland Border," reliefweb, 2 Dec 21, https://reliefweb.int/report/belarus/acaps-briefing-note-belaruspoland-migration-crisis-belarus-poland-border-2-dcember#:~:text=CRISIS

144. Alesia Rudnik, "Belarus dictator turns hybrid war into humanitarian crisis," The Atlantic Council, 15 Nov 21, https://www.atlanticcouncil.org/blogs/belarusalert/belarus-dictator-turns-hybrid-war-into-humanitarian-crisis/

145. "Russia-Ukraine border: NATO warning over military build-up," BBC, 15 Nov 21, https://www.bbc.com/news/world-europe-59288181

146. "Russia flies nuclear-capable bombers over Belarus as migrant crisis escalates," *Jerusalem Post*, 10 Nov 21, https://www.jpost.com/breaking-news/russian-bombers-sent-to-patrol-belarus-airspace-684576

147. "Russian planes intercepted by Belgian jets over North Sea-Netherlands," *U.S. News and World Report*, 12 Nov 21, https://www.usnews.com/news/world/articles/2021-11-12/russian-planes-over-north-sea-intercepted-by-belgian-f-16s-dutch-defence-ministry#:~:text=Russian%20Planes%20Intercepted%20by%20Belgian%20Jets%20Over%20North%20Sea%20%2D%20Netherlands,-By%20Reuters&text=By%20Reuters-,Nov.,2021%2C%20at%2010%3A39%20a.m.&text=AMSTERDAM%20(Reuters)%20%2D%20Two%20Russian,the%20Dutch%20Defence%20Ministry%20said

148. RivetJoint Twitter posts, 1 Nov 21, 2 Nov 21, 5 Nov 21, 7 Nov 21, 8 Nov 21, 10 Nov 21, 13 Nov 21, 15 Nov 21, 17 Nov 21, and 18 Nov 21.

149. Ida Louise Rostad et al., "Russia towards NATO: This is our country, and our waters," NRK.NO, 9 Nov 21, https://www.nrk.no/tromsogfinnmark/lavrov -advarer-norge-mot-a-la-nato-slippe-til-i-arktis-1.15499576

150. "Russia-Ukraine border: NATO warning over military build-up," BBC, 15 Nov 21, https://www.bbc.com/news/world-europe-59288181

151. Brian Weeden, "History of Anti-Satellite Tests in Space," 22 Feb 23, see spreadsheet at https://docs.google.com/spreadsheets/d/1e5GtZEzdo6xk41i2_ei3c8jRZDjvP4X wz3BVsUHwi48/edit#gid=1252618705; Ankit Panda, "The Dangerous Fallout of Russia's Anti-Satellite Missile Test," Carnegie Endowment for International Peace, 17 Nov 21, https://carnegieendowment.org/2021/11/17/dangerous-fallout-of-russia- s-anti-satellite-missile-test-pub-85804; "The Ministry of Defence reported on tests with the destruction of a satellite in orbit," RBC.ru, 16 Nov 21, https://www.rbc .ru/politics/16/11/2021/619393c39a79475f58e82c70

152. Assessment of ADS-B Exchange historical data, 15–16 Nov 21.

153. "Two Tu-95MS long-range missile carriers unexpectedly appeared off the coast of the United States having worked the delivery of nuclear strikes," Avia.pro, 19 Nov 21, https://avia.pro/news/dva-dalnih-raketonosca-tu-95ms-neozhidanno-poyavilis -u-poberezhya-ssha-otrabotav-nanesenie

154. RCAF operations Twitter post, 19 Nov 21.

155. Assessment of ADS-B Exchange historical data, 18 Nov 21.

156. Dylan Malyasov, "Portuguese F-16s intercept Russian bombers near NATO airspace," Defence Blog, 23 Nov 21, https://defence-blog.com/portuguese-f-16s -intercept-russian-bombers-near-nato-airspace/

157. "Press briefing notes on Poland/Belarus border, 21 December 2021," ReliefWeb, https://reliefweb.int/report/poland/press-briefing-notes-polandbelarus-border- 21-december-2021; "Is the Belarus migrant crisis a 'new type of war'? A conflict expert explains," The Conversation, 16 Nov 21, https://theconversation.com /is-the-belarus-migrant-crisis-a-new-type-of-war-a-conflict-expert-explains -171739; Michal Fiszer et al., "Manufactured Migrant Crisis on Polish Border Is a Form of Hybrid Warfare," Discourse Magazine, 18 Nov 21, https://www .discoursemagazine.com/politics/2021/11/18/manufactured-migrant-crisis-on -polish-border-is-a-form-of-hybrid-warfare/

158. RivetJoint Twitter post, 25 Nov 21.

159. Shane Harris and Paul Sonne, "Russia planning massive military offensive against Ukraine involving 175,000 troops, U.S. intelligence warns," Washington Post, 3 Dec 21.

160. RivetJoint Twitter post, 3 Dec 21.

161. "At the strategic exercise 'Thunder' it was decided to test hypersonic weapons," Izvestia, 1 Dec 21, https://iz.ru/1257657/2021–12–01/na-strategicheskikh-ucheniiakh -grom-reshili-oprobovat-giperzvukovoe-oruzhie

162. Paul Labbe et al., "Current and future hypersonic threats, scenarios, and defence technologies for the security of Canada," Defence Research and Development Canada Scientific Report, March 2022, https://cradpdf.drdc-rddc.gc.ca/PDFS /unc385/p814591_A1b.pdf

163. Kolja Brockmann and Dmitry Stefanovich, "Hypersonic Boost-Glide Systems and Hypersonic Cruise Missiles," SIPIRI, April 2022, https://www.sipri.org/sites /default/files/2022–04/2204_hgvs_and_hcm_challenges_for_the_mtcr.pdf, 9.

164. Izvestia, https://iz.ru/tag/giperzvukovoe-oruzhie; RIA Novosti, https://ria.ru /product_giperzvukovoe-oruzhie-rossii/

165. Seth Jones et al., "Moscow's Continuing Ukrainian Buildup," CSIS, 17 Nov 21, https://www.csis.org/analysis/moscows-continuing-ukrainian-buildup; "Russian Military Buildup along the Ukrainian Border," Congressional Research Service, 7 Feb 22, https://crsreports.congress.gov/product/pdf/IN/IN11806

166. "Something Was Badly Wrong: When Washington Realized Russia Was Actually Invading Ukraine," Politico, 24 Feb 23, https://www.politico.com/news/magazine /2023/02/24/russia-ukraine-war-oral-history-00083757

167. Shane Harris and Paul Sonne, "Russia planning massive military offensive against Ukraine involving 175,000 troops, U.S. intelligence warns," Washington Post, 3 Dec 21.

168. RivetJoint Twitter posts, 9 Dec 21.

169. FR 24 screenshot, 9 Dec 21.

170. FR 24 screenshots, 11 Dec 21, 13 Dec 21, and 17 Dec 21.

171. FR 24 screenshot, 16 Dec 21.

172. DS-B Exchange screenshots, 21 Dec 21.

173. "Russian Tu-22Me bombers worked out the destruction of Kyiv?" Avia.pro, 25 Dec 21, https://avia.pro/news/rossiyskie-bombardirovshchiki-tu-22m3 -otraboтali-unichtozhenie-kieva

174. FR 24 screenshots, 24 Dec 21, 27 Dec 21, and 29 Dec 21.

175. "Mission-ready nuclear submarines of Russian Pacific Navy Urgently leave port," Pravda, 26 Dec 21, https://english.pravda.ru/news/russia/149739-russian _submarine/

176. Robert-Gabriel Ticalau, "The 'Allied Resolve' military exercise between Belarus and Russia: A new threat to Ukraine?" Romanian Centre for Russian Studies, 1 Feb 22, https://russianstudiesromania.eu/2022/02/01/the-allied-resolve-military -exercise-between-belarus-and-russia-a-new-threat-to-ukraine/

177. Roger McDermott, "Russia's Military Exercise in Belarus Prepares for War," Jamestown Foundation, 26 Jan 22, https://jamestown.org/program/russias-military-exercise-in-belarus-prepares-for-war/

178. "Background Press Call by a Senior Administration Official on President Biden's Call with President Putin of the Russian Federation," 30 Dec 21, https://www.whitehouse.gov/briefing-room/press-briefings/2021/12/30/background-press-call-on-president-bidens-call-with-president-putin-of-the-russian-federation/

179. ADS-B Exchange screenshot, 30 Dec 21.

180. Peter Hennessy and James Jinks, *The Silent Deep: The Royal Navy Submarine Service since 1945* (London: Alan Lane, 2015), 643, 667.

181. The 2010 review of the British strategic deterrent asserted that the number of functional SLBM tubes would be reduced from "twelve to eight" and the number of warheads aboard topping off at forty. There is no reason to believe that these numbers cannot be altered as required, so the exact number of RVs aboard the Vanguards is unknown at this time. See H.M. Government, "Securing Britain in an Age of Uncertainty: The Strategic Defence and Security Review," October 2010, https://web.archive.org/web/20101222022127/http://www.direct.gov.uk/prod_consum_dg/groups/dg_digitalassets/%40dg/%40en/documents/digitalasset/dg_191634.pdf, 5.

182. Iain Cameron Twitter post, 1 Jan 22. For comparative purposes, Peter Hennessy and James Jinks explain the great pains the Royal Navy took during the Cold War to prevent Soviet acquisition of Royal Navy ballistic missile submarines as they departed their base. See Hennessy and Jinks, *The Silent Deep*, 283, 330, 342–346.

183. Nancy Shute, "For Russian, New Years Eve Remains the Superholiday," NPR, 27 Dec 11, https://www.npr.org/sections/thesalt/2011/12/27/144326826/for-russians-new-years-eve-remains-the-super-holiday

Chapter 5. Triumph of Disbelief

1. "Something Was Badly Wrong: When Washington Realized Russia Was Actually Invading Ukraine," *Politico*, 24 Feb 23, https://www.politico.com/news/magazine/2023/02/24/russia-ukraine-war-oral-history-00083757

2. Olivia Gazis, "U.S. says Russia is creating possible 'pretext for invasion' of Ukraine," CBS News, 14 Jan 22, https://www.cbsnews.com/news/russia-ukraine-invasion-pretext-united-states/; U.S. State Department, Press Briefing, 3 Feb 22, https://www.state.gov/briefings/department-press-briefing-february-3–2022/; Natasha Bertrand, "US alleges Russian planning false flag operation against Ukraine using 'graphic' video," CNN, 3 Feb 22, https://www.cnn.com/2022/02/03/politics/us-alleges-russian-false-flag-ukraine/index.html; "US warns Putin might launch false flag chemical weapons attack before invading Ukraine," *Daily Mail*, 17 Feb

22, https://www.dailymail.co.uk/news/article-10522545/Ukraine-crisis-Fears
-Russian-false-flag-operation-underway-justify-invasion.html; "What are false flag
attacks—and did Russia stage any to claim justification for invading Ukraine?" *The
Conversation*, 18 Feb 22, https://theconversation.com/what-are-false-flag-attacks
-and-did-russia-stage-any-to-claim-justification-for-invading-ukraine-177879

3. Confidential communication with the author. Of note, on 12 Jan 22, fake bomb
   threats were made to hundreds of Ukrainian *and* Russian schools. The purpose
   behind this was likely some form of information operations preparation for the
   false flag event when it took place. See "Russian Hybrid Threats Report: Troops
   Arrive in Belarus as Propaganda Narratives Heat Up," Atlantic Council, 21 Jan 22,
   https://www.atlanticcouncil.org/blogs/new-atlanticist/russian-hybrid-threats
   -report-troops-arrive-in-belarus-as-propaganda-narratives-heat-up/

4. UK Defence Intelligence Twitter post, 14 Mar 22.

5. Zabrodskyi et al., *Preliminary Lessons*.

6. Dominic Nicholls, "Russa deploys mobile crematoriums to follow its troops into battle," *The Telegraph*, 23 Feb 22, https://www.telegraph.co.uk/world-news/2022/02/23
   /russia-deploys-mobile-crematorium-follow-troops-battle/

7. One of these units, led by colonel Konstantine Ogiy, was ambushed and liquidated
   in toto. "Konstantin Ogiy, Colonel from the SOBR detachment, killed in Ukraine,"
   Zhytomyr Journal, 28 Feb 22, https://hindustannewshub.com/russia-ukraine-news
   /commander-of-kuzbass-sobr-killed-in-military-operation-in-ukraine-the-moscow
   -times/

8. Liveuamap Twitter feed depicting these events throughout the day before the plug
   was pulled on connectivity by the Kazakh government.

9. Alex Khrebet Twitter post, 5 Jan 22.

10. Rene D. Kanayama, "Events in Kazakhstan's Almaty of January 2022: Grass-root
    Revolt of Terrorism Inspired Insurgency?," *Sicurezza Terrorisimo Societa* Issue
    1/2022, 123–148; Raushan Zhandayeva and Rachael Rosenberg, "Kazakhstan's
    Bloody January: Digital Repression on the 'New Silk Road,'" Toda Peace Institute,
    Policy Brief No. 140, November 2022, https://toda.org/policy-briefs-and-resources
    /policy-briefs/kazakhstans-bloody-january-digital-repression-on-the-new-silk
    road.html

11. Julia Emtseva, "Collective Security Treaty Organization: Why Are Russian
    Troops in Kazakhstan?," EJIL Talk!, 13 Jan 22, https://www.ejiltalk.org/collective
    -security-treaty-organization-why-are-russian-troops-in-kazakhstan/

12. "The State Duma called events in Kazakhstan an attempt at a color revolution,"
    RIA Novosti, 6 Jan 22, https://ria.ru/20220106/kazakhstan-1766775249.html;
    "To the events in Kazakhstan, the decline of the American 'color revolutions'
    strengthens Russia," Ukrainia.ru, 7 Jan 22, https://ukraina.ru/20220107/1033025734

.html.; "Signs of 'color revolution' in Kazakhstan discovered," Lenta.ru, 8 Jan 22, https://lenta.ru/news/2022/01/08/canadakzh; "Kazakhstan: Color revolution mixed with Islamic jihad," Zvezda Weekly, 10 Jan 22, https://zvezdaweekly.ru /news/202218131-tip1B.html; "Medvedev said that there was an attempt of 'colour revolution in Kazakhstan,'" 28 Jan 22, https://www.interfax.ru/world/818902

13. Radioskaner Twitter post, 6 Jan 22; Gerjon Twitter post, 6 Jan 22; "The peacekeeping mission in Kazakhstan is accompanied by the expected cries of the Pozner-Goz-mans," Zavtra.ru, 6 Jan 22, https://zavtra.ru/events/mirotvorcheskaya_missiya_v _kazahstane_soprovozhdaetsya_ozhidaemimi_voplyami_poznero-gozmanov; RIA Novosti Twitter post, 7 Jan 22; Liveuamap Twitter post, 7 Jan 22; see also Kanayama, "Events in Kazakhstan's Almaty of January 2022."

14. Svyatoslav Kaspe, "Kazakhstan: What it was, what it means," GIS Reports, 3 Mar 22, https://www.gisreportsonline.com/r/kazakhstan-protests/

15. Kanayama, "Events in Kazakhstan's Almaty of January 2022: Grass-root Revolt of Terrorism Inspired Insurgency?" *Sicurezza Terrorisimo Societa* Issue 1/2022, 123–148. See also "Fugitive Police General Kudebaev Apprehended in Turkey, Returned to Kazakhstan," RFE/RL, 27 Apr 23, https://www.rferl.org/a/kazakhstan-kudebaev-apprehended-nazarbaev/32382111.html#0_8_10089_8766_2710_247803822

16. Victoria Hudson, "The impact of Russian soft power in Kazakhstan: Creating an enabling environment for cooperation between Nur-Sultan and Moscow," *Journal of Political Power*, Vol. 15, No. 3 (2022), 469–494.

17. Estonia in Ukraine Twitter post, 8 Jan 22.

18. Initial reports were that the cable break occurred on 9 January, but later information emerged that it occurred on 7 January. "Damage to SvalSat cable proves Russia further upping stakes," The Lansing Institute, 13 Jan 22, https://lansinginstitute .org/2022/01/13/damage-to-svalsat-cable-proves-russia-further-upping-stakes/

19. Atle Staalesen, "Human activity behind Svalbard cable disruption," *The Barents Observer*, 11 Feb 22, https://thebarentsobserver.com/en/security/2022/02/unknown -human-activity-behind-svalbard-cable-disruption

20. "Damage to SvalSat cable proves Russia further upping stakes," The Lansing Institute, 13 Jan 22.

21. Staalesen, "Human activity behind Svalbard cable disruption," *The Barents Observer*.

22. "Mysterious Atlantic cable cuts linked to Russian fishing vessels," EU Observer, 26 Oct 22, https://euobserver.com/nordics/156342

23. "ESA ends efforts to recover Sentinel-1B," Space News, 3 Aug 22, https://spacenews.com /esa-ends-efforts-to-recover-sentinel-1b/

24. ADS-B Exchange screen shot, 10 Jan 22. In addition to the regular "racetrack" lobes, this flight had additional lobes so it could observe in several directions.

25. Joseph Trevithick, "FAA's Statement on Mysterious Air Traffic Halt Leaves More Questions than Answers," The Drive/The Warzone, 31 Mar 22, https://www.thedrive.com/the-war-zone/43840/faas-statement-on-mysterious-air-traffic-halt-leaves-more-questions-than-answers

26. ADS-B Exchange screenshots, 11 Jan 22.

27. William Pugh, "Executing mission, fostering innovation, making history: Barksdale's first ACE exercise," 2nd Bomb Wing Public Affairs, 24 Jan 22, https://www.barksdale.af.mil/News/Article/2910004/executing-mission-fostering-innovation-making-history-barksdales-first-ace-exer/; "From Barksdale to Blytheville: Exercise Proves B-52s Can Provide Global Deterrence Anytime, Anywhere," Bossier Now, 26 Jan 22, https://bossiernow.com/from-barksdale-to-blytheville-exercise-proves-b-52s-can-provide-global-deterrence-anytime-anywhere/

28. "317th AW provides Agile Combat Employment during exercise Patriot Fury," 7th Bomb Wing Public Affairs, 8 Sep 22, https://www.mcchord.af.mil/News/Article-Display/Article/3160983/317th-aw-provides-agile-combat-employment-during-exercise-patriot-fury/

29. ADS-B Exchange screenshots, 12 Jan 22.

30. Oriana Pawlyk, "FAA briefly halted West Coast flights amid North Korean missile scare," Politico, 11 Jan 22, https://www.politico.com/news/2022/01/11/faa-west-coast-flights-north-korea-526894

31. Confidential communications with the author.

32. U.S. Department of State press statement, "United States Designates Entities and Individuals Linked to the Democratic People's Republic of Korea's (DPRK) Weapons Programs," 12 Jan 22, https://www.state.gov/united-states-designates-entities-and-individuals-linked-to-the-democratic-peoples-republic-of-koreas-dprk-weapons-programs/

33. Press conference by NATO Secretary General Jens Stoltenberg following the meeting of the NATO-Russia Council, 12 Jan 22, https://www.nato.int/cps/en/natohq/opinions_190666.htm

34. Liveuamap Twitter post, 11 Jan 22; "Russia Launches Military Drills Near Ukraine as Invasion Fears Remain High," Moscow Times, 12 Jan 22, https://www.themoscowtimes.com/2022/01/11/us-says-too-early-to-tell-if-russia-serious-on-security-talks-a76016

35. FR24 screenshots, 12 and 13 Jan 22.

36. FR24 screenshot, 13 Jan 22.

37. Press Briefing by Press Secretary Jen Psaki and National Security Advisor Jake Sullivan, 13 Jan 22, https://www.whitehouse.gov/briefing-room/press-briefings

/2022/01/13/press-briefing-by-press-secretary-jen-psaki-and-national-security
-advisor-jake-sullivan-january-13–2022/
38. ADS-B Exchange, 13 Jan 22.
39. "Russian Navy adds pressure on Ukraine and NATO," Navy Lookout, 21 Jan 22,
https://www.navylookout.com/russian-navy-adds-to-pressure-on-ukraine-and
-nato/
40. Anna Ringstrom, "Sweden boost patrols on Gotland amid Russia tensions,"
Reuters, 14 Jan 22, https://www.reuters.com/world/europe/sweden-boosts-patrols
-gotland-amid-nato-russia-tensions-2022–01–13/
41. Minna Alander and Michael Paul, "Moscow Threatens the Balance in the High
North," SWP Comment, No. 2, March 2022, German Institute for International and
Security Affairs, https://www.swp-berlin.org/publications/products/comments
/2022C24_Balance_HighNorth.pdf
42. "Authorities Confirm Sightings of Mysterious Drones over Swedish Nuclear Facilities," The Debrief, 17 Jan 22, https://thedebrief.org/authorities-confirm-sightings
-of-mysterious-drones-over-swedish-nuclear-facilities/
43. OSINTtechnical Twitter post, 16 Jan 22.
44. "U.S. Accuses Russia of Preparing 'False Flag' Operation in Ukraine," RFE/RL,
14 Jan 22, https://www.rferl.org/a/russia-false-flag-ukraine-accusations-invasion
/31654852.html
45. Vladimir Isachenlov and Yuras Karmanau, "Russia denies US claim it seeks
'false flag' pretext to invade Ukraine," Associated Press via Military Times, 17
Jan 22, https://www.militarytimes.com/flashpoints/2022/01/17/russia-denies
-us-claim-it-seeks-false-flag-pretext-to-invade-ukraine/
46. "Pushilin said that Kiev can arrange a staged chemical attack in the Donbass," Arguments & Facts, 27 Jan 22. Note that this article has been removed from the Argument
& Facts website at AIF.ru, but there are screenshots in the author's possession.
47. RT clip embedded in Sergej Sumlenny's Twitter post, 14 Jan 22.
48. USNavyCNO Twitter post, 15 Jan 22; USSTRATCOM Twitter post, 15 Jan 22.
49. Aircraft Spots Twitter post, 15 Jan 22.
50. redandblackattack Twitter post, 17 Jan 22.
51. U.S. Naval Forces Europe-Africa tweet, 15 Jan 22.
52. XSovietNews Twitter post, 16 Jan 22; "Russian Legislator Fedorov Recommends
Nuking Nevada to Convince U.S. that Russia Is Serious," MEMRI, 21 Feb 22,
https://www.memri.org/reports/russian-legislator-fedorov-recommends-nuking
-nevada-convince-us-russia-serious
53. The Lookout Twitter posts, 16 Jan 22.
54. "Russian cargo plane takes unexpected detour over Finland," YLE News, 18 Jan 22.

55. Robert-Gabriel Ticalau, "The 'Allied Resolve' military exercise between Belarus and Russia: A new threat to Ukraine?" Romanian Centre for Russian Studies, https://russianstudiesromania.eu/2022/02/01/the-allied-resolve-military-exercise-between-belarus-and-russia-a-new-threat-to-ukraine/

56. "Taking up combat duty!" *Red Star*, 17 Jan 22, http://redstar.ru/est-zastupit-na-boevoe-dezhurstvo/; Liveuamap Twitter post, 17 Jan 22; Pavel Podvig, "Strategic Rocket Forces," https://russianforces.org/missiles/

57. "Russian Deploys Nuclear Ballistic Missile in Arctic Military Base," Global Defense Corp, 17 Jan 22, https://www.globaldefensecorp.com/2022/01/17/russia-deploys-nuclear-ballistic-missile-in-arctic-military-base/

58. Screenshots of Russian Ministry of Defence website, 18 Jan 22.

59. "Russian Hybrid Threats Report: Troops Arrive in Belarus as Propaganda Narratives Heat Up," Atlantic Council, 21 Jan 22, https://www.atlanticcouncil.org/blogs/new-atlanticist/russian-hybrid-threats-report-troops-arrive-in-belarus-as-propaganda-narratives-heat-up/

60. This is based on an exhaustive examination of ADS-B Exchange historical databases from January 2021 to December 2022.

61. ADS-B Exchange screenshot, 18 Jan 22.

62. ADS-B Exchange screenshot, 18 Jan 22.

63. Radioskaner Twitter post, 20 Jan 22.

64. "Two strategic missile carriers TU-160 performed a planned flight in air space over the neutral waters of the Arctic Ocean, the Barents, and the White Sea," 20 Jan 22, Russia Ministry of Defence website (not accessible: screenshot available); Tatyana Pashkova, "Russian Tu-160 missile carriers made a training flight over the Arctic Ocean," Poliexpert.net, 21 Jan 22 https://politexpert.net/22630095-rossiiskie_raketonostsi_tu_160_sovershili_trenirovochnii_polet_nad_severnim_ledovitim_okeanom

65. NORAD Twitter post, 20 Jan 22.

66. Anna Clara Arndt and Liviu Horovitz, *Nuclear Rhetoric and Escalation Management in Russia's War against Ukraine: A Chronology* (Berlin: German Institute for International and Security Affairs, 2022), 10.

67. ADS-B Exchange and FR24 screenshots, 22 Jan 22. The author observed this activity while it was in progress. It was spoofing on Mode-S.

68. The author screenshot images of the 2S7s from the footage, but the original site it was on ceased to exist after 24 Feb 22.

69. Radioskaner Twitter post, 24 Jan 22; "Russian Navy adds pressure on Ukraine and NATO," Navy Lookout, 21 Jan 22, https://www.navylookout.com/russian-navy-adds-to-pressure-on-ukraine-and-nato/

70. H.I. Sutton. "Russian Navy Live Firing off the Irish Coast during Tensions in Europe," Naval News, 24 Jan 22, https://www.navalnews.com/naval-news/2022/01/russian-navy-live-firing-off-irish-coast-during-tensions/

71. Screenshots of footage from Russian Ministry of Defence website (now blocked), 22 Jan 22.

72. FR24 screenshots, 25 Jan 22.

73. "Russian PGRK YARS held and unexpected exercise near the borders of NATO," Avia.pro, 26 Jan 22, https://avia.pro/news/rossiyskie-pgrk-yars-proveli-neozhidan nye-ucheniya-vblizi-granic-nato; Russian Defense Policy Twitter post, 25 Jan 22, which is a retweet of an Interfax post; this is deliberate information.

74. "Mighty YARS conquer Siberia," VPK News, 26 Jan 22, https://www.vpk-news.ru/65532 (site is sporadically accessible; author has screenshots)

75. FR24 screenshot, 27 Jan 22.

76. Russian Ministry of Defence tweet and embedded video, 27 Jan 22. Note that Russia also ran a Tu-142 exercise over the Sea of Okhotsk at the same time: "The crews of anti-submarine aircraft of the Pacific Fleet performed training flights over the Sea of Okhotsk," Interfax, 27 Jan 22.

77. At least one aircraft squawked accidentally as it took off and swiftly gained altitude over Lake Baikal before it shut off its transponder. Belaya is a large Backfire base. FR24 screenshot, 27 Jan 22.

78. "Russia completed a check of the combat readiness of the troops of the Western Military District," Interfax, 29 Jan 22, https://www.interfax.ru/russia/819112

79. Arndt and Horovitz, *Nuclear Rhetoric and Escalation Management in Russia's War against Ukraine*, 11.

80. ADS-B Exchange screenshots, 24 Jan 22.

81. Dirk Kurbjuweit, "Putin's poker with the bomb," *Der Spiegel*, 28 Oct 22, https://www.spiegel.de/politik/wladimir-putin-und-sein-poker-mit-der-bombe-furcht-vor-einem-nuklearkrieg-a-d60389b5-fa83-4138-bb40-a417e5f25e0a. See also Hans von der Burchard Twitter post, 28 Oct 22 and TPYXA News Twitter post, 29 Oct 22.

82. Evergreen Twitter post, 26 Jan 22. The author observed the exercise on ADS-B Exchange while it was in progress.

83. "Russia warns of NATO nuclear threat: The military bloc is training its members to unleash atomic hellfire," RT, 27 Jan 22, https://www.rt.com/russia/547552-nato-prepare-nuclear-strikes/

84. Sergey Ketonov, "Tomahawks in the Black Sea," *Military Industrial Courier*, 10 Jan 22, https://www.google.com/search?q=multiferious&oq=multiferious&aqs=chr ome..69i57j0i10i433i512j0i10i512l8.2508j1j4&sourceid=chrome&ie=UTF-8

85. "The Northern Fleet began exercises in the Barents Sea," Lentra.ru, 26 Jan 22.

86. "Russian military base in Venezuela: Pros and cons," K-politika, 27 Jan 22, https://k-politika.ru/rossijskaya-voennaya-baza-v-venesuele-za-i-protiv/?utm_source =warfiles.ru; Viktor Sokirko and Dmitry Mayorov, "The Russian nuclear submarine with the Zircon near New York is a nightmare for the United States," Gazeta .ru, 30 Jan 22, https://www.gazeta.ru/army/2022/01/30/14478433.shtml; "The Russians are coming? What Moscow could do to make life difficult for the US in Latin America," RT, 17 Jan 22, https://www.rt.com/russia/546256-moscow-latin -american-relationship/; "Russia unveils military plans in Cuba and Latin America," RT, 26 Jan 22, https://www.rt.com/russia/547417-putin-military-plans-cuba/; Vladimir Vasiliev, "America's Vietnam syndrome could lead Russia and NATO to a second Caribbean crisis," *Military Industrial Courier*, 25 Jan 22, https://vpk-news. ru/articles/65519

87. ADS-B Exchange screenshot, 28 Jan 22.

88. ADS-B Exchange screenshot, 25 Jan 22.

89. FR24 screenshot, 28 Jan 22.

90. "Russia completed a check of the combat readiness of the troops of the Western Military District," Interfax, 29 Jan 22, https://www.interfax.ru/russia/819112

91. "Colonel Knutov stated the defenselessness of the United States and NATO in front of 'Putin's new weapon,'" InfoReactor, 31 Jan 22, https://inforeactor.ru/22597855-polkovnik_knutov_konstatiroval_bezzaschitnost_ssha_i_nato_pered_novim _oruzhiem_putina_; "The best weapon in Russia: Baranets pointed out the merits of the Russian rocket 'Satan,'" InfoReactor, 30 Jan 22, https://inforeactor. ru/407265-luchshee-oruzhie-v-rossii-baranec-ukazal-na-dostoinstva-rakety-rf-satana; "Rockets 'Skif',—a universal weapon. How the secret complex works," InfoReactor, 29 Jan 22, https://inforeactor.ru/22598603-universal_noe_i_ molnienosnoe; Oksana Volgina, "US needs three steps to defeat Russia—Russia only needs one to destroy the US," Politros, 30 Jan 22, https://politros.com/22622563 -netease_ssha_nuzhno_tri_shaga_dlya_pobedi_nad_rossiei_rossii_dostatochno _odnogo_chtob_unichtozhit_ssha; "The commander of the Northern Fleet checked the readiness of the crew of the nuclear submarine Knyaz Oleg to carry out the assigned tasks," Russian Ministry of Defence web page, 31 Jan 22.

92. ADS-B and FR24 Exchange screenshots, 27, 28, and 29 Jan 22.

93. ADS-B Exchange screenshot, 28 Jan 22.

94. U.S. Fleet Forces Twitter post, 29 Jan 22.

95. George Allison, "American 'nuke sniffer' aircraft arrives in the UK," UK Defence Journal, 30 Jan 22, https://ukdefencejournal.org.uk/american-nuke-sniffer-aircraft-arrives-in-the-uk/?fbclid=IwAR2qVtvx9CsTimOlcZ KWx_TYeoIaDVuMRfOiJfhQDFbjo96VdgNTGE

96. Sebastian Steinke, "US sends tracer aircraft over the Baltic Sea," FlugReview, 6 Aug 21, https://www.flugrevue.de/militaer/suche-nach-nuklearen-partikeln -vor-bornholm-usaf-schickt-spuerflugzeug-ueber-die-ostsee/

97. Russian Ministry of Defence announcement on 31 Jan 22 (unavailable; author has screenshots).

98. Russian Ministry of Defence announcement on 30 Jan 22 (unavailable; author has screenshots).

99. ADS-B Exchange screenshots, 31 Jan 22.

100. Communications Security Establishment, *Cyber Threat Bulletin: Cyber Activity related to the Russian Invasion of Ukraine*, https://www.cyber.gc.ca/sites/default/ files/cyber-threat-activity-associated-russian-invasion-ukraine-e.pdf

101. Alex Boutillier and Mercedes Stephenson, "Global Affairs Canada suffers cyber attack amid Russia-Ukraine tensions: Sources," Global News, 24 Jan 22, https://globalnews.ca/news/8533835/global-affairs-hit-with-significant -multi-day-disruption-to-it-networks-sources/

102. David A. Broniatowski et al., "Weaponized Health Communication: Twitter Bots and Russian Trolls Amplify the Vaccine Debate," *American Journal of Public Health*, 12 Sep 18, https://ajph.aphapublications.org/doi/10.2105/AJPH.2018.304567; see also Broniatowski, "Foreign and Domestic Online Manipulation of the Vaccine Debate," lecture at George Washington University, https://healthequity.ucla.edu/sites /default/files/Dr.%20Broniatowski_UCLA%20%281%29.pdf; Bruce Y. Lee, "That Anti-Vaccination Message May Be from a Russian Bot or Troll," *Forbes*, 25 Aug 18, https://www.forbes.com/sites/brucelee/2018/08/25/that-anti-vaccination-message- may-be-from-a-russian-bot-or-troll/?sh=625e5d45ff77; Julian Cardillo, "Social media is feeding the anti-vaccination movement," BrandeisNOW, 3 Nov 20, https:// www.brandeis.edu/now/2020/november/social-media-vaccine-disinformation .html; "Russian troll farms aiming disinformation war at Canadian anti-vaxxers: Global affairs expert," *National Post*, 19 Mar 22, https://nationalpost.com/news /canada/russian-troll-farms-aiming-disinformation-war-at-canadian-anti-vaxxers -global-affairs-expert; "Is anti-vaccination campaign the new Operation Infektion?" Digital Forensic Center, https://dfcme.me/en/is-anti-vaccination-campaign-the -new-operation-infektion/; Grant LaFleche, "How vaccination status might predict views on the Russian invasion of Ukraine," *Toronto Star*, 19 Mar 22, https://www .thestar.com/news/investigations/2022/03/19/how-vaccination-status-might -predict-views-on-the-russian-invasion-of-ukraine.html

103. The "Trucker Convoy" protests were a complex phenomenon and the subject remains under debate. This information is distilled from the 2,000 pages of the five-volume *Report of the Public Inquiry into the 2022 Public Order Emergency*, which can be found at https://publicorderemergencycommission.ca/final-report/

104. Confidential discussions with the author.

105. Caroline Orr Bueno, "Russia's Role in the Far-Right Truck Convoy: An Analysis of Russian State Media Activity Related to the 2022 Freedom Convoy," *Journal of Intelligence, Conflict, and Warfare,* Vol. 5, Issue 3 (2022), https://journals.lib.sfu.ca/index.php/jicw/article/view/5101

106. George Allison Twitter post, 2 Feb 22; "Two Aerospace forces aircraft completed a planned flight over the neutral waters of the Atlantic," *Arguments and Facts,* 3 Feb 22, https://aif.ru/society/army/dva_samoleta_vks_vypolnili_planovyy_polet_nad_neytralnymi_vodami_atlantiki

107. Thomas Nilsen, "Norway scrambles jets as group of Russian aircraft flies from the north," *Barents Observer,* 2 Feb 22, https://thebarentsobserver.com/en/2022/02/norway-scrambles-jets-russian-bombers-flies-towards-western-europe

108. Russian Ministry of Defence Twitter post, 2 Feb 22.

109. Liveuamap Twitter post, 2 Feb 22.

110. Liveuamap Twitter post, 2 Feb 22.

111. FR24 screenshot, 2 Feb 22.

112. ADS-B Exchange screenshots, 2 Feb 22.

113. Thomas Nilsen, "NATO jets scrambled for second day in a row," *Barents Observer,* 3 Feb 22, https://thebarentsobserver.com/en/security/2022/02/nato-jets-scrambled-second-day-row; Russian Defence Policy Twitter post, 3 Feb 22; "British fighters climbed to escort Russian strategic bombers-Russian Defense Ministry," Interfax, 3 Feb 22, http://militarynew.ru/story.asp?rid=1&nid=565958&lang=RU; The Lookout Twitter post, 3 Feb 22; Russian Ministry of Defence, "Two Tu-95 long-range strategic missile carriers carried out a planned flight in the airspace over the neutral waters of the Barents Sea, Norwegian Sea, and the north Atlantic Ocean," 3 Feb 22 [RU MoD link broken; screenshot on file]. "In the Ivanovo region, missilemen carry out a complete special treatment of YARS launchers," Russian Ministry of Defence, 4 Feb 22, https://function.mil.ru/news_page/country/more.htm?id=12407198@egNews [link severed]; "Two Tu-22M3 long-range bombers of the Aerospace Forces performed patrols in the airspace of the airspace of the Republic of Belarus," Russian Ministry of Defence, 5 Feb 22 [Ru MoD website blocked; screenshot on file]; "Autonomous launchers of the YARS mobile missile system changed positions during an exercise in the Ivanovo region," 6 Feb 22 [Ru MoD website blocked; screenshot on file]; Liveuamap Twitter post, 8 Feb 22.

114. FR24 screenshots, 2, 7, 8, and 9 Feb 22.

115. ADS-B Exchange screenshots, 4 Feb 22.

116. Alexander Sitnikov, "MiG-31 with Daggers in Kaliningrad took aim at NATO headquarters," *Free Press,* 9 Feb 22, https://svpressa.ru/war21/article/324661/?utm_source=warfiles.ru; see also "Chinese media called the use of the Kinzhal complex

by Russia a slap in the face of NATO," *Red Spring Information Service*, 22 Mar 22, https://rossaprimavera.ru/news/ce5c2e08

117. "RAF fails to intercept unidentified aircraft near Kingdom's borders," Politi expert, 3 Feb 22, politeexpert.net/275061-britanskie-vvs-poterpeli-fiasko-pri -perekhvate-neizvestnogo-samoleta-u-granic-korolevstva [link severed]; "Survival expert explains where to hide from nuclear strikes on Russia," News.ru, 3 Feb 22, https:// news.ru/society/specialist-po-vyzhivaniyu-rasskazal-gde-pryatatsya-pri-yadernom -vzryve/; "Survival expert Makeev voiced the main advice in case of a nuclear war," Politros, 2 Feb 22, https://politros.com/22621173-ukrainskii_aktivist_ganul _zastavil_organizatorov_otmenit_kontsert_basti_v_odesse; "Survival expert reveals how to survive a nuclear war," News.ru, 2 Feb 22, https://news.ru/lifestyle /ekspert-po-vyzhivaniyu-rasskazal-kak-spastis-vo-vremya-yadernoj-vojny/; "Honored pilot Popov: Tu-160M bomber will leave no chance for enemy fighters," Poliexpert, 4 Feb 22, https://politexpert.net/22622280-rls_zalog_uspeha; "Denis Manturov inspected Tu-160M strategic missile carrier in Kazan," *Argument and Facts*, 4 Feb 22, https://kazan.aif.ru/politic/person/denis_manturov _osmotrel_v_kazani_strategicheskiy_raketonosec_tu-160m; Knutov Yuri, "The revival of the White Swan," *Military Industrial Courier*, 7 Feb 22, https://vpk-news .ru/articles/65705

118. Arndt and Horovitz, *Nuclear Rhetoric and Escalation Management in Russia's War against Ukraine*, 11.

119. ADS-B Exchange and FR24 screenshots, 2–9 Feb 22.

120. ADS-B Exchange screenshot, 9 Feb 22

121. ADS-B Exchange screenshots, 9–10 Feb 22.

122. Nathan Hodge, "Russia and Belarus hold joint military exercises as diplomatic talks ramp up," CNN, 10 Feb 22, https://www.cnn.com/2022/02/10/europe/ukraine -russia-news-thursday-military-exercises-intl/index.html

123. "Ukraine tension: Russia stages military drills with Belarus," BBC, 10 Feb 22, https://www.bbc.com/news/world-europe-60327930

124. Violetta Khaneneva, "After the incident with the submarine, the Russian Defense Ministry call the US military attache," Gazeta.ru, 12 Feb 22, https://www.gazeta .ru/army/news/2022/02/12/17281201.shtml

125. Vera Zherdeva, "6–7 NATO destroyers that approach Crimea will be immediately destroyed from the shore," *Free Press*, 12 Feb 22, https://svpressa.ru/war21 /article/324940/

126. FR24 screenshots, 12 and 14 Feb 22.

127. Margaret Brennan, "New detail about Russian 'false flag' plan prompts U.S. to prepare for the worst in Ukraine," CBS News, 12 Feb 22, https://www.cbsnews .com/news/ukraine-russias-false-flag-plan-new-detail/

128. "Russia's Tu-160 strategic bombers perform scheduled flight over Barents, Norwegian seas," TASS, 14 Feb 22, https://tass.com/defense/1576451

129. Viktor Sokirko, "Super-sniffer from the United States began the hunt for the Russian missile Birevestnik," Gazeta.ru, 7 Feb 22, https://www.gazeta.ru/army/2022/02/07/14506351.shtml

130. TASS Twitter post, 11 Feb 22.

131. ADS-B Exchange, 14 Feb 22, and yes, I screenshot this while it was underway.

132. Thomas Nilsen, "Russia issues largest ever warning zone in Norwegian part of the Barents Sea," *Barents Observer*, 15 Feb 22, https://thebarentsobserver.com/en/security/2022/02/largest-ever-russian-notam-warning-norwegian-sector-barents-sea; "Fishermen despair over Russian large-scale exercise," *NRK*, 15 Feb 22, https://www.nrk.no/tromsogfinnmark/havfiskeflaten-reagerer-pa-russisk-storovelse-_-fiskebat-mener-ovelsen-er-en-overkjoring-av-norge-1.15855910

133. Christian W., "Russia alarmed about foreign troops on Bornholm," *CPH Post*, 18 Feb 22, https://cphpost.dk/2022-02-18/news/russia-alarmed-about-foreign-troops-on-bornholm/

134. "Deploying US rocket systems at Bornholm to lead to escalation with Russia, envoy says," TASS, 1 Jun 22, https://tass.com/politics/1459349?ref=hermes-kalamos.eu

135. FR 24 screenshot, 15 Feb 22.

136. "Hypersonic Daggers Withdrawn from the Kaliningrad Region," *The Reporter*, 15 Feb 22, https://topcor.ru/24039-giperzvukovye-kinzhaly-vyvedeny-iz-kaliningradskoj-oblasti.html?utm_source=warfiles.ru

137. Alex Kokcharov Twitter post, Timothy Ash Twitter post, Rgip Soylu Twitter post, Charlotte Lawson Twitter post, NetBlocks Twitter post and graphic, 15 Feb 22.

138. Col McGowan Twitter post, 15 Feb 22. This information came in on ACARS and was retransmitted by one of the flights.

139. Ekaterina Zaitseva, "Russian Tu-22M3 bombers and MiG-31K fighters arrived in Syria," Gazeta.ru, 15 Feb 22, https://www.gazeta.ru/army/news/2022/02/15/17294827.shtml; "Russia sends hypersonic missiles to Syria," RT, 15 Feb 22, https://www.rt.com/russia/549458-hypersonic-missiles-syria-drills/

140. Barbara Starr and Oren Liebermann, "US Navy aircraft had an 'extremely close' encounter with multiple Russian military jets over the Mediterranean," CNN, 16 Feb 22, https://www.cnn.com/2022/02/16/politics/us-navy-russia-jets-mediterranean/index.html

141. Sergy Sumlenny Twitter post, 15 Feb 22.

142. FR 24 screenshots, 16 Feb 22.

143. Irina Alshaeva and Dmitry Mayorov, "Replace the Soviet pyramids: What is the Voronezh radar capable of?" Gazeta.ru, 16 Feb 22, https://www.gazeta.ru/army/2022/02/16/14543557.shtml; "Aerospace-Under Control," *Red Star*, 16 Feb

22, http://redstar.ru/vozdushno-kosmicheskoe-prostranstvo-pod-kontrolem/; "Russian missile attack warning system to be updated by 2030," InfoReactor, 16 Feb 22, https://inforeactor.ru/408996-sistemu-preduprezhdeniya-o-raketnom -napadenii-v-rf-obnovyat-k-2030-godu; "The level of world standards: How the missile attack warning systems works in the Russian Federation," InfoReactor, 16 Feb 22, https://inforeactor.ru/409157-kedmi-ocherednaya-antirossiiskaya -provokaciya-mozhet-obernutsya-dlya-vms-ssha-poterei-korablei

144. FR 24 screenshots, 15 and 16 Feb 22; ADS-B Exchange screenshots, 15–16 Feb 22.

145. See The Sky Kings Twitter posts, 15–17 Feb 22.

146. This message is similar in format to Cold War–era EAMs and appeared on the #haveglass Twitter feed at 0318 h 17 Feb 22. It is impossible to determine whether this was an authentic EAM with existing information.

147. CANUK78 Twitter post, 17 Feb 22; ADS-B Exchange screenshot, 17 Feb 22.

148. The Lookout Twitter post, 17 Feb 22.

149. The Lookout Twitter post, 17 Feb 22.

150. FR 24 screenshots, 17 Feb 22; ADS-B screenshots, 17 Feb 22.

151. ADS-B screenshot, 17 Feb 22.

152. NORAD Twitter post, 17 Feb 22.

153. Russian MoD Twitter post, 17 Feb 22.

154. FR 24 screenshots, 17 Feb 22.

155. Note: The DM used a four-line paragraph for an article title. I am not putting all that here. Daily Mail, 17 Feb 22, https://www.dailymail.co.uk/news/article-10522545 /Ukraine-crisis-Fears-Russian-false-flag-operation-underway-justify-invasion.html

156. "Our hand will not waver to carry out the last order," The Reporter, 17 Feb 22, https://topcor.ru/24082-u-nas-ne-drognet-ruka-vypolnit-poslednij-prikaz-podryv -zaporozhskoj-ajes-kak-otvet-na-vtorzhenie-rf.html?utm_source=warfiles.ru

157. Владислав Демченко Twitter post, 15 Feb 22.

158. These included TV ZEVEZDA, Gazeta.ru, TASS, and RIA Novosti.

159. Daria Klester, "Training Process: Russia to Launch Ballistic Missiles," Gazeta.ru, 18 Feb 22, https://www.gazeta.ru/army/2022/02/18/14550781.shtml

160. Rob Lee Twitter post, 18 Feb 22; IgorGirkhin Twitter post, 18 Feb 22; OSINTtechnical Twitter post, 18 Feb 22.

161. ADS-B screen shots, 18 Feb 22.

162. marqs Twitter post, 18 Feb 22; see also "NATO Response Force," https://www .nato.int/cps/en/natohq/topics_49755.htm

163. ADS-B screenshots, 18 Feb 22.

164. FR 24 screenshots, 18 Feb 22.

165. "Putin launches nuclear war drills," RT, 19 Feb 22, https://www.rt.com/russia /549957-putin-launches-nuclear-war-drills/

166. Ropey, Trent Telenko, and From the Static Twitter posts, 19 Feb 22. Non-cognoscenti misinterpreted the fact that the OTHR is always "on" and that it cycles its frequencies depending on the atmospheric conditions. Some observers misinterpreted the sudden increase in frequency band as some kind of signal that Russia was about to launch a nuclear strike. See "Russian OTHR 29B6 Konteyner analysis," planesandstuff blog, 25 Feb 20, https://planesandstuff.wordpress.com/2020/02/25/russian-othr-29b6-konteyner-analysis/

167. ADS-B screenshots, 19 Feb 22; Thenewarea51 Twitter post, 19 Feb 22.

168. The Lookout Twitter post, 20 Feb 22; Benedikt Franke Twitter post, 20 Feb 22; Max Seddon Twitter post, 20 Feb 22; FR 24 screenshots, 20 Feb 22.

169. Russian Federation Security Council Meeting, Kremlin, 21 Feb 22, http://en.kremlin.ru/events/president/news/67825

170. Russian Federation Security Council Meeting.

171. Russian Federation Security Council Meeting.

172. Address by the President of the Russian Federation, 21 Feb 22, http://www.en.kremlin.ru/events/president/transcripts/67828

173. Address by the President of the Russian Federation.

174. "OTR-21 Tochka (SS-21)," CSIS Missile Defense Project, https://missilethreat.csis.org/missile/ss-21/

175. ADS-B screenshots, 21 Feb 22.

176. Christopher Miller Twitter post of Interfax post, 21 Feb 22.

177. OSINTtechnical Twitter post, 21 Feb 22.

178. Footage embedded in Russian Ministry of Defence Twitter post, 21 Feb 22.

179. FR 24 screenshots, 21 Feb 22.

180. ADS-B screenshots, 21 Feb 22.

181. Scott Ritter, "Ukraine's nuclear fantasy is dangerous," RT, 22 Feb 22, https://www.rt.com/russia/550057-zelensky-dangerous-nuclear-ukraine/

182. Viktor Sorkirko, "Ukraine is only capable of a dirty atomic bomb," Gazeta.ru, 22 Feb 22, https://www.gazeta.ru/army/2022/02/22/14566363.shtml

183. ADS-B and FR 24 screenshots, 22 Feb 22.

184. ADS-B screenshots.

185. ADS-B screenshots, 23 Feb 22.

186. ADS-B screenshots, 23 Feb 22.

187. Aircraft Spots Twitter post, 23 Feb 22; World Events Live Twitter post with embedded footage from Saratov, 23 Feb 22; Thenewarea51 Twitter post, 23 Feb 22. See also SoundCloud, https://soundcloud.com/tomteej/russian-bomber-hf-voice-net-24th-feb-2022-81310khz

188. Derived from The Lookout Twitter posts for 19, 23, 27, and 28 Feb 22 and 3 and 6 Mar 22. The Lookout conducts analysis of Northern Fleet operations using the Sentinel-1 satellite.

189. Oliver Alexander Twitter post, 23 Feb 22; Net Blocks Twitter post, 23 Feb 22.

190. Net Blocks Twitter post, 23 Feb 22;Dustin Voltz Twitter post citing Symantec and the Baltic countries.

191. Communications Security Establishment, "Cyber Threat Bulletin: Cyber Threat Activity Related to the Russian Invasion of Ukraine," https://www .cyber.gc.ca/sites/default/files/cyber-threat-activity-associated-russian-invasion -ukraine-e.pdf; "How the war in Ukraine is affecting space activities," OECD, 15 Nov 22, https://www.oecd.org/ukraine-hub/policy-responses/how -the-war-in-ukraine-is-affecting-space-activities-ab27ba94/

192. TJ (amateur radio enthusiast) Twitter post, 24 Feb 22.

193. Grady McGregor, "Map shows how civilian planes cleared Ukrainian airspace hours before Russia launched invasion," *Fortune*, 24 Feb 22, https://fortune. com/2022/02/24/russia-ukraine-invasion-airspace-putin-planes-flights-map/

194. Justin Bronk, *Russian Combat Air Strengths and Limitations: Lessons from Ukraine*, Center for Naval Analyses, 2023, https://www.cna.org/reports/2023/04/Russian -Combat-Air-Strengths-and-Limitations.pdf

195. "Full text: Putin's declaration of war on Ukraine," *The Spectator*, 24 Feb 22, https:// www.spectator.co.uk/article/full-text-putin-s-declaration-of-war-on-ukraine/

196. ADS-B screenshots, 24 Feb 22.

197. FR24 screenshots, 24 Feb 22.

198. "France says Putin needs to understand NATO has nuclear weapons," Reuters, 24 Feb 22, https://www.reuters.com/world/europe/france-says-putin-needs-understand-nato -has-nuclear-weapons-2022-02-24/

Chapter 6. Abyss Creep

1. Jack Watling and Nick Reynolds, *Operation Z: The Death Throes of an Imperial Delusion*, Royal United Services Institute, 22 Apr 22, https://static.rusi.org/special- report-202204-operation-z-web.pdf

2. Tom McMillan, "Know No Mercy: The Russian Cops Who Tried to Storm Kyiv by Themselves," *The Debrief*, 20 May 22, https://thedebrief.org/know -no-mercy-the-russian-cops-who-tried-to-storm-kyiv-by-themselves/

3. Watling and Reynolds, *Operation Z*.

4. TJ Twitter post, 24 Feb 22.

5. Watling and Reynolds, *Operation Z*.

6. Ashleigh Stewart, "The Battle of Hostomel: How Ukraine's unlikely victory changed the course of the war," Global News, 18 Feb 23, https://globalnews.ca/news/9491396/ ukraine-hostomel-battle-antonov-airport/; Tom Cooper, "Russian Heliborne Assault on Antonov/Hostomel Airport Seems to Have Failed," The Aviation Geek Club, 25 Feb 22, https://theaviationgeekclub.com/russian-heliborne-assault-on-antonov- hostomel-airport-seems-to-have-failed/; Sebastien Roblin, "Pictures: In Battle for

Hostomel, Ukraine Drove Back Russia's Attack Helicopters and Elite Paratroopers," *1945*, 25 Feb 22, https://www.19fortyfive.com/2022/02/pictures-in-battle-for-hostomel-ukraine-drove-back-russias-attack-helicopters-and-elite-paratroopers/

7. Roblin, "Pictures."

8. Andrew McGregor, "Russian Airborne Disaster at Hostomel Airport," *Aberfoyle International Security*, 8 Mar 22, https://www.aberfoylesecurity.com/?p=4812

9. Christo Grozev reported on Twitter at 1732 hours EET 24 Feb 22 that the Il-76s were in the air.

10. MtotolkoHelp Twitter post, 25 Feb 22.

11. Illia Ponomarenko Twitter post, 24 Feb 22; Stewart, "The Battle of Hostomel."

12. "Weapons & equipment seized from alleged Russian saboteurs in Ukraine," *The Hoplite*, 25 Mar 22, https://armamentresearch.com/weapons-equipment-seized-from-alleged-russian-saboteurs-in-ukraine-2022/?fbclid=IwAR3qxwUzwns5y1Vtm1onxLJpu77dYck5QJ; Roblin, "Pictures"; Illia Ponomarenko Twitter post, 24 Feb 22; Ukrainian Secretary of the Security Council Twitter post, 1 Mar 22; Riho Terras (former Estonian defense minister) Twitter post of Ukrainian intelligence analysis, 26 Feb 22. Note that former GRU Spetsnaz belonging to the RSB PMC may also have been involved. Intelligenceonline.com Twitter post, 1 Mar 22.

13. Glasnost Gone Twitter post, 27 Feb 22; UK Defence Intelligence tweet, 27 Feb 22.

14. McMillan, "Know No Mercy."

15. Josh Lederman, NBC News from Ukrainian Ambassador to the United States, Twitter post, 24 Feb 22.

16. Ukrainian SBU via Glasnost Gone Twitter post, 1 Mar 22.

17. Screenshot in author's collection.

18. TASS, 25 Feb 22.

19. Dmitry Rogozhin Twitter post, 25 Feb 22; Riho Terras Twitter post, 26 Feb 22.

20. President of Russia, "Meeting with Security Council permanent members," 25 Feb 22, http://en.kremlin.ru/events/president/news/67851

21. Andrew Roth, "It's not rational: Putin's bizarre speech wrecks his once pragmatic images," *The Guardian*, 25 Feb 22.

22. FR 24 data, 25 Feb 22.

23. Russian Ministry of Defence Twitter post, 25 Feb 22.

24. Russian Ministry of Defence Twitter post, 25 Feb 22; Status-6 Twitter post, 25 Feb 22.

25. "Topol-M advanced in an unknown direction," Avia.pro, 26 Feb 22, https://avia.pro/news/v-podmoskove-zametili-vydvinuvshiesya-v-neizvestnom-napravlenii-strategicheskie-kompleksy-topol

26. ADS-B Exchange screenshots, 25–27 Feb 22.

27. Euromaidan Press Twitter post, 1 Mar 22.

28. Bruce G. Blair, *The Logic of Accidental Nuclear War* (Washington, D.C.: Brookings Institution, 1993), ch. 4.

29. Anna Clara Arndt and Liviu Horovitz, *Nuclear Rhetoric and Escalation Management in Russia's War against Ukraine*, 14.

30. President of Russia, "Meeting with Sergei Shoigu and Valery Gerasimov," 27 Feb 22.

31. There are too many to list here.

32. "Putin orders 'special service regime' in Russia's deterrence forces," TASS, 27 Feb 22.

33. Maloney, *Deconstructing Dr. Strangelove*, 139–140.

34. ABC News This Week transcript, 27 Feb 22, https://abcnews.go.com/Politics/week-transcript-27–22-white-house-press-secretary/story?id=83130361

35. Translated by Max Seddon, Moscow bureau chief of the *Financial Times* and posted on Twitter, 27 Feb 22.

36. Shannon Bugos, "Putin Orders Russian Nuclear Weapons on Higher Alert," Arms Control Association, March 2022, https://www.armscontrol.org/act/2022–03/news/putin-orders-russian-nuclear-weapons-higher-alert

37. Arndt and Horovitz, *Nuclear Rhetoric*, 17.

38. ADS-B Exchange screenshots, 28 Feb 22; The Sky Kings Twitter posts, 28 Feb 22.

39. Marco Rubio Twitter post, 28 Feb 22.

40. Mike McFaul Twitter post, 28 Feb 22.

41. TJ (amateur radio enthusiast) Twitter post, 28 Feb 22.

42. Tvzvezda News, 28 Feb 22, https://tvzvezda.ru/news/20222281432–5bqzh.html

43. ADS-B Exchange screenshots, 28 Feb 22.

44. Pentagon briefing, 28 Feb 22, https://www.c-span.org/video/?518248–1/defense-department-briefing

45. Tvzvezda Twitter post, 1 Mar 22.

46. FR 24 screenshots, 1 Mar 22.

47. ADS-B Exchange, 1 Mar 22.

48. Finnish defense commentator Petri Makela Twitter post, 1 Mar 22.

49. Il-114LL Fact Sheet, https://www.ilyushin.org/en/aircrafts/special_aircraft/1180/

50. Il-114LL/Il-114FK/PR, Global Security, https://www.globalsecurity.org/military/world/russia/il-114fk.htm

51. Tom Bateman, "Planes and smartwatches near Finland's Russian border had GPS issues, and not for the first time," Euronews.Next, 9 Mar 22, https://www.euronews.com/next/2022/03/16/planes-and-smartwatches-near-finland-s-russian-border-had-gps-issues-and-not-for-the-first

52. "Remarks by Ambassador Linda Thomas-Greenfield at a UN General Assembly Emergency Special Session," 2 Mar 22, https://usun.usmission.gov/remarks-by-ambassador-linda-thomas-greenfield-at-a-un-general-assembly-emergency-special-session-on-ukraine/

53. Charles Szumski, "Russian jets carrying nuclear weapons violated Swedish airspace," Euracitiv, 31 Mar 22, https://www.euractiv.com/section/politics/short_news/russian -jets-carrying-nuclear-weapons-violated-swedish-airspace/

54. "Did Russian jets carry nuclear weapons into Swedish airspace?" *The Local*, 31 Mar 22, https://www.thelocal.se/20220331/were-the-russian-jets-that -violated-swedish-airspace-carrying-nuclear-weapons

55. "Kärnvapenbestyckade ryska plan kränkte svenskt luftrum," TV4, 31 Mar 22, https://www.tv4.se/artikel/6cNV5sPAaxdIgAsnItdVsK/kaernva-ryska-plan -kraenkte-svenskt-luftrum

56. Screenshots of Vladislav Demchenko Twitter post, 2 Mar 22. This Twitter page, established in January 2022, posted Cold War–era Soviet nuclear material but embedded current Russian nuclear weapons material. The page ceased to operate later in 2022.

57. TikTok post @svetylia97, screenshot by the author, 4 Mar 22. The footage depicts a 12th GUMO convoy almost identical in route march order as the convoy near Mozarhaisk-10.

58. Russia's Nuclear Activity Twitter post, 2 Mar 22 retweet of RIA Novosti Twitter post.

59. ADS-B Exchange screenshots, 2 Mar 22.

60. Ellen Mitchell, "Pentagon postpones missile test launch to deescalate Russia tensions," *The Hill*, 2 Mar 22, https://thehill.com/policy/defense/596603-pentagon -postpones-missile-test-launch-to-deescalate-russia-tensions/

61. Arndt and Horovitz, *Nuclear Rhetoric*, 20.

62. FR 24 screenshots, 3 Mar 22.

63. ADS-B Exchange screenshots, 3 Mar 22.

64. U.S. Strategic Command Twitter post, 3 Mar 22.

65. Frank Andrews, "Submarine re-supply ship makes first ever mooring at naval base on Okinawa," *Stars and Stripes*, 20 Dec 21, https://www.stripes.com/theaters/asi a_pacific/2021–12–20/uss-frank-cable-submarine-tender-us-navy-okinawa-4032312 .html

66. "Trident Support of the Guided Missile Submarine Fleet," *Navy Supply Corps Newsletter*, 2 Aug 13, https://scnewsltr.dodlive.mil/Latest-Issue/Article-Display /Article/2612169/trident-support-of-the-guided-missile-submarine-fleet/; Randall Guttery, "The Submarine's Secret Weapon: A Tender Tale," *Undersea Warfare*, Summer 2002, http://www.tendertale.com/ttp10.html

67. Jade McGlynn, "Who poisoned Roman Abramovich?" *The Spectator*, 29 Mar 22, https://www.spectator.co.uk/article/who-poisoned-roman-abramovich/; Shaun Walker and Pjotr Sauer, "Abramovich and Ukrainian MP have been poisoned this month," *The Guardian*, 28 Mar 22, https://www.theguardian.com/world/2022/mar /28/abramovich-and-ukrainian-mp-may-have-been-poisoned-this-month

68. Anonymous TV Twitter post, 3 Mar 22.

69. "Cyberattacks on Russian satellites are casus belli—Rogozin," Interfax, 2 Mar 22, https://interfax.com/newsroom/top-stories/75057/. See also "Russia space agency head says satellite hacking would justify war—report," Reuters, 2 Mar 22, https://www.reuters.com/world/russia-space-agency-head-says-satellite-hacking-would-justify-war-report-2022-03-02/; "Cyberattacks against Russian Satellites Are 'Casus Belli,'" Sputnik News, 3 Mar 22, https://sputnikglobe.com/20220302/roscosmos-chief-warns-cyberattacks-against-russian-satellites-are-casus-belli-1093513258.html

70. Mark Krutov and Sergei Dobrynin, "In Russia's War on Ukraine, Effective Satellites Are Few and Far Between," RFE/RL, 11 Apr 22, https://www.rferl.org/a/russia-satellites-ukraine-war-gps/31797618.html#:~:text=Russia; Pavel Luzin, "Russia's Military Space Program: 2022 Results," Eurasia Daily Monitor, 15 Dec 22, https://jamestown.org/program/russias-military-space-program-2022-results/; Pavel Luzin, "Satellites of Stagnation," Riddle Russia, 15 Mar 23, https://ridl.io/satellites-of-stagnation/

71. C4H10FO2P Twitter post, 9 Mar 22 pictures of captured GPS equipment; Boyko Nikolov, "UK: Russian pilots use civil Garmin GPA devices in the cockpit," BulgarianMilitary.com, 20 Jun 22, https://bulgarianmilitary.com/2022/06/20/uk-russian-pilots-use-civil-garmin-gps-devices-in-the-cockpit/#:~:text=An%20attached%20Garmin%20GPS%20navigator,devices%20on%20its%20fighter%20jets.%E2%80%9D; Dylan Malyasov, "Russian pilots still use non-military navigation equipment," DefenceBlog, 20 Jun 22, https://defence-blog.com/russian-pilots-still-use-non-military-navigation-equipment/

72. Tracey Cozzens, "Russia is expected to ditch GLONASS for LORAN in Ukraine invasion," GPS World, 17 Feb 22, https://www.gpsworld.com/russia-expected-to-ditch-glonass-for-loran-in-ukraine-invasion/

73. The Lookout Twitter post, 3 Mar 22.

74. President of Russia, "Meeting with Security Council permanent members," 3 Mar 22.

75. FR 24 screenshots, 3 Mar 22.

76. FR 24 screenshots, 4 Mar 22; ADS-B Exchange screenshot, 4 Mar 22; Russia's Nuclear Activity Twitter post, 4 Mar 22, repost of Russian Ministry of Defence information release.

77. FR 24 screenshots, 4 Mar 22; ADS-B Exchange screenshot, 4 Mar 22; Status-6 Twitter post, 4 Mar 22.

78. Screenshots of Vladislav Demchenko Twitter post, 5 Mar 22. See also https://planet4589.org/space/lvdb/launch/R-14

79. "Intercontinental missile system RT-2PM2 'Topol-M,'" Army Today, https://army-today.ru/tehnika/topol-m

80. Pavel Luzin, "Nuclear de-escalation option," Riddle Russia, 7 Mar 22, https://ridl.io/ru/optsiya-yadernoj-deeskalatsii/

81. Olga Lautman Twitter post, 6 Mar 22; Nexta Twitter post, 6 Mar 22; Shaun Walker Twitter post, 6 Mar 22.

82. Alexandra Vishnevskaya, "The Baltic Fleet began tracking two US missile destroyers in the Baltic Sea," Gazeta.ru, 7 Mar 22, https://www.gazeta.ru/army/news/2022/03/07/17394295.shtml; "US Navy destroyers with guided missiles enter the Baltic Sea," Lenta.ru, 7 Mar 22, https://lenta.ru/news/2022/03/08/esmincy

83. FR 24 screenshots, 7 Mar 22, 8 Mar 22.

84. ADS-B Exchange screenshots, 8 Mar 22.

85. U.S. Senate Select Committee on Intelligence, 10 Mar 22, https://www.intelligence.senate.gov/hearings/open-hearing-worldwide-threats-2#

86. Arndt and Horovitz, *Nuclear Rhetoric*, 20–23; U.S. Senate Select Committee on Intelligence, 10 Mar 22, https://www.intelligence.senate.gov/hearings/open-hearing-worldwide-threats-2#

87. Katherine Lawlor and Lateryna Stepanenko, "Warning Update," Critical Threats, 9 Mar 22, https://www.criticalthreats.org/analysis/warning-update-russia-may-conduct-a-chemical-or-radiological-false-flag-attack-as-a-pretext-for-greater-aggression-against-ukraine

88. Carol L. Lee and Teaganne Finn, "US warns Russia could use chemical weapons in false-flag operation in Ukraine," 9 Mar 22, https://www.nbcnews.com/politics/national-security/us-warns-russia-use-chemical-weapons-false-flag-operation-ukraine-rcna19391

89. Sam Fossum and Betsy Klein, "Biden warns Russia will pay a 'severe price' if it uses chemical weapons in Ukraine," CNN, 11 Mar 22, https://www.cnn.com/2022/03/11/politics/joe-biden-warning-chemical-weapons/index.html; Connor Finnegan, "Russia escalates false chemical weapons claims about US, Ukraine, by bringing them to the US," ABC News, 22 Mar 22, https://abcnews.go.com/Politics/russia-escalates-false-chemical-weapons-claims-us-ukraine/story?id=83366504

90. Chuprin Konstantin, "The Solution to Stalingrad," VPK, 1 Mar 22, https://vpk-news.ru/articles/65988

91. Defence Intelligence of Ukraine Facebook post, 11 Mar 22.

92. ADS-B Exchange screenshots, 10–11 Mar 22.

93. FR 24 screenshots, 9–11 Mar 22.

94. Sam Fossum and Betsy Klein, "Biden warns Russia will pay a 'severe price' if it uses chemical weapons in Ukraine," CNN, 11 Mar 22, https://www.cnn.com/2022/03/11/politics/joe-biden-warning-chemical-weapons/index.html

95. ADS-B Exchange and FR 24 screenshots, 12 Mar 22.

96. Christo Grozev Twitter posts, 11 and 17 Mar 22; Tim McMillan, "Deputy Commander of the Russian National Guard Ousted," The Debrief, 18 Mar 22, https://

thedebrief.org/deputy-commander-of-the-russian-national-guard-ousted/; "Putin purges 150 FSB officers," TVP World, 11 Apr 22, https://tvpworld.com/59577307 /putin-purges-150-fsb-officers; Olga Lautman, Russia's War on Democracy blog, 26 Mar 22, https://olgalautman.substack.com/p/not-only-shoigu-disappeared -from?r=k8nwv&s=w&utm_campaign=post&utm_medium=web

97. Patrick Reilly, "Russian elites planning to overthrow Putin: Ukrainian intelligence," *New York Post*, 20 Mar 22, https://nypost.com/2022/03/20/russian-elites -planning-to-overthrow-putin/

98. Christo Grozev, "The Remote Control Killers behind Russia's Cruise Missile Strikes on Ukraine," Bellingcat, 24 Oct 22, https://www.bellingcat.com/ news/uk-and-europe/2022/10/24/the-remote-control-killers-behind-russias -cruise-missile-strikes-on-ukraine/

99. This was a program called *NewsPower* on Tvzvezda aired on 14 Mar 22, based on screenshots of the presentation with maps.

100. "Operation NOBLE DEFENDER," NORAD, 31 Mar 22, https://www.dvidshub .net/video/837704/operation-noble-defender-march-2022; RCAF Operations Twitter post, 15 Mar 22.

101. NORAD Twitter post, 14 Mar 22.

102. Kathleen Barrios, "USS Forrest Sherman (DDG 98) Arrives in Stockholm," USNAVFOREUR and AF public affairs, 14 Mar 22, https://www.c6f.navy.mil/Press -Room/News/News-Display/Article/2965423/uss-forrest-sherman-ddg-98 -arrives-in-stockholm/

103. UK Defence Intelligence Twitter post, 14 Mar 22.

104. "The Kremlin is preparing bloody false-flag special operations in Russia," InformNapalm, 16 Mar 22, https://informnapalm.org/en/the-kremlin-is- preparing-bloody-false-flag-special-operations-in-russia/; Alexander Kovalenko, "Russia kills dozens to back fake story about Ukrainian shelling of Donetsk," 14 Mar 22, https://site.ua/alexander.kovalenko/russia-kills-dozens-to-back-fake -story-about-ukrainian-shelling-of-donetsk-i7zp9jk

105. Mikhail Alexandrov, "Expert: Russia has the right to destroy US biological laboratories in Georgia," REGNUM, 15 Mar 22, https://regnum.ru/amp/3533153

106. FR 24 and ADS-B Exchange screenshots, 14 Mar 22.

107. Kanishka Singh, "U.S. warns Russia any use of chemical weapons would have consequences," Reuters, 16 Mar 22, https://www.reuters.com/world/us-warns-russia -consequences-any-possible-russian-use-chemical-weapons-white-2022–03–16/

108. Senator Marco Rubio tweet, 16 Mar 22.

109. ADS-B Exchange screenshots, 16 Mar 22.

110. FR 24 screenshots, 16 Mar 22.

111. FR 24 screenshots, 16 Mar 22.

112. Russian Ministry of Foreign Affairs Twitter post, 17 Mar 22; "Cyberattack on Polish railways," RailTarget, 17 Mar 22, https://www.railtarget.eu/passenger/cyberattack -on-polish-railways-rail-transport-is-paralyzed-systems-manufactured-by-alstom -are-incapacitated-2041.html; "Polish soldier who defected to Belarus found dead from hanging in Minsk apartment," THEfirstNEWS, 17 Mar 22, https://www .thefirstnews.com/article/polish-soldier-who-defected-to-belarus-found-dead -from-hanging-in-minsk-apartment-28871

113. FR 24 screenshots, 17 Mar 22.

114. Christo Grozev Twitter post, 17 Mar 22.

115. ADS-B Exchange screenshots, 17 Mar 22.

116. Berlkium, https://www.lenntech.com/periodic/elements/bk.htm

117. U.S. Department of State, *Compliance with the Convention on the Prohibition of the Development, Production, Stockpiling and Use of Chemical Weapons and on Their Destruction: Condition (10)(C) Report*, April 2021, 17.

118. Alicia Sanders-Zakre, "Russia Destroys Last Chemical Weapons," *Arms Control Association*, Nov 17; U.S. GAO-03-482, "Weapons of Mass Destruction: Additional Russian Cooperation Needed to Facilitate U.S. Efforts to Improve Security at Russian Sites," March 2003.

119. CIT (en) Twitter post, 17 Jan 22; Tony Seward Facebook post, 9 Mar 22; information held by the author.

120. FR 24 screenshots, 18 and 21 Mar 22; ADS-B Exchange screenshots, 18–20 Mar 22.

121. "More than 20 Russian strategic bombers spotted at Engels airbase," Avia.pro, 20 Mar 22, https://avia.pro/news/bolee-20-rossiyskih-strategicheskih-bombard irovshchikov-zamecheny-na-aviabaze-engels

122. TASS Twitter post, 19 Mar 22.

123. ABC News Twitter post, 18 Mar 22.

124. George Galloway Twitter post, 22 Mar 22.

125. Samantha Locket et al., "Clear sign Putin is weighing up use of chemical weapons in Ukraine, says Biden," *The Guardian*, 22 Mar 22, https://www.theguardian .com/world/2022/mar/22/clear-sign-putin-is-weighing-up-use-of-chemical -weapons-in-ukraine-says-biden

126. U.S. Strategic Command Twitter post, 23 Mar 22.

127. Visegrad 24 Twitter post, retweeting Russian Pravda, 16 Mar 22.

128. Mark Kozubal, "MOD: We do not confirm the violation of Polish airspace," Wojsko, 21 Mar 22, https://www.rp.pl/wojsko/art35905901-mon-nie-potwierdzamy -naruszenia-przestrzeni-powietrznej-polski

129. OSINTdefender Twitter post, 22 Mar 22.

130. "Russian diplomats burning files in Warsaw, prepared to leave: Report," Polski-eRadio, 23 Mar 22, https://www.polskieradio.pl/395/7785/Artykul/2925661,Russian -diplomats-burning-files-in-Warsaw-preparing-to-leave-report

131. Aki Heikkinen Twitter post, 22 Mar 22.

132. Arndt and Horovitz, *Nuclear Rhetoric*, 29.

133. Bruno Tertrais, *French Nuclear Deterrence Policy, Forces, and Future: A Handbook* (Paris: Fondation pour la Recherche Stratégique, 2020), 55–57; "Intercontinental ballistic missile M51," Missilery.info, https://en.missilery.info/missile/m51

134. "France and the Ukraine 2022 Crisis: France Sends 2nd SSBN to Sea," Second Line of Defense, 11 Mar 22, https://sldinfo.com/2022/03/france-and-the-ukraine-2022 -crisis-france-sends-2nd-ssbn-to-sea/

135. Stephane Jezequel, "Why did France sail three submarines from Ile-Longue?" *Le Télégramme*, 21 Mar 22, http://letelegramme.fr/France/pourquoi-la-france -a-t-elle-fait-appareiller-trois-sous-marins-nucleaires-au-depart-de-l-ile-longue -21-03-2022-12954544.php

136. Thomas Brent, "France ups defence and deploys three of its four nuclear submarines," *Connexion France*, 25 Mar 22, https://www.connexionfrance.com/article/French-news /France-ups-defence-and-deploys-three-of-its-four-nuclear-submarines

137. "France claims successful testing of advanced ASMPA nuke missile," Anadolu Agency, 24 Mar 22, https://www.aa.com.tr/en/europe/france-claims-successful -testing-of-advanced-asmpa-nuke-missile/2545156; Richard Scott, "Successful flight test of upgraded ASMPA missile paves way for refurbishment," *Janes*, 30 Mar 22, https://www.janes.com/defence-news/news-detail/successful -flight-test-of-upgraded-asmpa-missile-paves-way-for-refurbishment

138. For details, see Bruno Tertrais, *French Nuclear Deterrence Policy, Forces and Future: A Handbook* (Paris: Fondation pour la Recherche Stratégique, 2020), https:// www.frstrategie.org/sites/default/files/documents/publications/recherches-et -documents/2020/202004.pdf

139. FR 24 screenshots, 23 Mar 22; ADS-B Exchange screenshots, 23 Mar 22.

140. Nexta retweeting *Washington Post*, 24 Mar 22.

141. Russian Ministry of Defence Twitter post, 26 Mar 22.

142. FR 24 screenshot, 24 Mar 22.

143. FR 24 screenshot, 24 Mar 22.

144. Christian Wilkie, " Biden says US would responds to Russia if Putin uses chemical or biological weapons," NBC News, 24 Mar 22.

145. Marco Rubio Twitter post, 25 Mar 22.

146. Air Power Twitter post, 25 Mar 22; see also "Eiselson Air Force Base completes two-year F-35A build-up," Overt Defense, 3 May 22, https://www.overtdefense .com/2022/05/03/eielson-air-force-base-completes-two-year-f-35a-build-up/, and Amy Hudson, "Eielson Days Away from Achieving Full Complement of F-35s," *Air and Space Forces*, 13 Apr 22, https://www.airandspaceforces.com/eielson -days-away-from-achieving-full-complement-of-f-35s/. Note that local Alaskan

media regularly mentioned the nuclear capability of the F-35. One of many examples: "Alaskan Delegation Celebrates Arrival of F-35s at Eielson AFB," *Alaska Business*, 22 Apr 20, https://www.akbizmag.com/industry/government /alaska-delegation-celebrates-arrival-of-f-35s-at-eielson-afb/

147. "The F-35 delivers unmatched global deterrence," *The Watch*, 30 Jan 23, https:// thewatch-magazine.com/2023/01/30/the-f-35-delivers-unmatched-global -deterrence/

148. Polina Ivanova Twitter posts, 25 Mar 22; "Russian Military Official Shifts Rhetoric, Says Army Now Focusing on 'Liberation' of Eastern Ukrainian Regions," RFE/ RL, 25 Mar 22, https://www.rferl.org/a/russia-ukraine-goals-scaled-back/31770879 .html

149. Paraphrased from "Remarks by President Biden on the United Efforts of the Free World to Support the People of Ukraine," The Royal Castle in Warsaw, 26 Mar 22, https:// www.whitehouse.gov/briefing-room/speeches-remarks/2022/03/26/remarks -by-president-biden-on-the-united-efforts-of-the-free-world-to-support-the-people -of-ukraine/

150. Arndt and Horovitz, *Nuclear Rhetoric*, 31.

151. Russian Defense Policy Twitter post, 31 Mar 22; "Strategic missile forces of the Russian Federation conduct an exercise," Interfax, 31 Mar 22, militarynews.ru /story.asp?rid=1&nid=571668&lang=RU; TASS Twitter post, 1 Apr 22; Russia's Nuclear Activity Twitter post, 1 Apr 22.

152. Heather Mongilio, "SECDEF Austin Extends Truman Deployment," USNI News, 31 Mar 22, https://news.usni.org/2022/03/31/secdef-austin-extends-truman -deployment-as-conflict-in-ukraine-continues

153. "Press Briefing by Press Secretary Jen Psaki, March 16, 2022," https://www .whitehouse.gov/briefing-room/press-briefings/2022/03/16/press-briefing -by-press-secretary-jen-psaki-march-16–2022/

154. Alyona Getmanchuk, "Putin will win unless the West sends Ukraine offensive weapons," The Atlantic Council, 31 Mar 22, https://www.atlanticcouncil.org/blogs /ukrainealert/putin-will-win-unless-the-west-sends-ukraine-offensive-weapons/; Luis Martinez, "What are Javelin missiles and why they're being mentioned repeat- edly during the impeachment hearings," ABC News, 15 Nov 2019, https://abcnews .go.com/Politics/javelin-missiles-ukraine/story?id=65855233#:~:text=In%20 March%202018%2C%20Trump%20approved,that%20fighting%20began%20in%20 2014; Jonathan Guyer, "What US weapons tell us about the Russia-Ukraine war," Vox, 29 Mar 23, https://www.vox.com/world-politics/2023/3/29/23652435/debate -weapons-ukraine-abrams-leopard-tanks-biden-zelenskyy; Anna Wieslander, "What I heard in Munich," The Atlantic Council, 22 Feb 23, https://www.atlantic council.org/blogs/new-atlanticist/what-i-heard-in-munich-four-fears-are-holding -the-west-back-from-quicker-bolder-support-for-ukraine/

155. FR 24 screenshots, 4 and 5 Apr 22.

156. ADS-B Exchange screenshots, 5–7 Apr 22.

157. Aki Heikkinnen Twitter post, 8 Apr 22.

158. Anish Dangwal, "Clash of NATO & Russian Fighters," *Eurasian Times*, 10 Apr 22, https://eurasiantimes.com/nato-jets-scrambled-four-times-to-intercept-russian-warplanes/; Kimberly Johnson, "NATO Fighters Scramble as Increase in Russian Aircraft Reported around Black Sea," *Flying*, 11 Apr 22, https://www.flyingmag.com/nato-fighters-scramble-as-increase-in-russian-aircraft-reported-around-black-sea/

159. FR 24 screenshots, 8 Apr 22.

160. ADS-B Exchange, 11 Apr 22.

161. Thomas Nilsen, "U.S. amphibious combat group deploys to the high north amid high tensions," *Barents Observer*, 12 Apr 22, https://thebarentsobserver.com/en/security/2022/04/us-amphibious-combat-group-moves-iceland-exercise-north-norway-winter-training; ADS-B Exchange and FR 24 screenshots, 7 Apr 22.

162. Hans Kristensen, "Lakenheath Air Base Added to Nuclear Weapons Storage Site Upgrades," Federation of American Scientists, 11 Apr 22, https://fas.org/publication/lakenheath-air-base-added-to-nuclear-weapons-storage-site-upgrades/; Julian Borger, "UK military vaults upgraded to store new US nuclear weapons," *The Guardian*, 12 Apr 22, https://www.theguardian.com/world/2022/apr/12/uk-military-vaults-upgraded-to-store-new-us-nuclear-weapons

163. FR 24 screenshots, 10–11 Apr 22.

164. Olena Roshchina, "Russia threatens chemical attacks on Mariupol," Pravda Ukraine, 11 Apr 22, reporting RIA Novosti on Telegram, https://www.pravda.com.ua/eng/news/2022/04/11/7338843/

165. *The Times of Moscow* Twitter post, 11 Apr 22; Carl Bildt Twitter post, 11 Apr 22; Ander Aslund Twitter post, 11 Apr 22.

166. Arndt and Horovitz, *Nuclear Rhetoric*, 41.

167. Alina Mykhailova Twitter post, 11 Apr 22; Petri Makela Twitter post, 11 Apr 22; Azov Twitter post, 11 Apr 22.

168. Alexander Ward and Jonathan Lemire, "Team Biden scrambles to respond to claims of Russia chemical weapon use," *Politico*, 4 Apr 22, https://www.politico.com/news/2022/04/12/biden-admin-russia-chemical-weapons-00024709

169. U.S. Department of Defense Twitter post, 11 Apr 22.

170. FR 24 screenshot, 11 Apr 22.

171. Arndt and Horovitz, *Nuclear Rhetoric*, 43.

172. Guy Falconbridge, "Russia warns of nuclear, hypersonic deployment if Swede and Finland join NATO," Reuters, 14 Apr 22, https://www.reuters.com/world/europe/russia-warns-baltic-nuclear-deployment-if-nato-admits-sweden-finland-2022-04-14/?fbclid=IwARostunxxcCOm5ocexnyIzzN

173. ADS-B Exchange screenshots, 14 Apr 22; FR 24 screenshots, 14 Apr 22.

174. "CIA Chief Says Threat Russia Could Use Nuclear Weapons Is Something U.S. Cannot 'Take Lightly,'" RFE/RL, 15 Apr 22, https://www.rferl.org/a/russia-nuclear-weapons-burns-cia/31804539.html

175. The Kyiv Independent Twitter post, 17 Apr 22.

176. FR 24 screenshots, 15 Apr 22.

177. Arndt and Horovitz, *Nuclear Rhetoric*, 45.

178. MFA Russia Twitter post, 16 Apr 22.

179. #haveglass Twitter post, 15 Apr 22.

180. Burak Bir, "NATO jets intercept Russian fighters, intelligence aircraft over Baltic Sea," AA World, 17 Apr 22, https://www.aa.com.tr/en/world/nato-jets-intercept-russian-fighters-intelligence-aircraft-over-baltic-sea/2874839

181. Visegrad 24 Twitter post, 16 Apr 22; "Most Poles fear Ukraine war threatens Poland security-poll," THEfirstNEWS, 15 Mar 22, https://www.thefirstnews.com/article/most-poles-fear-ukraine-war-threatens-polands-security—-poll-28781

182. "Poland/Belarus: New evidence of abuses," Amnesty International, 11 Apr 22, https://www.amnesty.org/en/latest/news/2022/04/poland-belarus-new-evidence-of-abuses-highlights-hypocrisy-of-unequal-treatment-of-asylum-seekers/

183. OSINTdefender Twitter post, 17 Apr 22.

184. "Belarusian paratroopers fight militants in tactical exercise," Belta, 20 Apr 22, https://eng.belta.by/society/view/belarusian-paratroopers-fight-militants-in-tactical-exercise-149659–2022/

185. Abraham Mahshie, "NATO Scrambling More Often in Response to Russian Jets Near Poland," *Air and Space Forces*, 14 Apr 22, https://www.airandspaceforces.com/nato-intercepts-of-russian-aircraft-near-polands-skies-increasing/?fbclid=IwAR3AB77qgeC1PojYtTXonoY2EJk_QwaNDXcxl3_SdPiv

186. "Top Belarus aide threatens Poland with 'destruction, death, and explosions,'" TVP World, 18 Mar 22, https://tvpworld.com/59703860/top-belarus-aide-threatens-poland-with-destruction-death-and-explosions?fbclid=IwAR2hoLdS3UxvKm7qnkT5f6V9q4cHdTZOVZwcg1em_h9oCv1xqFsl_Qbpe5M

187. RT tweet, 17 Apr 22.

188. FR 24 screenshots, 18 Apr 22.

189. Oliver Trapnell, "Britain's most powerful submarine arrives in Gibraltar hours after Putin's threats," *Express*, 17 Apr 22, https://www.express.co.uk/news/world/1596978/ukraine-news-britain-hms-audacious-nuclear-submarine-gibraltar-putin

190. "UGM/BGM/RGM-109 Tomahawk," Seaforces-online, https://www.seaforces.org/wpnsys/SURFACE/BGM-109-Tomahawk.htm

191. ADS-B Exchange, 18 Apr 22.

192. Radina Gigova, "Russian foreign minister says new phase of Ukraine operation beginning," CNN, 19 Apr 22, https://www.cnn.com/europe/live-news/ukraine-russia-putin-news-04–19–22/index.html

193. FR 24 screenshots, 19 Apr 22.

194. "Transcript: Senior Defense Officials Hold a Background Briefing," 19 Apr 22, https://www.defense.gov/News/Transcripts/Transcript/Article/3004245/senior -defense-officials-hold-a-background-briefing-april-19–2022/

195. ADS-B Exchange, 19 Apr 22.

196. Anthony Ruggiero et al., "Russia's Sarmat test underscores need to modernize US nuclear triad," *DefenseNews*, 28 Apr 22, https://www.defensenews .com/opinion/commentary/2022/04/28/russias-sarmat-test-underscores-need -to-modernize-us-nuclear-triad/; Timothy Wright, "Russia's new stratgic nuclear weapons: A technical analysis and assessment," IISS, 16 Jun 22, https://www. iiss.org/online-analysis/online-analysis/2022/06/russias-new-strategic -nuclear-weapons-a-technical-analysis-and-assessment/

197. Barbara Starr, "US defense secretary being regularly briefed," CNN, 20 Apr 22, https:// www.cnn.com/europe/live-news/ukraine-russia-putin-news-04–20–22/index.html

198. FR 24 screenshots, 20 Apr 22.

199. ADS-B Exchange screen shots and FR 24 screenshots, 20 Apr 22.

200. FR 24 screenshot, 21 Apr 22; The Barents Observer Twitter post, 21 Apr 22.

201. ADS-B Exchange screenshots, 21 Apr 22.

202. Ed Browne, "Russia's Dmitry Rogozin Delivers Nuclear Threat about Boris Johnson's Hair," *Newsweek*, 22 Apr 22, https://www.newsweek.com/russia-dmitry -rogozin-boris-johnson-hair-comment-sarmat-nuclear-missile-1699812

203. "Statistics: Cyber attacks April 2022," KON Briefing, 14 Sep 22, https://konbriefing .com/en-topics/cyber-attacks-2022–04.html

204. Whiz Security Twitter post, 21 Apr 22.

205. John Besley, "Five Eyes alert warns of heightened risk of Russian cyber attacks," *Belfast Telegraph*, 21 Apr 22, https://www.belfasttelegraph.co.uk/news/uk/five eyes-alert-warns-of-heightened-risk-of-russian-cyber-attacks/41576277.html

206. "Five Eyes nations warn against impending Russian cyber attacks," TechCentral.ie, 21 Apr 22, https://archive.org/details/a2_cple_Atc_B1_Bomber_Midway_Nukewar

207. Besley, "Five Eyes alert warns of heightened risk of Russian cyber attacks."

208. Julia Davis Twitter post, 21 Apr 22.

209. Pomerantsev, *This Is Not Propaganda*, 85.

Chapter 7. Ramp-Up

1. FR 24 screenshots, 22 Apr 22.

2. "Pentagon Press Secretary John F. Kirby Holds a Press Briefing," 22 Apr 22, https:// www.defense.gov/News/Transcripts/Transcript/Article/3008806/pentagon -press-secretary-john-f-kirby-holds-a-press-briefing/

3. ADS-B Exchange screenshot, 22 Apr 22.

4. RIA Novosti Twitter post, 23 Apr 22.

5. Russia's Nuclear Activity Twitter post, retweeted from TASS, 23 Apr 22.

6. OSINTdefender Twitter post, 23 Apr 22.

7. Sam Ramani retweeting Russian Ministry of Defence briefing, 23 Apr 22.

8. FR 24 screenshots, 24 and 25 Apr 22.

9. Zaini Majeed, "NATO jets scramble 'multiple times' to intercept Russian aircraft near alliance's airspace over past few days," Republicworld.com, 29 Apr 22, https://www.republicworld.com/world-news/russia-ukraine-crisis/nato-jets-scramble-multiple-times-to-intercept-russian-aircraft-over-baltic-and-black-sea-articleshow.html

10. FR 24 screenshots, 25 and 26 Apr 22.

11. President of Russia, "Meeting with Council of Lawmakers," 27 Apr 22, http://en.kremlin.ru/events/president/news/68297

12. FR 24 screenshots, 27 Apr 22.

13. "Pentagon Press Secretary John F. Kirby Holds a Press Briefing," 27 Apr 22, https://www.defense.gov/News/Transcripts/Transcript/Article/3012823/pentagon-press-secretary-john-f-kirby-holds-a-press-briefing/

14. Sam Ramani retweet Simonyan RT tweet, 28 Apr 22.

15. Def Mon Twitter post, 30 Apr 22; TASS Twitter post, "Retired Canadian army general may be holed up in Azovstal steel plant—DPR militia," 28 Apr 22.

16. ADS-B Exchange screenshots, 29 Apr 22.

17. "Angara-1.2 files its first mission," Russian Space Web, 29 Apr 22, https://www.russianspaceweb.com/angara1-flight1.html

18. CNN, 29 Apr 22, https://www.cnn.com/europe/live-news/russia-ukraine-war-news-04-29-22/index.html

19. FR 24 screenshots, 29 Apr 22.

20. NASA Space Flight Forum, 29 Apr 22, https://forum.nasaspaceflight.com/index.php?topic=54797.0

21. ADS-B Exchange screenshots, 29 Apr 22.

22. "Pentagon Press Secretary John F. Kirby Holds a Press Briefing," 29 Apr 22, https://www.defense.gov/News/Transcripts/Transcript/Article/3015652/pentagon-press-secretary-john-f-kirby-holds-a-press-briefing/

23. "Senate Armed Services Committee Hearing: Nuclear Weapons Council," 4 May 22, https://www.stratcom.mil/Media/Speeches/Article/3022885/senate-armed-services-committee-hearing-nuclear-weapons-council/

24. Mateusz Sobieraj Twitter post, 1 May 22; Team Luftwaffe Twitter post, 3 May 22; U.S. Marines Twitter post, 19 Apr 22; BAE Systems Maritime Twitter post, 26 Apr 22.

25. Arndt and Horovitz, Nuclear Rhetoric, 55; FR 24 screenshots, 1–4 May 22.

26. The Daily Telegraph Twitter post, 4 May 22 footage of Ben Wallace.

27. ADSB-exchange historical data, 4 May 22.

28. Tom O'Connor, "Russia Ambassador the U.S. Says NATO Not Taking Nuclear War Threat Seriously," *Newsweek*, 5 May 22, https://www.newsweek.com/russia -ambassador-us-says-nato-not-taking-nuclear-war-threat-seriously-1703968?fbcl id=IwAR1SAZFHJjnpu56AaBtsdFAfQthn_7vGXRRXes7DOBxOHBA1sEZ0 H6H9gZ0

29. FR 24 screenshots, 5 May 22.

30. Paul D. Shinkman, "Fears of False Flag Operation Grow as Russia Claims Ukraine Poised for Chemical Weapons Attack," *U.S. News and World Report*, 6 May 22, https:// www.usnews.com/news/world-report/articles/2022–05–06/russia-claims-ukraine -poised-for-chemical-weapons-attack-as-analysts-fear-false-flag-operation

31. "U.S. Senate Select Committee on Intelligence Hearings," 10 Mar 22, https://www .intelligence.senate.gov/hearings/open-hearing-worldwide-threats-2#

32. "U.S. Senate Select Committee on Intelligence Hearings."

33. Caitlin Doornbos, "Milley speaks with top Russian general days after post-Ukraine invasion call between Austin and Russian defense minister," *Stars and Stripes*, 19 May 22, https://www.stripes.com/theaters/us/2022–05–19/milley-russia-general -military-talk-ukraine-war-6059037.html

34. Ryan Chan Twitter post, 8 and 17 May 22; Antti Eskelinen Twitter post, 8 May 22; U.S. Embassy Riga Twitter post, 12 May 22; Warship.com Twitter post, 14 May 22.

35. Thenewarea51 Twitter posts, 11 May 22; ADS-B Exchange screenshots, 12 May 22.

36. Gerjon Twitter post, 13 May 22.

37. ADS-B Exchange screenshots, 13 May 22.

38. FR 24 screenshots, 7–15 May 22.

39. "Russian submarine launched torpedoes to pierce the Arctic ice," Newsnpr, 14 May 22, https://www.newsnpr.org/russian-submarine-launched-torpedoes -to-pierce-the-arctic-ice/

40. "Top US, Russian generals speak for the first time since Ukraine war began: Pentagon," AFP via *Times of India*, 19 May 22, https://timesofindia.indiatimes.com world/europe/top-us-russian-generals-speak-for-first-time-since-ukraine-war began-pentagon/articleshow/91671979.cms

41. Defender 22 Fact Sheet, https://www.europeafrica.army.mil/Portals/19 /documents/Infographics/DE22%20Factsheet%20.pdf?ver=Sz9mrhoGxIMTAd1i —z0OQ%3D%3D

42. "Deploying US rockets systems at Bornholm to lead to escalation," TASS, 1 Jun 22, https://tass.com/politics/1459349; Russia's Nuclear Activity, retweeting Russian Ministry of Defence social media.

43. Arndt and Horovitz, *Nuclear Rhetoric*, 69.

44. Julia Davis Twitter post of Solovyev on Rossiya1, 5 Jun 22.

45. "Belarus to conduct military mobilization exercises near Ukraine Border—BelTA," Reuters, 30 May 22, https://www.reuters.com/world/europe/belarus-conduct -military-mobilisation-exercises-near-ukraine-border-belta-2022–05–30/; "Brest Paratroopers to Cross River Located Near Border with Ukraine," Charter97, 1 Jun 22, https://charter97.org/en/news/2022/6/1/500744/; "Belarus has resumed its military exercises again," Pravda Ukraine, 16 Jun 22, https://www.pravda.com.ua /eng/news/2022/06/16/7352918/

46. Russia's Nuclear Activity retweet of TASS and Russian Ministry of Defence posts, 6 Jun 22.

47. FR 24 screenshots, 6 Jan 22.

48. Grzegorz Kuczynski, "Russia Violates Baltic Airspace and Waters Sending Warning to Sweden and Finland," *Russia Monitor*, 20 Jun 22, https://warsawinstitute .org/russia-violates-baltic-airspace-waters-sending-warning-sweden-finland/

49. "Why Russia-Lithuanian Tensions Are Rising," VOA Explainer, 22 Jun 22, https:// www.voanews.com/a/explainer-why-russia-lithuania-tensions-are-rising/6628802 .html

50. FR 24 screenshots, 21–23 Jun 22.

51. Arndt and Horovitz, *Nuclear Rhetoric*, 76.

52. "US Navy Tests Unarmed Trident II Strategic Weapon System," *Dialgo Americas*, 27 Jun 22, https://dialogo-americas.com/articles/us-navy-tests-unarmed-trident -ii-strategic-weapon-system/#.YrnU9uzMK9s

53. Lindsey Heflin, "B-2 executed new nuclear tactic in B61–12 JTA capstone test," 53rd Wing Public Affairs, 8 Jul 22, https://www.53rdwing.af.mil/News /Article/3087025/b-2-executes-new-nuclear-tactic-in-b61–12-jta-capstone-test /fbclid/b-2-executes-new-nuclear-tactic-in-b61–12-jta-capstone-test/

54. Delia Martinez, "Always Ready: Barksdale Airmen Execute ACE Capabilities," 2nd Bomb Wing Public Affairs, 7 Jul 22, https://www.barksdale.af.mil/News /Article/3086309/always-ready-barksdale-airmen-execute-ace-capabilities/fbclid /always-ready-barksdale-airmen-execute-ace-capabilities/

55. Arndt and Horovitz, *Nuclear Rhetoric*, 78.

56. "Russia threatens retaliation against Norway over access to Arctic islands," Reuters, 29 Jun 22, https://www.reuters.com/world/europe/russia-threatens -retaliation-against-norway-over-access-arctic-islands-2022–06–29/

57. FR 24 screenshot, 29 Jun 22.

58. Oliver Moody, "Putin's cyber-army hits Norway after Arctic blockade," *Times*, 3 Jun 22, https://www.thetimes.co.uk/article/putins-cyber-army-hits-norway -after-arctic-blockade-qwbvdq3cl

59. U.S. Naval Forces Europe-Africa Twitter post, 1 Jul 22.

60. George Allison, "More American nuclear submarines arrive in Scotland," *UK Defence Journal*, 6 Aug 22, https://ukdefencejournal.org.uk/more-american-

nuclear-submarines-arrive-in-scotland/?fbclid=IwAR23iTgKYDOAY89–8z4eXq
-jGk7KRrgcsaVpjGAdmzgmZcmkVzaB-dAp1Yw

61. Russia's Nuclear Activity retweet of Russian Ministry of Defence Twitter post; FR
    24 screenshots, 6 Jul 22; Arndt and Horovitz, *Nuclear Rhetoric*, 81.
62. Congressional Research Services, "Russia's Nuclear Weapons: Doctrine, Forces,
    and Modernization," 21 Apr 22, https://crsreports.congress.gov/product/pdf/R
    /R45861, 28.
63. Thomas Nilsen, "World's longest nuclear submarine handed over to the Russian
    Navy," *Barents Observer*, 8 Jul 22, https://thebarentsobserver.com/en/security
    /2022/07/worlds-longest-nuclear-submarine-handed-over-russian-navy
64. Stephanie M. Smith, *U.S. Air Force Operational Test & Evaluation Center Year in
    Review 2022*, 2, https://www.afotec.af.mil/Portals/69/HO_YIR_2022.pdf
65. Arndt and Horovitz, *Nuclear Rhetoric*, 83.
66. FR 24 screenshots, 12 Jul 22.
67. Russia's Nuclear Activity retweet Russian Ministry of Defence Twitter post.
68. NATO Air Command Twitter post, 15 Jul 22.
69. "Two Russian Tu-160 strategic bombers perform scheduled flight over Barents Sea,"
    TASS, 19 Jul 22, https://tass.com/defense/1481855?fbclid=IwARobm-5QESB9X
    eYprs2HeR3Z7FOoe884S4CM7mOAcbNIu4yL7auXnSrq_Gk
70. Thomas Nilsen, "Northern Fleet sabotage-reconnaissance group eliminated claims
    Ukrainian intelligence," Radio Canada International, 19 Jul 22, https://www.rcinet
    .ca/eyc on the-arctic/2022/07/19/northern-fleet-sabotage-reconnaissance-group
    -eliminated-claims-ukraine-intelligence/
71. Timo R. Stewart, *FIIA Briefing Paper 330: Russian Blackmail and the Black Sea Grain
    Initiative*, April 2023, Finnish Institute for International Affairs, https://www.fiia
    .fi/en/publication/russian-blackmail-and-the-black-sea-grain-initiative
72. Arndt and Horovitz, *Nuclear Rhetoric*, 89–92.
73. Transcript of French statement, 1 Sep 22, https://totalwonkerr.net/2022/09/17
    /france/, through Ankit Panda Twitter post, 23 Sep 22.
74. Ministry of Defence Lithuania, "Data on interceptions of aircraft completed near the
    Baltic States' border on August 1–7, 2022," https://kam.lt/en/data-on-interceptions
    -of-aircraft-completed-near-the-baltic-states-borders-on-august-1–7-2022/
75. Barbara Starr, "US postponed missile test," CNN, 5 Aug 22, https://www.cnn.com
    /2022/08/04/politics/us-postpones-missile-test/index.html
76. FR 24 screenshots, 8 Aug 22; "American military aircraft RC-135W RIVET JOINT spot-
    ted near Novaya Zemlya," Avia.pro, 10 Aug 22, https://avia.pro/news/amerikanskiy
    -voennyy-samolyot-rc-135w-rivet-joint-zamechen-v-rayone-novoy-zemli
77. ADS-B Exchange screenshots, 9 Aug 22.
78. FR 24 screenshots, 14 Aug 22.

79. "Britain says it has received approval from air traffic controllers to fly its spy plane near Russia's borders," Avia.pro, 17 Aug 22, https://avia.pro/news/velikobritaniya -zayavila-chto-poluchila-ot-dispetcherov-odobrenie-na-prolyot-svoego-samolyota

80. Arndt and Horovitz, *Nuclear Rhetoric*, 96.

81. FR 24 screenshots, 16–17 Aug 22.

82. Corey Bouchard, "Historic moment: B-52 lands at Loring AFB for first time in 29 years," WABI5 CBC, 12 Aug 22, https://www.wabi.tv/2022/08/12/historic-moment-b-52-lands-loring-afb-first-time-29-years/?fbclid=IwAR2iQo-j-WDhFoOMwW MeWsblzSMd9MYVbdlYr5LmdoHTUrGKkOxTobLRSdU; Scott Wakefield, "AFGSC's most recent Minuteman III test occurs on historic date," 377 Air Base Wing Public Affairs, 17 Aug 22, https://www.afgsc.af.mil/News/Article-Display /Article/3132112/afgscs-most-recent-minuteman-iii-test-occurs-on-historic-date /fbclid/afgscs-most-recent-minuteman-iii-test-occurs-on-historic-date/; "B-52 Bombers Complete Deployment to Europe," *Dialogo Americas*, 6 Oct 22, https:// dialogo-americas.com/articles/b-52-bombers-complete-deployment-to-europe -strengthen-ties-with-nato-allies-and-partners/

83. "American B-52 bombers strike for the first time with JDAM munitions 400 kilometres from the Russian border in the Arctic," Avia.pro, 19 Aug 22, https:// avia.pro/news/amerikanskie-bombardirovshchiki-b-52-vpervye-nanesli-udary -boepripasami-jdam-v-400-kilometrah; Henri Vanhanen Twitter post, 18 Aug 22; Petri Makela Twitter post, 18 Aug 22.

84. Fr 24 screenshot, 23 Aug 22; Russia's Nuclear Activity retweet of Russian Ministry of Defence, 23 Aug 22

85. ADS-B Exchange screenshots, 22 Aug 22.

86. ADS-B Exchange screenshots, 26Aug 22.

87. Team Barksdale Twitter post, 25 Aug 22.

88. Ryan Chan Twitter post, 27 Aug 22.

89. Imagery located at https://www.dvidshub.net/image/7439453/705th-muns -movement with a gallery depicting said activity.

90. "Slava-class guided missile cruiser," http://www.military-today.com/navy/slava _class.htm

91. "Russia Sends Slava-Class Cruiser Ustinov in the Mediterranean," Naval News, 7 Feb 22, https://www.navalnews.com/naval-news/2022/02/russia-sends-slava -class-cruiser-ustinov-in-the-mediterranean/; Sam LaGrone, " Sister Ship of Sunken Russian Cruiser *Moskva* Departs Mediterranean," USNI News, 24 Aug 22, https://news.usni.org/2022/08/24/sister-ship-of-sunken-russian-cruiser-moskva -departs-mediterranean-u-s-destroyers-follow-behind; H.I. Sutton, "New Intelligence: Russia Sends Nuclear Submarine to Mediterranean," Naval News, 2 Sep 22, https://www.navalnews.com/naval-news/2022/09/new-intelligence -russia-sends-nuclear-submarine-to-mediterranean/

92. Soraya Ebrahimi, "Russian warships seen acting unusually off Irish coast," *National News*, 30 Aug 22, https://www.thenationalnews.com/world/europe/2022/08/31/russian-warships-seen-acting-unusually-off-irish-coast/; Anita McSorley and Sara Rountree, "Russian submarine and destroyer in dramatic Irish Sea chase as Defence Forces 'monitor' situation," *Irish Mirror*, 31 Aug 22, https://www.irishmirror.ie/news/irish-news/russian-submarine-destroyer-dramatic-irish-27877776

93. Analysis conducted for the author by Robert Silliman based on frequency analysis of SSBN deployments to and from Faslane.

94. H.I. Hutton Twitter post, 30 Aug 22; "Royal Navy forces Russian vessel to U-turn after 'unusual transit' through Irish Sea," Navy Leaders, 31 Aug 22, https://www.navyleaders.com/news/royal-navy-forces-russian-vessel-u-turn-unusual-transit-through-irish-sea; "Military operation underway of south coast as Russian missile cruisers transit through area," *The Journal*, 30 Aug 22, https://www.thejournal.ie/russian-navy-cork-coast-british-navy-irish-defence-forces-5852924-Aug2022/; "UK Shadows Russian Warships Returning from Black Sea," *Maritime Executive*, 7 Sep 22, https://maritime-executive.com/article/photos-uk-shadows-russian-warships-returning-from-black-sea; "The appearance of the cruiser 'Marshal Ustinov' in the English Channel excited France," RG.RU, 8 Sep 22, https://rg.ru/2022/09/08/poiavlenie-krejsera-marshal-ustinov-v-la-manshe-vzbudorazhilo-franciiu.html

95. And the RT videos based on Western media videos were re-rebroadcast by Western media outlets. See *The Express*, 30 Aug 22, embedded video segment from RT, https://www.express.co.uk/news/uk/1662076/royal-navy-hms-lancaster-russian-cruiser-Irish-sea-transit-britain-ireland. Perusal of Russian "media" for the period 30 Aug 7–Sep 22 shows that practically every Russian outlet carried the story.

96. FR 24 screenshots, 30–31 Aug 22.

97. Seth G. Jones et al., "Mapping Ukraine's Military Advances," CSIS, 22 Sep 22, https://www.csis.org/analysis/mapping-ukraines-military-advances; Congressional Research Service, "Russia's War in Ukraine: Military and Intelligence Aspects," 13 Feb 23.

98. FR 24 screenshots, 1 Sep 22.

99. President of Russia schedule, 1 Sep 22, http://www.en.kremlin.ru/events/president/news/69247

100. ADS-B Exchange screenshots, 2 Sep 22.

101. USSTRATCOM Twitter post, 2 Sep 22.

102. Dougie Coull Photography Twitter post, 2 Sep 22.

103. Russia's Nuclear Activity retweet of Russian Ministry of Defence post, 5 Sep 22.

104. Department of Defense Twitter post, 6 Sep 22; Hans Kristensen Twitter post, 7 Sep 22; MatchlessMan410 Twitter post, 4 Sep 22, on the Marshal Krylov.

105. ADS-B Exchange screenshot, 8 Sep 22.

106. FR 24 screenshots, 8 Sep 22.

107. FR 24 screenshot, 12 Sep 22.

108. ADS-B Exchange screenshots, 8 Sep 22.

109. NORAD Twitter post, 12 Sep 22; The Lookout retweeting Russian Ministry of Defence post, 12 Sep 22.

110. Dmitry Twitter post with screenshots of Medvedev's rant on Telegram in Russian.

111. FR 24 screenshots, 13 Sep 22.

112. ADS-B Exchange screenshots, 13 Sep 22.

113. "SOCNORTH augments Operation Noble Defender with SOF assets," *Tip of the Spear*, 8–9 Oct 22. https://www.socom.mil/TipOfTheSpear/USSOCOM%20Tip%20of%20the%20Spear%20October%202022%20(Web).pdf. See also Wikipedia entry on St. Lawrence Island.

114. NORAD Twitter posts, 14–14 Sep 22.

115. USSTRATCOM Twitter posts, 23 Sep 22; Rusty Frank, "Prairie Vigilance keeps warbirds sharp," 5th Bomb Wing Public Affairs, 23 Sep 22.

116. FR 24 screenshots, 15 Sep 22; "Allies Work Out an Integrated Air Defense System Takedown with Bombers and Fighters," NATO press release, 16 Sep 22, https://ac.nato.int/archive/2022/allies-work-out-an-integrated-air-defense-system-takedown-with-bombers-and-fighters

117. Anton Gerashchenko Twitter post, 15 Sep 22.

118. Translated by Julia Davis and posted on Twitter, 16 Sep 22.

119. Liviu Horovitz and Anna Clara Arndt, *One Year of Nuclear Rhetoric and Escalation Management in Russia's War against Ukraine: An Updated Chronology*, German Institute for International and Security Affairs, February 2023, 112.

120. "Russia conducts military exercise with cruise missiles in front of Alaska," *Odessa Journal*, 16 Sep 22, https://odessa-journal.com/russia-conducts-military-exercises-with-cruise-missile-launches-in-front-of-alaska/; "Russia with Military Exercises in Ocean vis-à-vis Alaska," *High North News*, 22 Sep 22, https://www.highnorthnews.com/en/russia-military-exercises-ocean-vis-vis-alaska

121. Missile data drawn from sources compiled by Missile Defense Advocacy Alliance, https://missiledefenseadvocacy.org/missile-threat-and-proliferation/todays-missile-threat/russia/ss-n-30a-kalibr/

122. FR 24 screenshot, 16 Sep 22.

123. Thomas Nilsen, "Nuclear subs made trans-Arctic under ice transfer from Barents to Pacific," *Barents Observer*, 28 Sep 22, https://thebarentsobserver.com/en/security/2022/09/nuclear-subs-made-trans-arctic-under-ice-transfer-barents-pacific; "Two Russian nuclear powered submarines arrive at home," *Naval Recognition*, 29 Sep 22, https://www.navyrecognition.com/index.php/naval-news/naval-news-archive/2022/september/12263-two-new-russian-nuclear-powered-submarines-arrive-at-home.html

124. "President Biden warns Vladimir Putin not to use nuclear weapons," *60 Minutes Overtime*, 16 Sep 22, https://www.cbsnews.com/news/president-joe-biden-vladimir-putin-60-minutes-2022-09-16/?fbclid=IwAR28LzA1WXSwOMWkg6tA8Fh1QqUI6MQuv4DXerEjpQ

125. ADS-B Exchange screenshot, 16 Sep 22.

126. "The Ministry of Defence conducted exercises with YARS missile systems in the Sverdlovsk region," Kommersandt, 18 Sep 22, https://www.kommersant.ru/doc/5569892

127. ADS-B Exchange screenshots, 16 Sep 22; TheIntelFrog Twitter post, 14 and 18 Sep 22; #haveglass Twitter post, 19 Sep 22.

128. H.I. Sutton Twitter post, 17 Sep 22; Francis Scarr Twitter post, 19 Sep 22.

129. Martin Kragh Twitter post, 17 Sep 22.

130. Horovitz and Arndt, *One Year of Nuclear Rhetoric*, 113; marqs retweet of Simonyan, 20 Sep 22; Russian Embassy Bangladesh Twitter post, 18 Sep 22.

131. Sergej Sumlenny Twitter post, 20 Sep 22. See "Vladmir Ionesya" Wikipedia entry.

132. Ali Makela Twitter post, 20 Sep 22.

133. ADS-B Exchange screenshots, 20 Sep 22.

134. President of Russia, "Address by the President of the Russian Federation," 21 Sep 22.

135. ADS-B Exchange screenshots, 21 Sep 22.

136. Aircraft Spots screenshot, 21 Sep 22.

137. ADS-B Exchange screenshots, 21 Sep 22.

138. FR 24 and ADS-B Exchange screenshots, 21 Sep 22.

139. Michael MacKay Twitter post, 21 Sep 22.

140. Rune Granbakk Twitter post, 21 Sep 22.

141. Brian Whitaker, "Photos show Russian ships were close to Nord Stream sabotage site four days before explosion," *Medium*, 18 Apr 23, https://brian-whit.medium.com/photos-show-russian-ships-were-close-to-nord-stream-sabotage-site-four-days-before-explosions-f5a4524be1a7; "Russian 'Special Vessel' Spotted before Nord Stream Blasts-Danish Media," *Moscow Times*, 28 Apr 22, https://www.themoscowtimes.com/2023/04/28/russian-special-vessel-spotted-before-nord-stream-blasts-danish-media-a80972

142. Horovitz and Arndt, *One Year of Nuclear Rhetoric*, 114–115.

143. Secretary Blinken at the United Nations, 22 Sep 22, https://www.state.gov/secretary-antony-j-blinken-at-the-united-nations-security-council-ministerial-meeting-on-ukrainian-sovereignty-and-russian-accountability/

144. Horovitz and Arndt, *One Year of Nuclear Rhetoric*, 117; Samuel Ramani Twitter post, 22 Sep 22.

145. ADS-B Exchange, 22 Sep 22.

146. The Lookout Twitter post, 22 Sep 22.

147. ADS-B Exchange and FR 24 screenshots, 23 Sep 22.

148. FR 24 screenshot, 23 Sep 22.

149. ADS-B Exchange and FR 24 screenshots, 24–25 Sep 22.

150. Diaplous Group, "White Paper: The EU's Energy Arteries at Knifepoint: Nord Stream Incident—What's Next?" 12 Oct 22, https://www.sceguk.org.uk/wp-content /uploads/sites/24/2022/10/DIAPLOUS-WHITE-PAPER-NORD-STREAM -INCIDENT-1.pdf; Hans Sanderson et al., "Environmental impact of sabotage of the Nord Stream pipelines," Biological Sciences, 10 Feb 23, https://assets.researc hsquare.com/files/rs-2564820/v1/befd8db2-e0d7-4d1a-9567-d77da3aea7db .pdf?c=1676844660

151. Pomerantsev, This Is Not Propaganda, 49.

152. Aura Sabadus, "Nord Stream Seep Dive Finds Putin's Fingerprints," Centre for European Policy Analysis, 29 Mar 23, https://cepa.org/article/nord-stream-deep dive-finds-putins-fingerprints/; James Sherr, "Pipes and Mirrors: The Nord Stream Explosions," International Centre for Defence and Security (Estonia), 28 Apr 23, https://icds.ee/en/pipes-and-mirrors-the-nord-stream-explosions/#_ftnref8

153. TASS, 27 Sep 22, https://tass.com/politics/1513883

154. Jason Jay Smart Twitter post, 27 Sep 22.

155. ADS-B Exchange screenshots, 27 Sep 22.

156. FR 24 screenshots, 27 Sep 22.

157. Horovitz and Arndt, One Year of Nuclear Rhetoric, 121–122.

158. ADS-B Exchange screenshots, 28 Sep 22.

159. "Russian jet 'released missile' near RAF aircraft during patrol over Black Sea," Sky News, 20 Oct 22, https://news.sky.com/story/russian-jet-released-missile-near-raf -aircraft-during-patrol-over-black-sea-12725414?fbclid=IwAR2-MEMGN80F57 M6u4wuWqWykRZnlN9toUnMqsDohdAYhkY556ENbTcmKZc

160. Vilius Petkauskas, "We breached Russian satellite network, say pro-Ukraine parti-sans," Cybernews, 10 Oct 22, https://cybernews.com/cyber-war/we-breached-russian -satellite-network-say-pro-ukraine-partisans/

161. Iain Cameron Twitter post, 29 Sep 22.

162. The Lookout Twitter post, 30 Sep 22 and 7 Oct 22.

163. Alexander Grigoriev, "Teachings in an adult way: Both in number and skill," Zvez-daTV, 29 Sep 22.

164. ADS-B Exchange screenshots, 29–30 Sep 22.

165. Samuel Ramani Twitter posts, 28 and 29 Sep 22.

166. Horovitz and Arndt, One Year of Nuclear Rhetoric, 123.

167. Annamarie Horndern Twitter post, 30 Sep 22; Sergej Sumlenny Twitter post, 30 Sep 22.

168. ADS-B Exchange screenshots, 30 Sep 22.

169. Aki Heikken Twitter post, 30 Sep 22; Thenewarea51 Twitter post, 30 Sep 22; Jamming Twitter post, 30 Sep 22; Forsvarsdepartment Twitter post, 30 Sep 22.

170. Anna Ahronheim Twitter post, 30 Sep 22.

Chapter 8. Why Do These Things Always Happen in October?

1. Konrad Muzyka Twitter post, 2 Oct 22; Masano Twitter post, 2 Oct 22; The Intel Crab, 3 Oct 22; "Russian nuclear military train," *Daily Mail*, 3 Oct 22, https://www.dailymail.co.uk/news/article-11274515/Russian-nuclear-military-train -seen-possible-warning-West.html; Haley Ott, "Rare video shows Russia moving equipment belonging to nuclear weapons unit," CBS News, 4 Oct 22, https://www .cbsnews.com/news/russia-nuclear-weapons-train-video/

2. "Who's behind Rybar?," *Meduza*, 18 Nov 22, https://meduza.io/en/feature/2022/11/18 /who-s-behind-rybar

3. Author's analysis of the Rybar footage.

4. Milton Leitenberge and Raymond A. Zilinskas, *The Soviet Biological Weapons Program: A History* (Cambridge, MA: Harvard University Press, 2012), 88–97.

5. Confidential discussion with the author.

6. Anton Gerashchenko Twitter post, 2 Oct 22; STRATMil Twitter post, 3 Oct 22; Shashank Joshi Twitter post, 3 Oct 22; Truha News Twitter post, 3 Oct 22; "NATO warns that Russia has moved the submarine that carries the weapon of the Apocalypse," 2 *El Confidencial*, 2 Oct 22, https://www.elconfidencial.com /mundo/2022-10-02/arma-apocalipsis-bomba-rusia-ucrania-submarino_3499 957/?fbclid=IwAR1nTapJzcEqwT6d0HRk-6HZi_1z0VY; Carly Bass, "Missing submarine hints at Russia's chilling plan: Radioactive tsunami," Yahoo! News, 3 Oct 22, https://au.news.yahoo.com/missing-submarine-russia-chilling-plan -radioactive-tsunami-015840437.html

7. DoubtMethod and The_Real_Fly Twitter posts, 3 Oct 22.

8. Dmitry Stefanovich Twitter post, 3 Oct 22.

9. Elon Musk Twitter post, 3 Oct 22.

10. Matthew Champion, "Elon Musk Spoke to Putin before Tweeting Ukraine Peace Plan: Report," *Vice*, 11 Oct 22, https://www.vice.com/en/article/ake44z/elon -musk-vladimir-putin-ukraine

11. Maura Reynolds, "Fiona Hill: Elon Musk Is Transmitting a Message for Putin," *Politico*, 17 Oct 22, https://www.politico.com/news/magazine/2022/10/17/fiona -hill-putin-war-00061894

12. Andrij Melnyk Twitter post, 3 Oct 22.

13. News Nation, 3 Oct 22, https://www.youtube.com/watch?v=AXs73q-3Xeg

14. ADS-B Exchange screenshots, 1–5 Oct 22.

15. FR 24 and ADS-B Exchange screenshots, 4 Oct 22; Faytuks News Twitter post, 4 Oct 22; Northern Sentry Twitter post, 4 Oct 22.

16. "F-35 fighter jets scrambles after detection of Russian aircraft," VG.NO, 5 Oct 22, https://www.vg.no/nyheter/innenriks/i/wAOX6P/f-35-kampfly-rykket-ut-etter-oppdagelse-av-russisk-fly; "Four Russian fighter jets intercepted after entering Polish airspace: Reports," PolskieRadio, 6 Oct 22, https://www.polskieradio.pl/395/9766/artykul/3048957,four-russian-fighter-jets-intercepted-after-entering-polish-airspace-reports

17. Hans Kristensen Twitter post, 5 Oct 22.

18. "Poland has discussed hosting nuclear weapons with US, says president," Notes from Poland, 5 Oct 22, https://notesfrompoland.com/2022/10/05/poland-has-discussed-hosting-nuclear-weapons-with-us-says-president/?fbclid=IwARonOGoAoyTvT1yz-IiviBYx2bBgR8C

19. "Remarks by President Biden at Democratic Senatorial Campaign Committee Reception," 6 Oct 22, https://www.whitehouse.gov/briefing-room/speeches-remarks/2022/10/06/remarks-by-president-biden-at-democratic-senatorial-campaign-committee-reception/

20. ADS-B Exchange screenshots, 6 Oct 22.

21. FR 24 screenshots, 6 Oct 22.

22. Horovitz and Arndt, *One Year of Nuclear Rhetoric*, 125.

23. Kathryn Watson, "John Bolton says Biden 'overstated' the situation with Armageddon remark," CBS News, 7 Oct 22, https://www.cbsnews.com/news/john-bolton-biden-overstated-the-situation-with-armageddon-remark/

24. As detailed in Maloney, *Deconstructing Dr. Strangelove* and *Emergency War Plan*.

25. The Lookout Twitter post, 22 Oct 22.

26. Samuel Ramani Twitter post, 9 Oct 22.

27. NEXTA Twitter post, 10 Oct 22.

28. Dmitry retweeting GUR (Ukrainian Military Intelligence) tweet, 10 Oct 22.

29. Horovitz and Arndt, *One Year of Nuclear Rhetoric*, 126–127.

30. ADS-B Exchange screenshots, 11 Oct 22.

31. RCAF Operations and NORAD Twitter posts, 11 Oct 22.

32. The Lookout Twitter post of Russian Northern Fleet post, 11 Oct 22.

33. FR 24 and ADS-B Exchange screenshots, 11 Oct 22.

34. Stephen Collinson, "Biden sends a careful but chilling new nuclear message to Putin in CNN interview," CNN, 12 Oct 22, https://www.cnn.com/2022/10/12/politics/joe-biden-nuclear-message-putin-cnntv-analysis/index.html

35. Truxha News Twitter post, 12 Oct 22.

36. FR 24 and ADS-B Exchange screenshots, 12 Oct 22; Horovitz and Arndt, *One Year of Nuclear Rhetoric*, 128; "Command-staff exercises of the Strategic Missile Forces are being held at Mari El," TASS, 12 Oct 22, https://tass.ru/armiya-i-opk/16038661

37. Emmanuel Macron Twitter post, 13 Oct 22; Soraya Ebrahimi, "France will not trade nuclear strikes with Russia over Ukraine," *National News*, 13 Oct 22, https://www.thenationalnews.com/world/europe/2022/10/13/france-wont-retaliate-with-nuclear-weapons-if-russia-uses-them-in-ukraine/

38. Horovitz and Arndt, *One Year of Nuclear Rhetoric*, 129; Edward Hunter Christie Twitter post reposting Josep Borrell remarks, 13 Oct 22.

39. CBC Power and Politics tweet, 13 Oct 22.

40. Horovitz and Arndt, *One Year of Nuclear Rhetoric*, 130.

41. FR 24 and ADS-B Exchange screenshots, 13 Oct 22.

42. USSTRATCOM Twitter post, 14 Oct 22.

43. FR 24 and ADS-B Exchange screenshots, 14–15 Oct 22.

44. ADS-B Exchange screenshots, 16 Oct 22.

45. Faytuks News Twitter post, 16 Oct 22.

46. ADS-B Exchange screenshots, 16 Oct 22.

47. Evergreen Intel Twitter post, 15 Oct 22; Aki Heikkinen Twitter post, 15 Oct 22; Viktor Kovalenko Twitter post, 15 Oct 22; Forsvaret Twitter post, 14 Oct 22; The Lookout Twitter post, 17 Oct 22.

48. Thomas Nilsen, "Russia escalates Arctic military posture ahead of NATO's nuclear drill," *Barents Observer*, 16 Oct 22, https://thebarentsobserver.com/en/security/2022/10/russia-escalates-arctic-military-poseur-ahead-natos-nuclear-drill

49. Tim Ellis, "Eielson-based F-16s intercept 2 Russian bombers near Alaska," *Alaskan Public Media*, 21 Oct 22, https://alaskapublic.org/2022/10/21/eielson-based-f-16s-intercept-2-russian-bombers-near-alaska/

50. David Cenciotti, "Let's Have a Look at This Year's NATO Nuclear Strike Exercise in Europe," *The Aviationist*, 28 Oct 22, https://theaviationist.com/2022/10/28/steadfast-noon-2022-report/; ADS-B Exchange screenshots, 17 Oct 22.

51. ADS-B Exchange screenshot, 17 Oct 22; Sheila Weir (photographer) Twitter post, 17 Oct 22.

52. FR 24 screenshot, 18 and 21 Oct 22.

53. FR 24 screenshots, 3–4 Feb 22 and 18 Oct 22.

54. Dan Sabbagh and Jessica Elgot, "Ben Wallace prepared to quit if PM drops pledge on defence spending," *The Guardian*, 18 Oct 22, https://www.theguardian.com/politics/2022/oct/18/james-heappey-defence-spending-quit-hints

55. Benjamin Lynch, "Fears group Putin could detonate nukes as Ben Wallace scrambles to Washington," *Mirror*, 18 Oct 22, https://www.mirror.co.uk/news/politics/fears-grow-putin-could-detonate-28273343

56. "No need for alarm, says UK defence minister about his U.S. visit," Reuters via EuroNews, 21 Oct 22, https://www.euronews.com/2022/10/20/britain-politics-wallace-u-s

57. FR 24 screenshots, 18 Oct 22.

58. ADS-B Exchange screenshots, 18 Oct 22.

59. J.J. Twitter post, 19 Oct 22 (planespotter reposting official photos).

60. FR 24 screenshot, 19 Oct 22; ADS-B Exchange screenshot, 19 Oct 22.

61. Bill Toulas, "Internet connectivity worldwide impacted by severed fiber cables," BleepingComputer, 20 Oct 22, https://www.bleepingcomputer.com/news/technology/internet-connectivity-worldwide-impacted-by-severed-fiber-cables-in-france/

62. Francis Scarr interpretation of Rossiya1 talk show, 19 Oct 22.

63. "Beware of Nuclear False Flag Blaming Russia. If Someone Sets Off a Nuke, It Will Be the US, Not the Russians," GlobalResearch, 10 Oct 22, https://www.globalresearch.ca/beware-nuclear-false-flag-blaming-russia-if-someone-sets-off-nuke-it-will-us-not-russians/5795895

64. Georgy Shamuev, "Doomsday machine: What is a nuclear strike and how does it work," @hi-tech, 19 Oct 22, https://hi-tech.mail.ru/review/59990-mashina-sudnogo-dnya-chto-takoe-yadernyy-udar-i-kak-on-rabotaet/#anchor379221

65. ADS-B Exchange screenshots, 20 Oct 22.

66. "The EKS Kupol network design," Russian Space Web, https://www.russianspaceweb.com/eks-network.html

67. ChristO Twitter 8 May 23 repost of Russian Telegram channel VChK-OGPU which details the scandal affecting Vympel Corporation, the Voronezh contractor.

68. "The decline of the nuclear triad? Ground and space echelons of the early warning system," Topwar.ru, 11 Jan 20, https://topwar.ru/166564-zakat-jadernoj-triady-nazemnyj-i-kosmicheskij-jeshelony-sprn.html

69. Maxim Starchak, "Sanctions further delay Russian missile early warning program in space," Defense, 12 Mar 23, https://www.defensenews.com/space/2023/03/12/sanctions-further-delay-russian-missile-early-warning-program-in-space/

70. Confidential discussion with the author.

71. Garda Crisis 24, 20 Oct 22, https://crisis24.garda.com/alerts/2022/10/europe-russia-implements-heightened-alert-levels-oct-20-extends-flight-restrictions-at-southwestern-airports-until-oct-28-update-89

72. "EAEU prime ministers to meet in Yerevan on 20–21 October," Belta, 11 Oct 22, https://eng.belta.by/politics/view/eaeu-prime-ministers-to-meet-in-yerevan-on-20-21-october-153754-2022/; FR 24 screenshots, 20 Oct 22.

73. Sam Ramani Twitter post, 20 Oct 22.

74. Will Stewart et al., "Mystery of Russia's 'missing' nuclear tests," Daily Mail, 20 Oct 22, https://www.dailymail.co.uk/news/article-11335665/Mystery-missing-Russian-nuclear-tests.html?fbclid=IwAR3h3FXXprV-yO6s4AEmb8fBNGID3wqAgsgJm55tq; "Known for wild conspiracy theories, political scientist Valery Solovey is now in police custody," Meduza, 20 Oct 22, https://meduza.io/en/news/2022/02/16/known-for-wild-conspiracy-theories-political-analyst-valery-solovey-is-now-in-police-custody

75. Institute for the Study of War Twitter post, 20 Oct 22.

76. ADS-B Exchange screenshots, 20 Oct 22.

77. FR 24 screenshots, 21 Oct 22.

78. Alex Horton, "U.S., Russian defence ministers hold first talks in months," *Washington Post*, 21 Oct 22, https://www.washingtonpost.com/national-security/2022/10/21/shoigu-austin-phone-call/

79. "Russian, US defense chiefs to clear the air-envoy," TASS, 21 Oct 22, https://tass.com/politics/1526013

80. Mike Wall, "Russia launches 4 satellites to orbit in 5th mission of last 2 weeks," Space.com, 24 Oct 22, https://www.space.com/russia-soyuz-launch-four-satellites-october-2022

81. ADS-B Exchange screenshots, 21 Oct 22.

82. Hans Kristensen Twitter post, 23 Oct 22.

83. "Kyiv Prepares Another Provocation," Fresh News Asia, 23 Oct 22, https://m.en.freshnewsasia.com/index.php/en/32550-2022-10-23-11-37-23.html

84. UK Ministry of Defence Twitter post, 23 Oct 22.

85. "Pentagon chief tells Russia's Shoigu he rejects any Russian pretext for escalation," Reuters, 23 Oct 22, https://www.reuters.com/world/europe/pentagon-chief-tells-russias-shoigu-he-rejects-any-russian-pretext-escalation-2022-10-23/; Austin Twitter post, 23 Oct 22.

86. Dave Lawler, "Dirty bomb fears prompt urgent US-Russia calls," Axios, 23 Oct 22, https://www.axios.com/2022/10/24/dirty-bomb-false-flag-russia-ukraine

87. Press release by the UK Foreign, Commonwealth & Development Office, 24 Oct 22.

88. ADS-B Exchange screenshots, 23 Oct 22.

89. Dave Lawler, "Dirty bomb fears prompt urgent US-Russia calls," Axios, 23 Oct 22, https://www.axios.com/2022/10/24/dirty-bomb-false-flag-russia-ukraine

90. "A pretext for escalation," *Meduza*, 25 Oct 22, https://meduza.io/en/feature/2022/10/25/a-pretext-for-escalation

91. Oliver Alexander Twitter post containing detailed map comparison, 24 Oct 22.

92. Flash News Twitter post retweeting data from the Belarusian Gayun Monitoring Group, 24 Oct 22.

93. ADS-B Exchange screenshots, 24 Oct 22.

94. FR 24 screenshots, 24 Oct 22.

95. David Martin, "Russia has notified the U.S. its annual nuclear exercise has begun, U.S. officials say," CBS News, 26 Oct 22, https://www.cbsnews.com/news/russia-nuclear-exercise-annual-nuclear-capable-missiles-launch-american-officials-say/

96. The Lookout Twitter post, 25 Oct 22; James Cameron Twitter post, 25 Oct 22.

97. Ruslan Trad retweeting TASS, 25 Oct 22.

98. Screenshot of RIA Novosti Telegram post, 26 Oct 22.

99. "Kremlin: Tasks of training strategic deterrent force have been fully completed," Interfax, 26 Oct 22, https://www.interfax-russia.ru/main/kreml-zadachi-trenirovki-strategicheskih-sil-sderzhivaniya-vypolneny-polnostyu

100. FR 24 screenshots, 25 Oct 22.

101. ADS-B Exchange screenshots, 26 Oct 22.

102. ADS-B Exchange screenshots, 26 Dec 22; Hans Kristensen Twitter post, 26 Oct 22.

103. Horovitz and Arndt, *One Year of Nuclear Rhetoric*, 139.

104. Horovitz and Arndt, 139–140.

105. "Russia warns West: We can target your commercial satellites," Reuters, 27 Oct 22, https://www.reuters.com/world/russia-says-wests-commercial-satellites-could-be-targets-2022-10-27/

106. Brett Tingley, "White House says US would respond if Russia targets commercial satellites," Space.com, 27 Oct 22, https://www.space.com/white-house-response-russia-target-commercial-satellites

107. Visegrad 24 Twitter post, 31 Oct 22; Charles Maynes, "Russia is suspending a Ukraine grain export deal," NPR, 29 Oct 22, https://www.npr.org/2022/10/29/1132608672/russia-says-it-is-suspending-a-grain-export-deal-with-ukraine

108. Hans Kristensen Twitter post, 27 Oct 22; Flash News Twitter post, 30 Oct 22.

109. Francis Scarr Twitter post, 1 Nov 22, with embedded clip from Rossiya1.

110. Dmitry Twitter post, translating Medvedev Telegram post, 1 Nov 22.

111. *Moscow Times* Twitter post, 4 Nov 22; Max Seddon Twitter post translating Medvedev Telegram post, 4 Nov 22.

112. Max Seddon Twitter post translating Dugin video, 4 Nov 22.

113. C4H10FO2P Twitter post translating commentary by Patriarch Kiril, 4 Nov 22; Sergey Radchenko Twitter post, 4 Nov 22.

114. UK Ministry of Defence Twitter post, 1 Nov 22.

115. USSTRATCOM Twitter post, 1 Nov 22.

116. COMSUBLANT Twitter post, 1 Nov 22.

117. FR 24 screenshots, 1 Nov 22.

118. Russian Ministry of Defence Twitter post, 2 Nov 22.

119. Russian Ministry of Defence social media, 7 Nov 22.

120. Horovitz and Arndt, *One Year of Nuclear Rhetoric*, 142; Starchak, "Sanctions further delay Russian missile early warning program in space."

121. ADS-B Exchange, 3 Nov 22.

122. FR 24 screenshots, 3 Nov 22.

123. Ryan Chan Twitter post, 7 Nov 22; SkyScanWorld Twitter post, 4 Dec 22; Snoopy Twitter post, 7 Nov 22; TheIntelFrog Twitter post, 7 Nov 22.

124. ADS-B Exchange screenshots, 8 Nov 22.

125. Ryan Chan Twitter post of imagery from Barksdale AFB Facebook page, 7 Nov 22.

126. ADS-B Exchange screenshots, 1 Nov 22, 20 Nov 22, and 29 Nov 22; Shashank Joshi Twitter post, 22 Nov 22.

127. Richard Wood Twitter post, 17 Nov 22.

128. Saturnax (submarine observer) Twitter post, 21 Oct 22.

129. The Lookout Twitter post, 9 Dec 22

130. U.S. Spec Ops Europe Twitter post, 9 Nov 22; ADS-B Exchange screenshot, 9 Nov 22.

131. USSTRATCOM Twitter post, 13 Nov 22; Dean Evans, "SPDE's Rapid Dragon capability demonstrated in Norway," Air Force Research Laboratory Public Affairs, 9 Nov 22, https://www.afmc.af.mil/News/Article-Display/Article/3215261 /sdpes-rapid-dragon-capability-demonstrated-in-norway/

132. NATO Twitter post, 12 Nov 22.

133. Horovitz and Arndt, *One Year of Nuclear Rhetoric*, 146.

134. FR 24 screenshots, 10 Nov 22.

135. ADS-B Exchange screenshots, 5–10 Nov 22.

136. FR 24 screenshots, 12–13 Nov 22; ADS-B Exchange screenshots, 12–13 Nov 22.

137. "Putin's spy chief says he discussed Ukraine with CIA director in Turkey," Alarabiya News, 30 Nov 22, https://english.alarabiya.net/News/world/2022/11/30 /Putin-s-spy-chief-says-he-discussed-Ukraine-with-CIA-director-in-Turkey

138. Natasha Bertrand, "CIA director Bill Burns met with Russian counterpart Monday," CNN, 14 Nov 22.

139. FR 24 screenshots, 12–13 Nov 22.

140. Warships IFR Twitter post, 15 Nov 22.

141. ADS-B Exchange, 15 Nov 22.

142. ADS-B screenshots, 15 Nov 22.

143. FR 24 screenshots, 15 Nov 22.

144. Author's analysis of the event in real time of 287 Twitter feeds and mainstream media, 15 Nov 22.

145. FR 24 screenshots, and ADS-B Exchange screenshots, 16 Nov 22.

146. ADS-B Exchange and RF 24 screenshots, 17–25 Nov 22.

147. Stephen Clark, "Russia launches two Soyuz rockets on military space missions," *Spaceflight Now*, 7 Dec 22, https://spaceflightnow.com/2022/12/07/russia -launches-two-soyuz-rockets-on-military-space-missions/

148. USSTRATCOM Twitter post, 28 Nov 22.

149. "Russia's YARS ICBM loaded into a silo in Central Russia," TASS, 13 Dec 22, https:// tass.com/defense/1550211

150. Evgeniy Maksimov Twitter post, 18 Dec 22.

151. FR 24 screenshots, 23 Dec 22.

152. USSTRATCOM Twitter post, 13 Dec 22.
153. USSTRATCOM Twitter post, 13 Dec 22.
154. William Morgan, "US Air Force moves HQ to RAF Fairford in strategic move to strengthen command and control," Cheltenham News, 14 Dec 22, https://www.gloucestershirelive.co.uk/news/cheltenham-news/usaf-moves-hq-raf-fairford-7925637?fbclid=IwAR2F5_B_D2J5qm8KH-s-O4BLfSgBTTeHV82EIHFByqA9BKO4Zt8UcvOJroM
155. Manu Gomez Twitter post, 13 Dec 22.
156. Massimo Frantarelli Twitter post, 14 Dec 22; ADS-B Exchange screenshot, 14 Dec 22.
157. UK Defence Intelligence Twitter post, 23 Dec 22.
158. ADS-B Exchange screenshots, December 2022.
159. Anton Gerashenko Twitter post, 23 Dec 22; Samuel Ramani Twitter post, 23 Dec 22.
160. NOEL Twitter post, 22 Dec 22.
161. Timothy Snyder Twitter post, 26 Dec 22.
162. MAKS22 Twitter post, 28 Dec 22.

SELECTED BIBLIOGRAPHY

Belton, Catherine. *Putin's People: How the KGB Took Back Russia and Then Took on the West*. New York: Farrar, Straus, and Giroux, 2020.

Carter, Ash. *Inside the Five-Sided Box*. New York: Dutton Publishing, 2019.

Clover, Charles. *Black Wind, White Snow*. New Haven: Yale University Press, 2022.

Conley, Heather A., and Caroline Rohlof. *The New Ice Curtain: Russia's Strategic Reach into the Arctic*. London: Rowman and Littlefield, 2013.

Conradi, Peter Conradi. *Who Lost Russia? How the World Entered a New Cold War*. London: Oneworld Publications, 2017.

Cooley, Alexander, and John Heathershaw. *Dictators without Borders: Power and Money in Central Asia*. New Haven: Yale University Press, 2017.

Cooper, Tom. *Moscow's Game of Poker: Russian Military Intervention in Syria, 2015–2017*. Warwick: Helion, 2018.

Dawisha, Karen. *Putin's Kleptocracy: Who Owns Russia?* New York: Simon and Schuster, 2014.

Durkalec, Jacek. *Nuclear-Backed "Little Green Men": Nuclear Messaging in the Ukraine Crisis*. Warsaw: Polish Institute of International Affairs, 2015.

Ekstrom, Markus. *Rysk operative-strategisk ovningsverksamhet under 2009 och 2010*. Stockholm: Swedish Defence Research Agency, 2010.

Esper, Mark T. *A Sacred Oath: Memoirs of a Secretary of Defense during Extraordinary Times*. New York: William Morrow, 2022.

Fridman, Ofer. *Russian Hybrid Warfare: Resurgence and Politicisation*. London: Hurst, 2018.

Galeotti, Mark. *Putin's Wars: From Chechnya to Ukraine*. Oxford: Osprey, 2022.

Gertz, Bill. *iWar: War and Peace in the Information Age*. New York: Threshold Editions, 2017.

Golts, Alexander, and Michael Korman. *Russia's Military: Assessment, Strategy, and Threat*. Washington, D.C.: Center on Global Interests, June 2016.

Haldeman, H.R. *The Ends of Power*. New York: Dell, 1978.

Hanebrink, Paul. *A Specter Haunting Europe: The Myth of Judeo-Bolshevism*. Cambridge, MA: Harvard University Press, 2018.

Kahn, Herman. *Thinking about the Unthinkable*. New York: Discus Books, 1962.

Kovalev, Andrei A. *Russia's Dead End: An Insider's Testimony from Gorbachev to Putin*. Lincoln, NE: Potomac Books, 2017.

Laqueur, Walter. *Black Hundred: The Rise of the Extreme Right in Russia*. New York: Harper Perennial, 1994.

Laruelle, Marelene. *Russian Eurasianism: An Ideology of Empire*. Baltimore: Johns Hopkins University Press, 2012.

Maloney, Sean. *Deconstructing Dr. Strangelove: The Secret History of Nuclear War Films*. Lincoln, NE: Potomac Books, 2020.

Maloney, Sean. *Emergency War Plan: The American Doomsday Machine, 1945–1959*. Lincoln, NE: Potomac Books, 2021.

Moore, Martin. *Democracy Hacked: Political Turmoil and Information Warfare in the Digital Age*. New York: OneWorld Publications, 2018.

Norberg, Johan. *Training to Fight: Russia's Military Exercises 2011–2014*. Stockholm: Swedish Defence Research Agency, 2015.

Pomerantsev, Peter. *This Is Not Propaganda: Adventures in the War against Reality*. New York: PublicAffairs, 2019.

Schweizer, Peter. *Victory: The Reagan Administration's Secret Strategy that Hastened the Collapse of the Soviet Union*. New York: Basic Books, 1994.

Schweizer, Peter. *Reagan's War: The Epic Story of His Forty-Year Struggle and Final Triumph over Communism*. New York: Doubleday, 2002.

Shekhovtsov, Anton. *Russia and the Western Far Right*. New York: Routledge, 2018.

Snyder, Timothy D. *The Road to Unfreedom: Russia, Europe, America*. New York: Crown, 2018.

Stein, Matthew. *Compendium of Central Asian Military and Security Activity*. Fort Leavenworth: Foreign Military Studies Office, 2021.

Streusand Douglas E., et al., *The Grand Strategy that Won the Cold War: Architecture of Triumph*. London: Lexington Books, 2016.

Stumpf, David K. *Minuteman: A Technical History of the Missile that Defined American Nuclear Warfare*. Fayetteville: University of Arkansas Press, 2020.

Watts, Clint. *Messing with the Enemy: Surviving in a Social Media World of Hackers, Terrorists, Russians, and Fake News*. New York: Harper Collins, 2018.

INDEX

ABOUT THE AUTHOR

SEAN M. MALONEY, PHD, is a professor of history at Royal Military College of Canada and served as the historical advisor to the Chief of the Land Staff during the war in Afghanistan. He previously served as the historian for 4 Canadian Mechanized Brigade, the Canadian Army's primary Cold War NATO commitment after the reunification of Germany and at the start of the Balkan conflicts. Dr. Maloney has extensive field experience in that region, particularly in Croatia, Bosnia, Kosovo, and Macedonia from 1995 to 2001, where he inadvertently observed the activities of the Al Qaeda organization and its surrogates. This work was interrupted by the 9/11 attacks. From 2001 Dr. Maloney focused nearly exclusively on the war against the Al Qaeda movement and its allies, particularly on the Afghanistan component of that war. He traveled regularly to Afghanistan from 2003 to 2014 to observe and record coalition operations in that country and was the first Canadian military historian to go into combat since World War II. After returning to Royal Military College, Dr. Maloney refocused on the Cold War, releasing *Deconstructing Dr. Strangelove: The Secret History of Nuclear War Films* in 2020 and *Emergency War Plan: The American Doomsday Machine, 1945–1960*, a reconstruction and analysis of nuclear war plans in the 1950s, in 2021. *Emergency War Plan* won the Air Force Historical Foundation's Air Power Book Prize in 2024. He is near completion of a history of Soviet nuclear weapons in Ukraine during the Cold War.

www.ingramcontent.com/pod-product-compliance
Lightning Source LLC
Jackson TN
JSHW081247130925
90895JS00001B/1